The Biology and Management of Capricornis and Related Mountain Antelopes

Edited by Hiroaki Soma

CROOM HELM
London · New York · Sydney

© 1987 Hiroaki Soma
Croom Helm Ltd, Provident House,
Burrell Row, Beckenham, Kent BR3 1AT
Croom Helm Australia, 44-50 Waterloo Road,
North Ryde, 2113, New South Wales

Published in the USA by
Croom Helm
in association with Methuen, Inc.
29 West 35th Street
New York, NY 10001

British Library Cataloguing in Publication Data

The biology and management of *Capricornis*
 and related mountain antelopes.
 1. Japanese serow — Ecology
 I. Soma, H.
 599.73′58   QL737.U53
 ISBN 0-7099-4458-6

Library of Congress Cataloging-in-Publication Data
ISBN 0-7099-4458-6

Typeset in Times Roman by Leaper & Gard Ltd, Bristol, England
Printed and bound in Great Britain by
Biddles Ltd, Guildford and King's Lynn

The Biology and Management of *Capricornis* and Related Mountain Antelopes

A sketch of a Japanese serow drawn by Ms Masako Takasaki

Capricornis and its related species include the mountain ungulates such as chamois, Rocky mountain goats, saigas, Mongolian gazelles, gorals, Japanese serows, Formosan serows and Sumatran serows. This book presents over thirty selected papers from a major international conference on these groups held in Japan in May 1986, with contributors from Europe, North America and Asia. It is one of the first books specifically to focus on these animals. Topics covered include distribution and breeding, behaviour, ecology and reproduction, pathology and veterinary aspects, and conservation of endangered species. The book is likely to interest zoologists, veterinarians and animal conservationists concerned with these groups of mammals.

A Japanese serow standing by Mt Gozaisho

Photograph taken by Japanese Serow Center

Contents

Acknowledgements

I would like to express my heartfelt thanks to all contributors in the planning of this book. In particular, I wish to thank the people of Komono-cho, workers at the Japan Serow Center, Gozaisho Ropeway, Mie Prefecture, Mie Kotsu Co. Ltd, Kinki Nippon Railway Co. Ltd, the Japan Environmental Agency, the Japan Cultural Agency, the Japan Shipbuilding Industry Foundation, Gifu University, Kyushu University, Mie University, Tokyo Medical College, the Japan Association of Zoological Gardens & Aquariums, the Mammalogical Society of Japan, the Society of Chromosome Research, the Genetic Society of Japan, and the Association of Museums in Mie Prefecture, for their generous support throughout the organisation of the International Symposium on *Capricornis* and its related species. I am also grateful to Mr T. Hardwick of Croom Helm for his help in the editing of this book.

Hiroaki Soma

Preface

The Japanese serow (*Capricornis crispus*) has been protected by law since 1955 in Japan, because it was becoming rarer and approaching extinction. Thereafter, the serow population has increased gradually. The Japanese serow is thought to be a primitive relict species on the islands of Japan, and the geographical range of the serow has retracted upwards into the mountain forests to avoid contact with humans. Little was therefore known about these animals. However, increasing losses of forest habitat due to exploitation of the mountain forests or expanding cultivation by local foresters have driven the Japanese serow back into the lowlands of Japan. Since then, complaints of damage to trees and other vegetation have accumulated against the serow. In some prefectures the shooting of Japanese serow was allowed in order to prevent damage to forests. The animals killed were taken for research by the Departments of the Environment and by universities.

The Japan Serow Center was set up at the summit of Mt. Gozaisho, Komono-cho, Mie Prefecture, in 1962 and has made a great effort to breed the serow and its related species in captivity. In addition, the International Studbook of *Capricornis crispus* in captivity was established in Japan, and the state of breeding of the Japanese serows is now reported annually. However, without detailed scientific research, it is impossible to conduct sensible protection, conservation or management of the serow in captivity or in the wild. Thus, the necessity for cooperative research into the unresolved problems regarding the biology and management of *Capricornis* and its related species that are also in danger of extinction has become apparent.

With this in mind, the International Symposium of *Capricornis* and its related species was held at Komono-cho, Mie Prefecture, from 11 to 13 May 1986 with the assistance of the people of Komono-cho and the Japan Serow Center. The opening ceremony was honoured by the attendance of their Imperial Highnesses Prince and Princess Hitachi. At the symposium, the biology, ecology, anatomy, pathology, bacteriology, reproduction and genetics of serows, gorals, chamois, Rocky Mountain goats, saigas and Mongolian gazelles were discussed by distinguished investigators from Europe, North America, east Asia and Japan, and thereby good scientific cooperation between West and East for the management of wild antelopes was maintained. Accordingly, this book is based on the proceedings of the symposium held at the foot of Mt. Gozaisho. I hope that it will facilitate an understanding of the status of the serow and other related antelopes and the preservation of these animals.

Hiroaki Soma

Part One:
Evolution and Breeding History
of the Rupicaprini

1

On the evolution of the Caprinae

Valerius Geist
Faculty of Environmental Design, University of Calgary, Calgary,
Alberta, Canada, T2N 1N4

INTRODUCTION

The evolution of the subfamily Caprinae follows much the same pattern as does that of other families whose species are distributed from the tropics to the Arctic (Geist 1971a,b, 1977, 1978a, 1983, 1985a, 1986b). A similar pattern is found in gallinaceous birds (Geist 1977). This very evolutionary analysis indicates that the segregation of rupicaprids from caprids is an artefact. In recent decades we have learned much about ungulates, yet taxonomic models were developed much earlier on the basis of morphological differences that were little understood. We are now gaining that understanding. I hope to show that caprids are an evolutionary progression out of the rupicaprids, and that their differences are a consequence of exploiting open habitats as opposed to those rich in cover, as well as a result of the colonisation of cold, seasonal climates.

I shall also exclude the tribe Saigini from the Caprinae (see Schaller 1977). The genera *Saiga* and *Panthalops* are highly specialised, cold-adapted gazelles as I pointed out earlier (Geist 1974a), a conclusion independently arrived at by Hofmann (submitted). Consequently I include in the Caprinae only three tribes, the Rupicaprini (goat-antelopes), the Caprini (sheep and goats), and the Oviboni (shrub-oxen, musk-oxen and euceratheres). The former is the original stem group, and the other two are independent radiations into open landscapes and cursorialism. I shall not give here a detailed taxonomic account; Schaller (1977) gave a good taxonomic overview of caprids which is acceptable with some revisions. I follow Groves and Grubb (1985) in uniting serow and goral into one genus (*Nemorhaedus*).

In the following pages I shall bring to bear a diversity of information on caprid biology. My goal is to devise a testable model of evolution, in particular of speciation.

THE ZOOGEOGRAPHIC PATTERN AND SPECIES TYPES

One finds in the caprids, as well as in other large Ice Age mammals, a geographic pattern of distribution in which social organs (horns, hair patterns, coat coloration, glands), differentiate progressively with latitude or altitude (Figure 1.1). A similar progression is found in social behaviour (visual, vocal, and olfactory signals). From the tropics to alpine and arctic regions, species evolve increasingly larger horns or accentuated coat patterns, larger rump patches, and shorter or bushier tails, terminating as 'grotesque giants' in climatically extreme, seasonal environments. In general, body size also increases with this progression.

In tropical latitudes primitive large mammals tend to have adaptive syndromes that centre on *material resource defence.* These are recognisable as weapons that allow a maximum of damage to the body's surface (Geist 1978a,b). Such weapons are found in tropical rupicaprids (*Nemorhaedus sumatrensis, N. goral*; see Figure 1.2), which can be identified as primitive by various features.

(1) they are less specialised in cheek teeth and limbs than other caprids (Dolan 1965);
(2) their social behaviour is conservative (Lovari 1985);
(3) zoogeographically they overlap with other mammalian species of great antiquity such as deer of the genera *Rusa*, and *Muntiacus*, the Sumatran rhino (*Dicerorhinus*), the sun bear (*Helarctos*), the mouse deer (*Tragulus*), the flatheaded cat (*Tatailurus*), etc. (Thenius and Hofer, 1960, Thenius 1969, Figure 1.3).
(4) *Nemorhaedus* or closely related genera appear earlier (late Pliocene) than other caprids in the fossil record (Kurten 1968).

In other mammalian lineages one finds in southern latitudes not only the most primitive members of their families (as well as *secondarily* primitive members such as the deer *Mazama* and *Pudu*, Krieg 1948 and others, see Bubenik 1966), but also ecological radiations based on this primitive body form. In Old World deer, for instance, one can identify at one extreme a small-bodied food generalist (*Muntiacus*) (Barette 1977a), and at the other a giant coarse-food specialist (*Rusa*), with several ecological specialists in between, all based on the primitive three-pronged antler plan (Geist 1971b, 1983). Analogously, the giraffe is the primitive giant food specialist, and the okapi a primitive forest species; giraffes advanced in body plan, with huge horn-like organs, were the extinct grazing giraffes from open landscapes, the sivatheres (W.R. Hamilton 1973, Churcher 1978).

In caprids the tropical radiation of primitives is poorly represented. One finds neither a 10 kg dwarf, nor a 300 kg giant. Nor are there fossils of such forms. However, the mainland serow (*Nemorhaedus sumatrensis*) is

Figure 1.1: Body size versus latitudinal/altitudinal position in the Caprinae (the positions are approximate only). Resource defenders are at low latitude; 'grotesque giants' are at high latitudes; most are of Pleistocene origin (Geist 1983)

Figure 1.2: (A) A 'war-coloured' monomorphic resource defender, the serow (*Nemorhaedus sumatrensis*), with dagger-like weapons (which allow the animal to inflict maximal surface damage or pain). (B) A gregarious, dimorphic, ornate caprid, the markhor (*Capra falconeri*), whose horns probably act as 'shields', and as 'luxury' organs. They function as symbols of the ability to procure food as well as to spare resources from the body towards horn growth (Geist 1986a)

in its wide tropical distribution usually as big as the largest cold-adapted rupicaprid, the mountain goat (*Oreamnos*) (Dolan 1965). Tropical grazer niches over 200 kg in weight are filled by bovines; cervids fill most small and medium herbivore niches in forest and savannah on level ground, leaving the steep slopes to rupicaprids and caprids.

Morphological evolution is tied to geographic dispersion, not only within a genus but also between genera. For instance, for each geographic and consequently evolutionary advance by Asiatic *Ovis*, there is a comparable advance by *Capra*. Therefore one finds a primitive sheep with a primitive goat, and an advanced sheep with an advanced goat (Geist 1985a). Similarly, one finds the more primitive, 'ornate' chamois (Lovari and Scala 1980, Masini 1985) with 'goats', and the 'black', more recent, chamois with 'ibex'. The same pattern can be seen in northern cervids (Geist 1986b), while *Cervus* closely paces the rupicaprids and caprids in their evolutionary radiations (Figure 1.3).

Geographic patterns also reveal the relative age of a genus. The tahr (*Hemitragus*) is split into three very distinct, widely separated species, one in the Himalayas (*H. jemlahicus*), one in the Nilgiri Hills of southern India (*H. hylocrius*), and one in Arabia across the Gulf of Oman (*H. jayakeri*). This implies considerable geological antiquity. So does the intermediate morphology and behaviour between rupicaprids and caprids (see Geist

Figure 1.3a: Rupicaprids and caprids are distributed geographically with Old World deer, and with one another, at comparable stages of evolutionary development. Primitive is found with primitive, advanced with advanced. The further one goes from the tropics, the more advanced or recent in time are the species. (a) Lower: tropical serow with rusa deer; upper: temperate zone Japanese serow with sika deer. (b) Lower: cold-zone Manchurian goral with Dybowski's sika and Izubr stag (primitive wapiti); upper, facing right; extreme eastern radiation, *Oreamnos* and wapiti in America; upper, facing left: extreme western radiation, *Rupicapra* and west European red deer. (c) Lower: the most primitive sheep (urial), goat (markhor), and red deer (Kashmir stag); middle: Armenian mouflon, wild goat and eastern red deer (This is the next step in the western radiation); upper: European mouflon, Spanish goat and west European red deer. The mouflon is not sympatric with the other two species and may be a human transplant, but it is advanced over the Armenian mouflon in its social organs. (d) Lower: bharal, Himalayan argali, Tibetan red deer; upper: Siberian ibex, argali and wapiti, the most advanced species of their respective lineages

Figure 1.3b

Figure 1.3c

Figure 1.3d

1971a, Schaller 1977, C. Rice, personal communication). Thirdly, *Hemitragus* occupies tropical latitudes and overlaps geographically with serow and goral, as would be expected from a primitive caprid. In the fossil record it appears at the beginning of the major glaciations in Europe concurrently with musk- and shrub-oxen, but without true sheep or goats; in its second appearance prior to the last (Würm) glaciation, the tahr is in the company of sheep, goats and chamois (Kurten 1968).

Neighbours in a geographic pattern are expected to be usually related, which indicates that *Hemitragus* and *Ammotragus* have an affinity (Geist 1971a).

Geographic dispersion is not only linear, but can also be radial, as discussed for American sheep (Geist 1971a, 1985a), and for ibex (Nievergelt 1981). Here the most primitive form is in the centre, and the evolved ones describe a circle about it; an evolutionary progression is expressed with distance from the centre. *Ovis, Capra, Rupicapra* and red deer (*Cervus elaphus*) appear to have the same centre of radiation which lies between the Himalayas and Asia Minor.

Island dwarfs are relatively few in caprids. Such dwarfs arise in the absence of predation on oceanic islands (Azzaroli 1982). A typical island dwarf is the extinct cave goat of Mallorca and Menorca (*Myotragus*) (Kurten 1968, Azzaroli 1982). The extinct postglacial Sardinian goral (*Capricornis melonii*, Kurten 1968, p. 175) may be another. The serow of Taiwan (*N. swinhoei*) is very small (18–20 kg. Zuh-Ming 1961) and appears paedomorphic; the Japanese serow (*N. crispus*) is also fairly small and, significantly, has enlarged cheek teeth. The wild goats (*Capra aegagrus*) of Crete in the Mediterranean are small compared with those from the mainland of Asia Minor (Schaller 1977). The natural origin of the small Cyprian mouflon (*Ovis orientalis ophion*) is in doubt, as is that of the Corsican and Sardinian mouflons (*O. o. musimon*). They may be early domestic sheep transplanted by humans (Bunch *et al.* 1978b, Valdez 1982, Davis 1984). The remains of sheep, goats, pigs and humans appear simultaneously in the lower Neolithic of Cyprus, while the endemic fauna disappears (Davis 1984); on Corsica and Sardinia mouflons do not appear earlier than 6000 years BP. These later mouflons share haemoglobin A with domestic sheep (Bunch *et al.* 1978b). Mouflons, introduced to the European mainland, are in some populations 6–17 per cent larger in linear dimensions than their island predecessors (Pfeffer 1967). This is a difference about as large as between island and mainland mouflons from Asia Minor and Persia (skull length in island rams, mean $=232 \pm 5.9$ mm SD $n=13$; 252 ± 5.9 mm in mainland rams, $n=7$; Pfeffer 1967). Therefore, mouflons are probably not permanently depressed in body size by their life on islands, a fact arguing in favour of the transplant hypothesis.

Paedomorphic dwarfs are mainland dwarfs. They are secondarily primitive, reduced forms. I suggest that the mouflons (*Ovis orientalis gemelini*) and 'blue sheep' (*Pseudois nayaur*) originate as paedomorphic dwarfs, derived from urial- and markhor-like sheep and goats, respectively (Geist 1987, Figure 1.4). They are remarkably similar in external appearance. Armenian mouflon rams (the most paedomorphic) have the coat patterns and short ruffs of juvenile urial rams, the primitive heteronym horn flare and flat tail of *Ammotragus*, and an advanced karyotype of $2n=54$ versus $2n=58$ for urials (Nadler *et al.* 1974). The $2n=54$ karyotype of *Pseudois* is not homologous with that of mouflons (Bunch *et al.* 1978a). Mouflon-like sheep appear late in the Pleistocene record of North Africa and Europe (Kurten 1968). They evolve enlarged ruffs and enhanced coat markings with dispersal south-east into Persia. The fallow deer (*Dama dama, D. mesopotamica*) parallel the evolution of mouflons geographically, including a former North African and European distribution (Halternorth 1961, Kurten 1968).

Hybrid species have not been identified in caprids, but the Caucasian turs (*Capra cylindricornis/caucasica*) appear to produce a hybrid zone (Heptner *et al.* 1961). So do various populations of sheep in Persia (Valdez *et al.* 1978); here mouflons (karyotype $n=54$) hybridise with more primitive urials ($n=58$) to give intermediate forms. In the Caucasus, turs and wild goats (*Capra aegagrus*) live sympatrically and do not interbreed, but did so in captivity and in reintroductions when the two forms first met; crosses between turs, ibex and wild and domestic goats are fertile (Heptner *et al.* 1961), but not those between *Pseudois* and *Capra*. The Chitlan goat, characterised by horns with a heteronym twist, lies geographically at the junction point of wild goats and markhor (*C. falconeri*) (Schaller 1977) and could well be of hybrid origin.

One can thus identify in large mammals at least *five* different types of species, which are also reflected in the caprid radiation.

(1) The *hypermorphic species* which is a product of altitudinal/latitudinal evolution, in which social organs and behaviour differentiate progressively with the climatic severity and the length of the productivity pulse of the habitat (Geist 1971a,b, 1977, 1978a, 1983, 1986b).

(2) The *paedomorphic species* which is a product also of altitudinal/latitudinal evolution, but in which dispersal is from a long to a short productivity pulse. There is thus apparently a symmetry in evolution, depending on the direction of colonisation by a speciating population.

(3) The *ecological (within-latitude) species* arises as a variant of a body plan typical of a given latitude or region.

(4) The *island dwarf species* which appears to be a product of severe efficiency selection (Geist 1978a, 1983) in the absence of predation (see below).

(5) The *hybrid species*, a product of interbreeding between populations with a distinct history.

Figure 1.4: Hypermorphs and paedomorphs in Asiatic sheep and goats. On the right the (northern) argali cline from Geist (1971a, Fig. 43). On the left are goats. C, Afghan urial, and G, markhor, are sympatric and the most primitive of their respective genera. B, Armenian urial, is a late-Pleistocene sheep and therefore a recent descendant of C, being its paedomorphic form. Reversing the markhor's characteristics to a more primitive level must of course generate a form very similar to the Armenian urial. This is the 'blue sheep' (*Pseudois*), which abuts on markhor distribution. Both paedomorphs are more 'ancestor-like' than the presumed parent species (Geist 1987)

Speciation does not appear to be a singular process, as depicted by conventional neo-Darwinism (Mayr 1966); rather, there appear to be several ways of speciating. The evolution of new social body plans is closely bound to geographic dispersion as evidenced by latitudinal/altitudinal clines; social evolution must be slower than ecological speciation, as was foreseen by Hofer (1972). After speciation there is considerable regional diversification at the subspecific level, as seen in serow and goral (Dolan 1965, Groves and Grubb 1985), in ibex (Nievergelt 1981), in urials and mouflons (Valdez *et al.* 1978), in markhor (Schaller 1977), argalis (*Ovis ammon*), snow sheep (*O. nivicola*) (Heptner *et al.* 1961) and American sheep (*O. dalli/canadensis*) (Cowan 1940). All regional variants of a species appear to have the same chromosome number; these races appear to be 'fine tuned' to regional environments. Goldschmidt (1940) raised this important point earlier.

Caprids have a great geographic radiation, but a rather limited ecological radiation within any given area. Their high species diversity is a function of their far-flung altitudinal and latitudinal distribution, not of their ecological plasticity. Old World primates have their greatest diversity in climatically benign tropical and subtropical climates, with but one genus radiating into cold-temperate, periglacial and arctic climates (*Homo*) to give off a typical Pleistocene 'grotesque giant'. Caprids are the obverse; they are most capable and abundant in extreme environments, and consequently they are more a Pleistocene than a Tertiary family. This conclusion is supported by the fossil record (Thenius 1969).

ADAPTIVE SYNDROMES

Behavioural, ecological and life historical characteristics tend to vary with morphological ones. Together with physiological and biochemical attributes they form an *adaptive syndrome*. This is a constellation of logically related adaptations which allow one to identify objectives, strategies, tactics and tactical requirements.

The most primitive adaptive syndrome is that of *resource defence*. Here the objective is to produce the smallest neonate compatible with survival, so as to maximise the resources available for defence of a resource territory (Geist 1978a). Probably the most primitive variant is a resource territory defended by one male and one female. This can occur only in very productive areas in which a small, 'defendable' territory can provide the requisites of life year round. In this syndrome we expect intense patrolling by the male, a check-off system of marking as Barrette (1977a,b) discovered for muntjacs, a combat strategy which confuses opponents via 'war-coloured monomorphism' of both sexes (Geist 1974a,b), fighting tactics that incorporate surprise and inflict a maximum of painful surface damage, and short and stout weapons that readily withdraw from the opponent's body (Geist 1978a). Male and female should be of the same size and shape, and are expected to mark one another. We also expect low neonatal investment, low birth numbers, long gestation periods, a postpartum oestrus, year-round reproduction and slow maturation. In brain morphology we expect relatively small size and simple corticular folding (Geist 1978a, p. 188). Food habits are expected to be diverse, and there ought to be great ability at detoxifying plant toxins; male and female ought to have the same food habits. We expect very little fattening since lipogenesis is calorically expensive (see Geist 1974a). These adaptations ought to be overlain by seasonality, as the primitive syndrome of resource defence is retained with latitudinal or altitudinal dispersal. Primitive rupicaprids and duikers ought to be similar through convergent evolution.

The four variations on resource defence are:

(1) Individual defence of territory with male and female territory overlapping. This is probably practised by the tropical serow (*Nemorhaedus sumatrensis*), judging from the limited information (Engelmann 1938, Schaller 1977), and appears to be the norm for the Japanese serow (Akasaka 1974, Akasaka and Maruyama 1977, Kishimoto 1981, and Chapter 7 this volume, Sakurai 1981).
(2) Single territory defence by individuals with large male territories overlapping smaller female territories, which appears to be an occasional strategy of the Japanese serow (Kishimoto, Chapter 7, this volume).
(3) Kin-defended group territories.
(4) Facultative, opportunistic territoriality alternating with gregarious existence. This is probably found in *Rupicapra* and *Oreamnos* (Kramer 1969, Schroeder 1971, Kuck 1977, Chadwick 1977, 1983, Knaus and Schroeder 1983, S. Lovari, pers. comm.).

It appears therefore that *obligatory* resource defence gives way to *facultative* resource defence as rupicaprids invade cold climates. This is probably due to a reduction in density and diversity of material resources, and thus a spread in home range size, with a concomitant reduction in its 'defendability'. Rupicaprids, typically, remain concentrated feeders (Hofmann 1973) with broad food habits (Heptner *et al.* 1961, Hibbs 1967, Akasaka and Maruyama 1977, Schaller 1977, Chadwick 1983, Knaus and Schroeder 1983) and retain cheek teeth of moderate height (Dolan 1965).

A very dark, if not black, body colour is associated with resource defence and willingness to face down predators (Figure 1.5). Black appears to be the colour of aggression, and in normally light-coloured species may be reserved for the rutting coat of males, or is the facial colour of males in the nuptial coat. This subject deserves further attention.

With invasion of open habitat goes the formation of the *selfish herd* as

Figure 1.5: Large mammals that confront predators tend to be black in body colour. (A) *Rusa unicolor*, (B) *Alces alces*, (C) *Bibos gaurus*, (D) *Okapia johnstoni*, (E) *Bubalus arnee*, (F) *Ovibos moschatus*, (G) *Sus scrofa*, (H) *Helarctos malayanus*

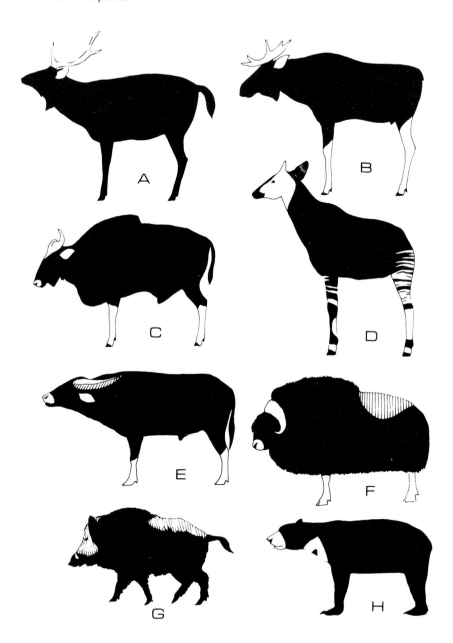

an anti-predator defence strategy (Hamilton 1971, Treisman 1975, Geist 1982). Such gregariousness is not compatible with surface damaging weapons (Geist 1978a,b), and caprids regress dagger-like horns and evolve parrying-type weapons that now function as shields, just as do other mammals. *Rupicapra* leads off with curved horns that basically 'blunt' the thrust; *Hemitragus* is more advanced, with larger horns which function as weapons and as defences. This trend accelerates in the genera *Capra* and *Ovis* (see Figure 1.2).

Deactivating 'dagger-like weapons' may have happened several times independently: once in *Rupicapra*, possibly in the Villafranchian *Procamptoceras*, whose short, very stout, closely spaced, frontward curving, sexually dimorphic horns (Kurten 1968) suggest parrying. It happened again in the Caprini, and again in the shrub- and musk-oxen. In the latter two lineages, horns and heads, in becoming parrying weapons, set the stage in extreme environments to become large luxury organs symbolic of the male's ability to procure resources (Geist 1977, 1978a, 1986a).

In the absence of resource defence, males usually compete for females within dominance hierarchies. This sets the stage for ecological segregation of males and females outside the mating season. This occurs because females must maximise *security* for the young, and males must maximise *body growth* in order to gain dominance. Therefore, females should stay on 'secure' areas and graze these closely. That would make such areas unsuitable for males since they require more feed per unit time. Consequently 'hungry' males leave for less secure, but richer areas. That is, females trade off security for food, and males trade off food for security (Geist 1982). This is possible since females can only *optimise* neonatal growth, since they must not produce too large a young as such are likely to die of dystocia (see Geist 1971a for review, pp. 284–7). Males, however, can *maximise* body growth, and thus can safely utilise more food. This predicts that males will forage further from escape terrain than females, that they will forage more opportunistically, that on average they will feed on better forage, that males will be more frequently killed by predators, and that males will be more sensitive to food shortages. The first, second and third predictions were verified by Shank (1982) and Francisi *et al.* (1985) for bighorn sheep and alpine ibex, and the fourth by Hornocker (1970) for wapiti (*Cervus elaphus nelsoni*). That males are sensitive to food shortages is indicated by the findings of Alados (1985) that Spanish goat males separate from females at *high*, but not at *low* densities.

Whether males segregate in winter from females so as to minimise competition with their offspring (either growing as fetuses in the females, or having been raised in the preceding summer), as suggested by Geist and Petocz (1977), is contentious. In line with this hypothesis is the finding that the older bighorn rams get, the less time they spend on female ranges (see Geist 1971a, Figure 24, p. 173).

The evolution of large horns accelerates as species leave cliffs to forage some distance beyond; small-horned caprids — excepting the chamois (Elsner-Schack 1985) — stick close to rocks, whereas large-horned ones roam well beyond cliffs. To reach safety, individuals must *sprint* to steep cliffs. The body form changes from a climbing to a saltatorial form, and ultimately to a cursorial one (shrub-ox/musk-ox type). However, with bounding and sprinting the young must become a capable runner. It must be large and well developed at birth, and have a supply of rich milk to grow rapidly into a speedy, enduring runner. In such circumstances the female must develop a great ability to spare resources from growth towards reproduction. This ability in males allows large horns, in that horns 'absorb' the surplus nutrients, being very sensitive to nutrition. Consequently, cursorialism, large neonatal weight, and rich (high milk solids) milk go together (Geist 1986a), reaching a maximum in species that rely almost purely on running, such as argali sheep (*Ovis ammon*). This subject requires urgent attention; limited, unpublished data suggest that in bovids the size of horn-like organs is indeed associated with high neonatal weight and high milk solids, as is the case in cervids (see Figure 1.8).

Large horns, secondarily, can become sledgehammer-like weapons. With that they can become symbols of dominance (Geist 1966), not just symbols of greater resource procurement (Geist 1971a, 1977, 1978a), or — where horns grow larger with age — symbols for living securely on home ranges relatively safe from predators. Consequently, in bighorns young males following large-horned males choose to follow to rich or secure ranges, or both. Large-horned males, however, should not encourage others to follow them because the followers would consume the rich forage resources of their limited home ranges. This the followers overcome by mimicking oestrous females (Geist 1971a), parasitising the large-horned males' inhibitions to be aggressive towards females. Female sheep in groups are related by maternal descent, and consequently need not rely on such complex behaviour to gain acceptability. This may be the reason for the great difference in the social behaviour of males and females.

One thus observes a latitudinal/altitudinal progression from resource defence, on steep, shrubby or forested areas, to gregariousness in open, steep landscapes, to gregariousness in less steep, open landscapes and a concomitant change from climbing to saltatorialism, and finally to cursorialism. This leads from 'battle-dress' monomorphism, to great dimorphism in body size and secondary sexual organs with saltatorialism, to a return to 'battle-dress' monomorphism in cursorial, open country forms (i.e. musk-oxen). In this caprids act like African antelope (Estes 1974). Battle-dress monomorphism is a function of intersexual aggression or of cooperation in defending a common territory (Geist 1974b). That is, either male and female have confrontations, as in *Oreamnos* (Geist 1965, Kuck 1977, Chadwick 1983), domestic horses (Berger 1986) or hyenas (Hamilton *et al.*

1986), or male and female cooperate in resource defence (Geist 1974b, 1977, 1978a).

Resource defence, climbing, saltatorialism and cursorialism follow a gradient of increasing visibility to predators as cover declines. We expect along this gradient increasing neonatal size, richness of milk and horn size (Figures 1.6–8), except for a decline in extreme and unproductive habitats where few predators are expected.

Olfactory marking is expected to change from the marking of localities and objects (i.e. localised dung deposition as in serow: Engelmann 1938, Yamamoto 1966; marking of branches or the female with preorbital or cornual glands as in serow, chamois, and mountain goat: Engelmann 1938, Geist 1965, Kramer 1969, Akasaka 1974, Knaus and Schroeder 1983, Lovari 1985) to marking of 'self' (i.e. rutting pits in mountain goats, urine spraying in goats: Shank 1972; and chamois: Kramer 1969, Lovari 1985) and marking of companions (i.e. self-marking by subordinate male bighorns on the preorbital glands of dominant males, Geist 1971a). This progression is also noticeable in cervids and in African antelopes.

The causes of body-colour variation remain largely to be explored, though W.J. Hamilton (1973) made a stimulating though contentious attempt at synthesis. His views on heat load explain why caprids from dry, warm climates are light, with consequently strong markings (wild and

Figure 1.6 Neonatal mass in ungulates (irrespective of phylogeny) as a function of 'openness' of their habitat. The largest young are born to species from open plains with a high biomass, where most predators are expected (Geist 1987)

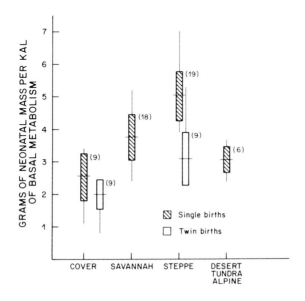

Figure 1.7: Neonatal mass as a function of relative antler mass in deer (Geist 1987)

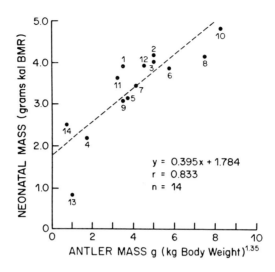

Figure 1.8: Per cent milk solids as a function of relative antler mass in cervids

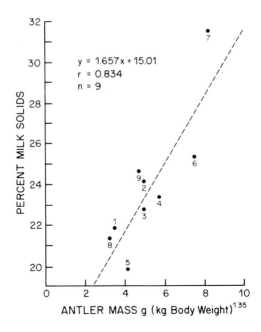

Spanish goats, Nubian and Walia ibex, most urials and continental mouflons, Abruzzo and Pyrenean chamois). Colour is most homogeneous in cold-climate caprids, usually dark (Alpine ibex and chamois, Siberian ibex and argali, turs, Japanese serow, Manchurian goral, most pachycerine sheep); very dark and white forms can exist side by side (Dall's and Stone's sheep, mountain goats and bighorns). This remains to be explained.

Ecologically, we expect a progression from concentrate to roughage feeding with intrusion into open landscapes, and a concomitant change in the morphology of teeth and digestive tracts. We expect a loss of ability to detoxify plant toxins with grazing.

We expect evolutionary age to follow approximately the same progression, with rupicaprids being the oldest forms, and the cursorial forms by and large the youngest. And that is found in the fossil record (Thenius and Hofer 1960, Kurten 1968, Thenius 1969, Kurten and Anderson 1980). Serow or goral-like species appear in the Villafranchian of Europe, where they survive the Pleistocene faunistic turmoil only as specialised island dwarfs (Sardinian goral, cave goat). The Formosan serow (*Nemorhaedus swinhoei*) and the Japanese serow (*N. crispus*), are likely results of that early radiation. The goral's zoogeography suggests that it appeared after Formosa and the Japanese islands were segregated from the mainland. The Caprini and Oviboni appear with the early glaciations and spread widely, even reaching North America. All early Oviboni (*Praeovibos priscus*, *Symbos cavifrons*, *Megalovis latifrons*, *Soergelia elisabetha*, *Euceratherium collinum*) are very large cursorial forms. The takin (*Budorcas taxicolor*), close to the geographic origin of the rupicaprids, may therefore be regarded as a primitive shrub oxen (Hofmann, submitted). Tahr appear early in the major glaciations. Ibex and large sheep appear in the middle of the large glaciations but as genera they may be earlier; *Capra camburgensis* is the first ibex in Europe and it combines features of other goats, including Spanish ibex. At the same time large sheep make their entry into Europe (Korobitsyna *et al.* 1974). These are supplanted in late glacial times, prior to the last extensive (Würm) glaciation, by modern ibex, mouflons and *Rupicapra*. Exact chronological ageing is problematic, but the sequence is evident. In North America *Ovis* appears to be present by 350 000 BP, but does not become abundant till late glacial times; *Oreamnos* has a poor fossil record, but it is present during the last glaciation and expands geographically just prior to the last glacial maximum (see Geist 1985a).

The order in which caprids appear in *time* is thus the same as the order they occupy in *space*; from their centre of origin in south-east Asia, caprids radiated outwards so that the most recent and advanced species are on the outward fringes. There is evidence of repeated radiation, in both the geographic and fossil records (Kurten 1968, Geist 1985a, Masini 1985). However, repeated evolutionary radiations from the same glacial refugia are predicted by the dispersal theory of speciation (Geist 1971a,b). Note also the parallel in evolutionary advancement and geographic position between caprids and *Cervus* (Figure 1.3), which is also duplicated in the fossil record. Primitive caprids arise with primitive cervids, and advanced caprids arise with advanced cervids. One could repeat this with other lineages.

Where predation is heavy, we expect exceptionally large-horned, cursorial caprids (Geist 1986a). The Rancholabrean of North America was characterised by many large bodied, cursorial predators, as well as a great diversity of predators. It thus follows that the euceratheres, as well as the musk-oxen, should be very large in the Rancholabrean, and very large-horned — as were bison, stagmoose (*Cervalces*), and the proboscidians, be they mastodonts or mammoths (see Kurten and Anderson 1980 for accounts of the Rancholabrean faunas), and this is found to be the case.

The evolution of behaviour within a species can be greatly affected by habitat, as is so well illustrated by Nievergelt's (1981) comparison of alpine ibex from Switzerland and Walia ibex from Ethiopia. In chamois the *rupicapra* subspecies from the Alps and the *ornata* subspecies from southern Italy differ noticeably (Lovari 1985). Conversely, there are also convergences; argalis and bighorns are very different sheep, but in social behaviour they are astonishingly similar (R.G. Petocz, pers. comm., films by Mongolian People's Republic). Here one should differentiate differences in the form of signals, as opposed to changes in social dynamics. The latter is likely to be purely circumstantial and not genetically fixed, but the former is most likely a product of natural selection, granted the great resistance of social signals to natural selection (Hofer 1972).

WEAPON DIFFERENTIATION

As indicated earlier, weapon morphology in ungulates can be divided into two principal types (Figure 1.2), one associated with the defence of material resources, and one with life in the gregarious 'selfish' herd (Geist 1978a,b). In essence, material resource defence hinges on weapons that allow the aggressor to inflict a maximum of severe surface damage (and pain) to the body of the opponent, without getting its weapons stuck or caught in the opponent's anatomy. Weapons in gregarious forms can be used to parry the blows of opponents, and often to hold on to the opponent's head as well. Secondarily, they may become rank display symbols as in American mountain sheep, and possibly organs of sexual selection (Geist 1966, 1971a, 1986a).

Weapons that identify resource defenders are sharp tusks (as in *Moschus*), lacerating dentition (*Hypertragulus*), short, sharp horns or antlers (*Oreamnos, Cephalophus, Mazama*), or short ossicones (*Okapia, Giraffa*). By this criterion, all rupicaprids proper should be resource defenders, be it obligatory or facultative. That is, at some time in the seasonal cycle, or in the lifetime of an individual, we expect them to defend a *resource territory.*

As indicated earlier, we also expect resource defenders to be monomorphic, to reduce neonatal mass, to hide the young, to extend the gestation period, and to reduce birth number the shorter the resources and the more frequently individuals must invoke resource defences. These expectations have now been met by recent research on the Japanese serow. With a birth weight of 2.5 kg (A. Komori, personal communication) and an adult weight of 40 kg for females, the neonatal investment is only 2.25 g/kcal of basal metabolic rate (BMR) (1.96 g/kcal BMR for the Formosan serow: Chen Pao-Chung, Chapter 13, this volume; 2.4 g/kcal BMR for mainland serow at 4.1 kg neonatal mass and 70 kg maternal weight: Meckvichai, this volume); for the long-tailed goral it is only 2.0 g/kcal BMR (see Geist 1981, p. 114). This compares with an average of 3.5 g/kcal BMR for ungulates (Geist 1981, p. 113). The gestation period of over 200 days (210–240 days for Formosan serow, Chen Pao-Chung, Chapter 13 this volume) is long for so small a ruminant, and as long as in the much larger mainland serow (210–230 days, Meckvichai, this volume). Reproduction may even be biannual in the Japanese serow (Miura, Chapter 23, this volume). Sexual maturation in captive serow is relatively slow, in that five out of 29 females gave birth at 4 years of age or older (A. Komori, personal communication). Sexes are identical in size and appearance, and are monomorphic in food habits (Suzuki and Takatsuki 1986). The brain has simple convolutions compared with the domestic goat (Atoji, this volume) and at 141 g (12.3 g/wt kg to the power of 0.66) is quite small. The serow thus fits the expectations of a resource defender.

As indicated earlier, caprids twice surged into open landscapes where the chief anti-predator strategy is to group into 'selfish herds' (Hamilton 1971). In these, surface-damaging weapons are selected against as incompatible with gregarious life; injury of conspecifics attracts predators and loss of herd members reduces the mathematical dilution effect, increasing the chances of predation on the remaining herd members. Selection is for weapons allowing ready parrying, which then evolved into clashing weapons in the Caprini and Oviboni. In general, short, sharp weapons are selected against in 'selfish' herds, whether long canines or incisors, short dagger-like horns, antlers or ossicones, and are replaced by parrying or grappling weapons (Geist 1978a,b). Such weapons in turn can evolve into 'luxury organs' via female choice (Geist 1977, 1978a, 1982, 1986a) or into rank symbols (Geist 1966, 1971a). Cursorialism selects for large young (Figure 1.6); horn size tracks the female's ability to produce large young (Figure 1.7) and rich milk (Figure 1.8).

In resource defenders, defences against weapons take two basic forms, evasion — so as to avoid injury, or catching the blow against body armour

ON THE EVOLUTION OF THE CAPRINAE

— so as to reduce injury (Geist 1966, 1971a, 1978a,b). Which option is followed depends on body size; small-bodied species (i.e. *Rupicapra*) can readily evade attacks due to the favourable force to mass ratio. Large-bodied species cannot readily accelerate away due to an unfavourable force to mass ratio, which sets the stage for the evolution of dermal armour (i.e. *Oreamnos*, Geist 1965, 1967, 1971a). Sheep and goats, which catch horn blows with their heads, have thick skins about their heads and necks, but not on their bodies (Geist 1971a). The thickness of skin in rutting serow, goral, tahr, takin or chamois has not been reported on.

We have thus two extremes in weaponry in caprids which are ecologically correlated. Where material resources are defendable on small, productive territories, there short, sharp, dagger-like weapons evolve in a syndrome of resource defence. Where material resources are widely and thinly scattered so as to be undefendable, males may advertise their resource-procuring ability via large horns that are sensitive to nutrition (Geist 1977).

ANTI-PREDATOR STRATEGIES

As indicated earlier, one can recognise three fundamental anti-predator strategies in caprids:

(1) hiding in thick vegetation or broken rock rubble and cliffs,
(2) placing obstacles in the way of pursuing predators, and
(3) maximising distance between predator and prey.

There are variations and combinations of these strategies, and in large bodied species confronting predators can become a feasible fourth option, such as the famous defensive ring of musk-oxen facing wolves, sled dogs and humans (Pedersen 1958, Gray 1974), or the attacks of serow on bears and hunting dogs (Engelmann 1938, Akasaka 1974).

The anti-predator strategies assumed are a function of the type of landscape exploited, so that browsers in highly productive areas may hide, climbers in open cliffs may specialise in rapid climbing or in jumping, species exploiting grasslands around cliffs mix jumping with sprinting (saltatorial), and species in open landscapes away from cliffs may become cursorial; they may also confront predators.

In the hiders in tropical areas such as *Nemorhaedus* one finds fairly primitive, that is, generalised, legs compared with highly specialised climbers such as *Oreamnos* and *Budorcas*; specialised climbing as opposed to jumping evolves where snow or vegetation obscures footholds, making exploration of each foothold mandatory. Climbing may also be mandatory

24

with big body size since footholds may give way under force of a rapidly decelerating, heavy body. In short, body size may constrain anti-predator strategy, or generate opportunities.

This point is well illustrated in saltatorial (New World) as opposed to cursorial (Old World) sheep. In saltatorial species, the focus of attention is on escape terrain, usually cliffs. Therefore an individual exploits forage resources within a narrow radius around a safe site. This distance is the 'safe distance' across which it can just barely sprint to safety, should it be surprised by a predator. Such species must be built to accelerate rapidly. However, since mass increases as the cube of linear dimensions, but muscle power only as the square of linear dimensions, the larger-bodied a species, the less power it has to accelerate body mass (Thompson 1961). Only in small species is the power to mass ratio favourable for rapid sprinting to secure cliffs.

The upper limit for 'accelerators' appears to be the upper weight of bighorn sheep or ibex (about 120 kg). That weight, however, appears to limit birth rates in cold climates. The survivable size of sheep neonates in cold climates is 3−4 kg. This is met by the smallest pachycerine sheep, the Dall's sheep of northern Canada and Alaska. Females weigh about 50 kg, males about 90 kg. To double the birth rate would be to produce two 4 kg lambs, or 8 kg of neonatal mass. This requires a female double the *metabolic mass*, that is, not $2 \times 50 = 100$ kg, but $2.54 \times 50 = 127$ kg (Geist 1981). The size of the female sheep now is that of a Siberian argali, which is at, or above, the safe size for 'accelerators'. In order to ingest twice the amount of food, the size of the home range has to be doubled; the minimum distance by which the home range must be increased is 1.42 (if the sheep forages about a point source) or 2.0 if the cliff is linear. That is, the large sheep must forage 1.42−2.0 times beyond the 'safe' distance from cliffs 50 per cent of the time. The large sheep would therefore be disadvantaged not only by reduced acceleration, but also by the greater distance to safety. Consequently sprinting to safety in cliffs becomes impossible for a sheep twice the metabolic size of a Dall's sheep. The large sheep must turn to cursorialism for safety.

Consider, if body mass increases 2.54 times, then linear dimensions increase 1.36 times and the individual has only $(1.36)^2 = 1.85$ times the power to push $(1.36)^3 = 2.52$ times the mass. This is a 26 per cent reduction in the power to accelerate.

The shift to cursorialism imposes within reason little constraint on body size, although body proportions do change with size in cursorial forms (V. Geist, in prep.). Cursorial Oviboni can be much larger than cursorial/saltatorial argalis, which can be larger than saltatorial sheep or ibex. Yet argalis are strictly the size needed to bear *two* 4 kg lambs, whereas mountain sheep are of a size needed to bear *one* such lamb (Geist 1981).

PREDATION EFFECT

As pointed out in Geist (1978a, 1983) and as I shall discuss later, we expect hypermorphs to evolve rapidly during a colonisation pulse, and to be 'fine-tuned' subsequently by *efficiency selection* under conditions of resource shortage (see Figure 1.11). Body size is expected to decline subsequent to the formation of the new form. How far can body size drop?

In the absence of predation, giant deer and elephants dwindled to tiny body sizes (Azzaroli 1982). In the presence of a large diversity of huge predators, Siberian species such as *Bison, Cervalces*, Mammutus, as well as the musk-oxen (*Symbos*) and euceratheres grew in North America to large size and carried very large horn-like organs (Kurten and Anderson 1980). With the extinction of the gigantic Rancholabrean predators and the ascent of the smaller Siberian predators, *Bison, Alces* and *Cervus* shrank to their current small body and horn/antler sizes (Wilson 1980, Guthrie 1984). It therefore appears that large predators generate hypermorphs, small predators generate small herbivores, and lack of predators generates paedomorphic dwarfs.

PRODUCTIVITY PULSE

Body size is a function of various factors, one of which is the duration of the annual productivity pulse in forage (Geist 1978a, 1986a, Guthrie 1984). In seasonal climates a vegetation pulse in summer allows individuals to escape from forage competition for a short time, so that they can ingest high-quality forage freely. In the Arctic this freedom from want is short, because the pulse is short. In the humid tropics there may be no productivity pulse at all; in the dry tropics the pulse may last the rainy season. In temperate latitudes the pulse should be the longest (Figure 1.9).

Margalef (1963) equated the *height* of the pulse with body size, but it appears to be the *length* that is related to size. If so, then body size becomes curvilinear with latitude, invalidating Bergmann's rule (Figure 1.10). This also predicts that in mountains, where migratory populations can exploit a large range of altitudes, body size will be larger than in populations which can exploit only a small range of altitudes (Shackleton 1973). Still, the height of the pulse (a function of quality or digestibility of the forage) may not be irrelevant, since within a latitude increased food abundance does lead to increased body size. This happens to deer on seral vegetation following a forest fire, in areas with severe predation, or during colonisation of vacant habitat as shown by chamois in New Zealand (Bauer 1985). We also expect in tropical areas that species exploiting plant communities along flooding rivers will be larger in size than close relatives on dry uplands. In view of the great plasticity of body size with environmental

Figure 1.9: Productivity pulse in relation to maintenance and growth levels of food intake (Geist 1986b). (A) This explains the general concept. (B) Productivity pulses at various latitudes, and their respective lengths

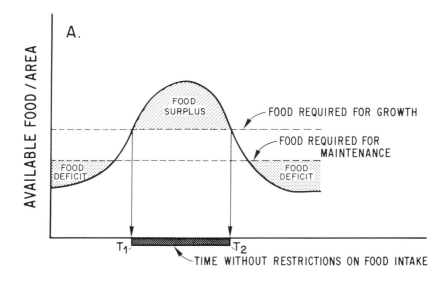

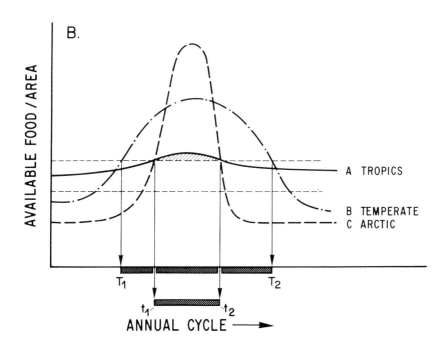

Figure 1.10: Relative body mass in cervids as a function of latitude

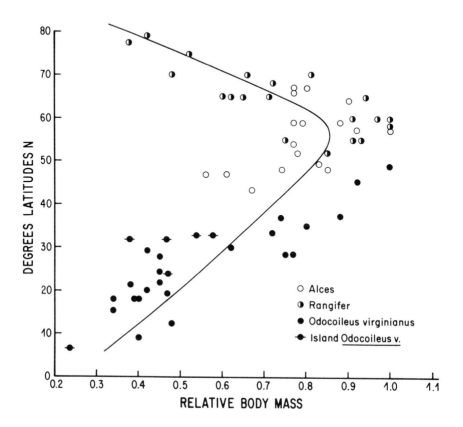

factors, and because body size is closely linked to shape, size and shape are weak criteria in taxonomy (Ingebrigtsen 1923).

PHENOTYPE PLASTICITY AND SPECIATION

To understand both hypermorphic and paedomorphic evolution, an understanding of phenotype plasticity is essential. It is becoming evident that genomes are reactive systems that interact with the environment through the phenotype in order to adjust the phenotype adaptively (Geist 1978a). Extremes in phenotypes have been identified by a number of investigators and related to environmental variation (Ingebrigtsen 1923, Beninde 1937, Gottschlich 1965, Geist 1971a,b, 1977, 1978a, 1983, Watson and Moss 1972, Shackleton 1973, Ellenberg 1978, Packard and Mech 1983).

Extremes in phenotype are related to extremes in availability of material resources, 'dispersal phenotypes' being related to abundance, 'maintenance phenotypes' to scarcity (Geist 1971a, 1978a, Ellenberg 1978). Beninde (1937) pointed out that red deer, reintroduced to a given region, soon adopt the phenotype typical of red deer formerly occupying that region. This conclusion was reached independently by James (1983) in experimental studies with birds. The earlier work of Ingebrigtsen (1923), Vogt (1936, 1948), Vogt and Schmidt (1951) and Beninde (1937), followed by the post-war studies of Szunyoghy (1963) and Gottschlich (1965), led Central European workers to abandon most subspecies of red deer, and regard the current geographic variation as ecotypic (Wagenknecht 1981, Vorreyer 1982). The phenotype plasticity discovered by Central European zoologists was independently confirmed and expanded upon by animal scientists, and formalised in the 'centripetal theory of growth' (Hammond 1960).

The foregoing research established that size and shape were linked, and fluctuated with material resource availability along a phenotypic hypermorph/paedomorph axis. Red deer, for instance, vary in mainland Europe five-fold in mass (which equals 1.5 times in linear dimensions: Wagenknecht 1981). What does not vary environmentally is the expression of social organs, namely rump patch shape and size, body colour and hair patterns, shape of antlers (a valid criterion only in phenotypic hypermorphs), vocalisations, and — probably — pheromones. Therefore, while size and shape are weak taxonomic criteria, the expression of social organs is not.

At present we are poorly informed about size plasticity, but a spread in size as large as that observed in red deer is unlikely in caprids, at least in saltatorial species. As shown above, size would affect acceleration. Bighorn sheep during late glacial colonisation of lower North America were larger than today's bighorns, but only by about 10 per cent linear dimensions (see Geist 1985b). Size variation in mouflons (Pfeffer 1967) and *Rupicapra* appears to be modest (Knaus and Schroeder 1983). A good comparison would be adult chamois from dispersing and stable populations in New Zealand.

A glance at the geographic dispersion of clinal variation in various mammalian families shows a hypermorph/paedomorph variation, *interspecifically*, with latitude/altitude. However, social evolution also occurs with increasingly open habitat, so that the hider/cursorial axis is also one of increasing 'ornateness'. The reason for this axis was given in Geist (1986a). In essence 'luxury organs' become adaptive as signals of ability to procure material resources by the male since this 'excess capacity' is adaptive to females to bear huge neonates and produce rich milk, which are essential to the survival of cursorial young. Sexual dimorphism in size and external appearance follow this hider/cursorial axis in curvilinear fashion (Estes

1974, Geist 1974b, 1977). That is, the hiders in thickets and the extremely social, cursorial forms of the open plains both display male-image monomorphism. This monomorphism in hiders is related to communal resource defence (Geist 1974b) whereas in the cursorial forms it is a function of intersexual competition (Geist 1974b, 1978a, Berger 1986, Hamilton *et al.* 1986).

In the geographic clines we thus have a confounding of two axes of directional evolution, namely, the hiders/cursorial axis, and the hypermorph/paedomorph axis related to the productivity pulse with altitude/latitude. One must therefore compare forms latitudinally/ altitudinally within the same habitat type.

Irrespective of this, the evolution of hypermorphs is described as a rapid, pulse-like change in body size and social organs as a consequence of colonising unexploited, rich habitat (Geist 1966, 1971a, 1971b, 1977, 1978a, 1983). In essence, colonisers, upon meeting superabundant forage, change to large dispersal phenotypes and switch from material resource competition to social competition. All individuals are expected to reach their potential genetic size due to resource abundance. This exposes the genetic variation in size to natural selection. Since body size is highly adaptive in combat (Clutton-Brock *et al.* 1984), this leads to sharp selection for body size, as well as to sturdy combat adaptations. There will also be selection for maximum reproduction, and efficiency of social communication. Intense selection for body size should lead to phenodeviants (Wilson 1975), and a broadening of variation. There should be an evolution of new adaptations, as a consequence of the dispersal phenotype's readiness to explore, experiment and innovate (Geist 1978a).

Once colonisers have filled the (large) area they colonised, they revert to maintenance phenotypes as resources become scarce. These are *efficiency* phenotypes that must spare material resources from growth and maintenance for reproduction. There is consequently continuous natural selection for the ability to reproduce successfully at smaller and smaller body sizes. Consequently, over geologic time, the species will decline in size and will also evolve continuously ever more efficient organs of food processing and procurement (i.e. larger teeth in *Bison*, specialised teeth as in *Myotragus* (Kurten 1968), or larger probosci as in tapirs, macrauchenids, oreodonts (Scott 1937).

The decline in body size is halted by predation, as indicated earlier. We thus expect a quick formation of a 'new design'; this new form is geographically removed from its parent species; the new form is largest in body size right at the outset; this is followed by a gradual, long decline in body size and social organs and a specialisation in food-procuring organs (Figure 1.11). In short, speciation is rapid in hypermorphs, but gradual in paedomorphic species. There may also be hybridisation with related species as part of the speciation process (Bartos and Zirovnicky 1981, 1982, Bartos *et*

Figure 1.11: A model of hypermorph evolution; rapid 'speciation' during colonisation is followed by shrinkage in size and 'fine tuning' of adaptations due to efficiency selection in maintenance phenotypes (Geist 1978a, 1983)

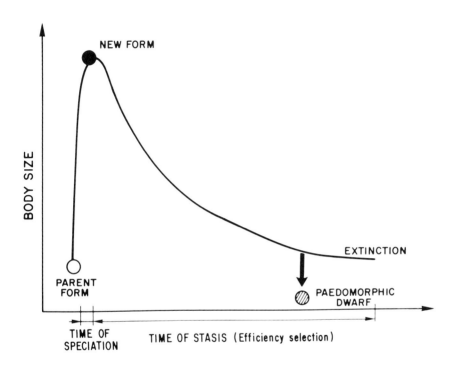

al. 1981). How sympatric species of the same social plan, but segregated by ecological adaptations, came to be is not readily discernible in large mammals; it may have followed the classical neo-Darwinian model (Mayr 1966).

It is vital to recognise that phenotype plasticity implies that epigenetic mechanisms protect the genome against natural selection, making Darwinian evolution (irreversible genetic change via natural selection) a rare event (Geist 1978a). Such evolution is expected only under extreme environmental circumstances. At the level of structural genes, change is expected to be largely random; phenotypic adaptability thus ensures the existence of DNA clocks.

31

TESTING WITH CYTOLOGY, IMMUNOLOGY AND MOLECULAR BIOLOGY

The pattern of social evolution with geographic dispersal described here has been subjected to detailed tests using cytogenetic, immunologic, electrophoretic and molecular techniques, best summarised by Hight and Nadler (1976). A series of important, international, collaborative studies have been undertaken, some of which are Lay *et al.* (1971), Nadler *et al.* (1973, 1974), Korobitsyna *et al.* (1974), Nadler and Bunch (1977) and Bunch *et al.* (1978).

At the protein level, *Capra* and *Ovis* are exceedingly similar, more so than most mammalian species. Cytogenetic studies of chromosome banding led to the same conclusion, but the pattern of Robertsonian fusions of chromosomes did differ within and between genera, and followed largely, but not entirely, the zoogeographic, morphological and behavioural pattern proposed by Geist (1971a). *Capra* had at $2n = 60$ the oldest karyotype; *Ammotragus* and urial sheep had the same, homologous $2n = 58$ karyotype, and were close serologically (Schmitt 1963), but distant from *Capra*; argalis differed by one extra fusion, having $2n = 56$; bighorns, Dall's sheep, mouflons and domestic sheep all had an additional fusion, and $2n = 54$; snow sheep from the Kolyma Highland were the surprise at $2n = 52$, implying derivation from *Ovis dalli*. *Pseudois* also had $2n = 54$, but its chromosome fusions were not homologous with those of the *Ovis* $2n = 54$ karyotype. *Pseudois* and *Capra* do not produce viable hybrids, neither do *Ammotragus* and *Ovis*, but *Ammotragus* and *Capra* occasionally do (Petzsch 1956, Bunch *et al.* 1977). Studies by Hight and Nadler (1976) using immunodiffusion techniques did not give as clear a pattern as the cytological studies. *Ammotragus* was more distant from *Ovis* and *Capra*, and each genus shared characteristics with one that it did not share with the other. The distance between bighorns and Dall's sheep was greater than between bighorns, mouflons and domestic sheep. It appeared that geographic neighbours were more distant immunologically than forms more distant in space. Does the chance to meet and hybridise enlarge immunologic distance?

These studies largely confirmed the pattern of Caprini evolution as proposed by Geist (1971a, pp. 313–46). *Ammotragus* is at the root of *Ovis* and *Capra* evolution; karyotypes evolved parallel with social characteristics showing geographic clines; caprids radiated from glacial refugia repeatedly showing convergences; as predicted, argalis were similar to bighorns due to convergence, and derived from urials; sheep and goats identified as primitive or advanced on morphological and behavioural criteria were with one exception judged so by cytogenetics as well. However, there were two surprises: mouflons, which appeared primitive, were a recently evolved form, albeit a throwback towards ancestral characteristics via paedomorphism. This was also confirmed palaeontologically. Secondly, snow sheep were found to have a karyotype of $2n = 52$. This implies that they are derived from Dall's sheep, and are late Pleistocene westward migrants from Alaska/Beringia into eastern Siberia. The snow sheep tested were from a region (Kolyma Range) where rather advanced snow sheep are found, as had been noted by Geist (1971a, pp. 344–5), and as was acknowledged by Korobitsyna *et al.* (1974). The most primitive pachycerine sheep is apparently the Kamchatka variety, which should have a $2n = 54$ karyotype. This remains to be determined. The cytogenetic studies of Soma (this volume) on rupicaprids confirm the trend in *Ovis* that chromosome fusion accompanies geographic dispersion.

Recently an interdisciplinary study of *Rupicapra*, combining morphology, zoogeography (Lovari and Scala 1980, Scala and Lovari 1984), palaeontology (Masini 1985), behaviour (Lovari 1985), and protein analysis (Nascetti *et al.* 1985) demonstrated that there was an earlier radiation (*R. pyrenaica*), followed by a later radiation (*R. rupicapra*). Both radiations came from the east, the second almost obliterating the first. All these results confirm that the analysis of speciation in Ice Age mammals as undertaken here does lead to verifiable conclusions.

The speciation of caprids is a problem that palaeontology, by itself, cannot solve for us; if protagonists in the quarrel about gradualism versus punctuated equilibrium agree on anything, it is that the fossil record is much too coarse to reveal the process of speciation (Vrba 1980). The fossil record is often also too fragmentary for reliable identification (Churcher 1984). We must therefore search for new methods accessible to neontology to investigate the dynamics of the evolutionary process, using the fossil record to test our hypotheses. That I have attempted to do.

POSTSCRIPT

Since this chapter was written, the discovery of a relict megafauna in southern Sapin has thrown new light on the origins of the hitherto mysterious Spanish ibex (*Capra pyrenaica*). Meunier (1984) reported the rediscovery of the *angulatus* stag in the Sierra Morena. This form of red deer is well known from the interglacial deposits of the Holstein interglacial of Central Europe dating back to about 230 000 years BP. It was replaced by west elaphine red deer with modern antler forms in the subsequent Riss glaciation. In the very same region in Spain was discovered a new form of roe deer (*Capreolus*, not yet named), which was identified as an Ice Age relict by Meunier, as was the *angulatus* stag.

Also found in southern Spain is the Spanish ibex. In the fossil record of Central Europe there appears, about 230 000 BP, along with the *angulatus* stag, the very first ibex (*Capra camburgensis*). This ibex, according to

Kurten (1968), shows affinities to the Spanish ibex. Since the latter occupies a glacial refugium along with two relict deer, it is likely that it too is a relict species, namely the first of the European ibex. Also, the presence of the 'ornate' form of chamois (*Rupicapra pyrenaica*, see Nascetti *et al.* 1985) in Spain is logical if this form is a relict of the first major chamois advance, as suggested by Lovari and Scala (1980). The ornate chamois thus *predates* the Riss glaciation, whereas the Alpine chamois (*R. rupicapra*) probably appeared with the advanced red deer in the Riss glaciations.

The above enables the following conclusions to be drawn.

(1) It allows a preliminary dating of the relict megafauna of Spain, and the establishment of the modern fauna of Central Europe.
(2) It confirms that faunas go together, so that *angulatus* deer go with Spanish ibex, ornate chamois and primitive roe deer, and west elaphine red deer go with Alpine ibex, Alpine chamois and advanced European roe deer.
(3) It shows that the European late Pleistocene megafauna did not cross via Gibraltar into Africa, but via Sicily. Otherwise the *angulatus* stag and the Spanish ibex would have been in the Atlas Mountains, rather than the Barbary stag (*Cervus elaphus barbarus*), a very close relative of today's continental red deer, and the audad *Ammotragus*.
(4) It suggests that the Alpine ibex spread radially from Central Europe in the Riss glaciation, explaining why the west Caucasian tur (*Capra caucasica*) is similar to the Alpine ibex and so very different from the east Caucasian tur (*C. cylindricornis*). It is also logical that the Caucasian chamois (*Rupicapra rupicapra caucasica*), which differs little from the Alpine chamois (Heptner *et al.* 1961), should be sympatric with the west Caucasian tur.
(5) It suggests that the east Caucasian tur is part of an old radiation of round-horned goats, supporting the conclusions of Geist (1985a) that *Pseudois* is an old form.

REFERENCES AND FURTHER READING

Akasaka, T. (1974) Japanese serow in the wild. *Wildlife 16* (10), 452–8
Akasaka, T. (1977) Food habits and feeding behavior of Japanese serow in Nibetsu Akita prefecture. *Wildlife Conservation in Japan* 67–80
Akasaka, T. and N. Maruyama. (1977) Social organisation and habitat use of Japanese serow in Kasabori. *J. Mammal. Soc. Japan 7* (2), 87–102
Alados, C.L. (1985) Group size and composition of the Spanish ibex (*Capra pyrenaica* Schinz) in the sierras of Cazorla and Segura. In S. Lovari (ed.) *The biology and management of mountain ungulates.* Croom Helm, London, pp. 134–47

Altmann, D. (1980) Verhalten's Studien am Mishmi-Takin (*Budorcas taxicolor taxicolor* Hodgson) im Tierpark, Berlin. *Milu, Berlin 5* (3), 342–58

Azzaroli, A. (1982) Insularity and its effects on terrestrial vertebrates: evolutionary and biogeographic aspects. In E.M. Gallitelle (ed.) *Paleontology, essentials of historic geology.* S.T.E.M. Mucci, Modena (Italy), pp. 193-213

Barrette, C. (1977a). The social behaviour of captive muntjacs *Muntiacus reevsi* (Ogilby 1839). *Z. Tierpsychol. 43*, 188–213

Barrette, C. (1977b) Some aspects of the behaviour of muntjacs in Wilpattu National Park. *Mammalia 41*, 1–34

Bartos, L. and J. Zirovnicky. (1981) Hybridisation between red and sika deer. II Phenotype analysis. *Zool. Anz. Jena 207*, 271–87

Bartos, L. and J. Zirovnicky. (1982) Hybridisation between red and sika deer. III Interspecific behaviour. *Zool. Anz. Jena 208*, 30–6

Bartos, L., J. Hyanck and J. Zirovnicky. (1981) Hybridisation between red and sika deer. I Craniological analysis. *Zool. Anz. Jena 207*, 260–70

Bauer, J.J. (1985) Fecundity patterns of stable and colonising chamois populations of New Zealand and Europe. In S. Lovari (ed.) *The biology and management of mountain ungulates.* Croom Helm, London, pp. 154–65

Beninde, J. (1937) *Naturgeschichte des Rothirsches.* P. Schoeps, Leipzig, 223 pp.

Berger, J. (1986) *Wild horses of the Breat Basin.* University of Chicago Press, Chicago, Ill., 326 pp.

Brandburg, S.M. (1955) Mountain goat in Idaho. Wildlife Bull. No. 2. State of Idaho, Dept. of Fish & Game, Boise, Idaho, 142 pp.

Bubenik, A. (1966) *Das Geweih.* Paul Parey, Berlin. 214 pp.

Bunch, T.D., A. Rogers and W.C. Foote. (1977) G-band and transferrin analysis of audad-goat hybrids. *J. Heredity 68*, 210–12

Bunch, T., C.F. Nadler and L. Simmons. (1978a) G-band patterns, hemoglobin, and tranferrin types of the bharal. *J. Heredity 69*, 316–20

Bunch, T.D., T.C. N'guyen and J.J. Lauvergne. (1978b) Hemoglobins of the *Corsico-Sardinian* mouflon (*Ovis musimon*) and their implications for the origin of Hb A in domestic sheep (*Ovis aries*). *Ann. Genet. Sel. anima. 10* (4), 503–6

Bunnell, F.L. (1978) Horn growth and population quality in Dall sheep. *J. Wildl. Mgmt. 42*, 764–75

Chadwick, D.H. (1977) The influence of mountain goat social relationships on population size and distribution. In W. Samuel and W.G. Macgregor (eds) *Proc. 1st International Mountain Goat Symposium.* Fish and Wildlife Branch, Province of British Columbia, Victoria, pp. 74–91

Chadwick, Ph. H. (1983) *A beast the color of winter.* Sierra Club Books, San Francisco, CA., 208 pp.

Churcher, C.S. (1978) *Giraffidae.* In V.J. Maglio and H.B.X. Cook (eds) *Evolution of African mammals.* Harvard University Press, Cambridge, Mass., pp. 509–535

Churcher, C.S. (1984) Sangamona: the furtive deer. In H.H. Grenoway and M.R. Dawson (eds) *Contribution in Quaternary vertebrate palaeontology.* Carnegie Museum of Natural History. Special publication No. 8. Pittsburgh, PA, pp. 316–31

Clutton-Brock, T.H., F.E. Guinness and S.D. Albon. (1982) *Red deer.* University of Chicago Press, Chicago, Ill., 378 pp.

Cowan, I. McT. (1940) Distribution and variation in the native sheep of North America. *Am. Midl. Natur. 24* (3), 505-80

Davis, S.J.M. (1984) *Kirokitia and its mammalian remains. A Neolithic Noah's ark. Fouilles recent à Khirokitia (Chypre) 1977–1981 (2 tomes).* Recherches sur les Civilisations, Paris

Dolan, J.M. (1965) Beitrag zur systematischen Gliederung des Tribus Rupicaprini, Simpson 1445. *Z. Zool. Syst. Evolutionsforsch. 1* (3–4), 314–407

Ellenberg, H. (1978) Zur Populations Okologie des Rehes (*Capreolus capreolus* L. Cervidae). In *Mitteleuropa spixiana* (Suppl. 2), 211 pp.

Elsner-Schack, T. von. (1985) What is good chamois habitat? In S. Lovari (ed.) *The biology and management of mountain ungulates.* Croom Helm, London, pp. 71–84

Engelmann, C. (1938) Verber die Grossauger Szetschwan, Sikongs und Osttibet. *Z. Säugetierkunde 13,* Sonderheft, 76 pp.

Estes, R. (1974) Social organisation of the African Bovidae. In V. Geist and F. Walther (eds) *The behaviour of ungulates and its relation to management.* IUCN Publ. (New Sciences) No. 24, Vol. 1, pp. 166–205

Francisi, F., S. Focardi and L. Boitana. (1985) Male and female alpine ibex: phenology of space use and herd size. In S. Lovari (ed.) *The biology and management of moutain ungulates.* Croom Helm, London, pp. 124–33

Geist, V. (1965) On the rutting behaviour of mountain goat. *J. Mammal. 45,* 551–68

Geist, V. (1966) On the evolution of horn-like organs. *Behaviour 27,* 177–214

Geist, V. (1967) On fighting injuries and dermal shields of mountain goats. *J. Wildl. Mgmt 31,* 192–4

Geist, V. (1971a) *Mountain sheep.* University of Chicago Press, Chicago, Ill., 383 pp.

Geist, V. (1971b) The relation of social evolution and dispersal in ungulates during the Pleistocene, with emphasis on the Old World deer and the genus *Bison. Quat. Res. 1,* 285–315

Geist, V. (1974a) On the relation of ecology and behaviour in the evolution of ungulates: theoretical considerations. In V. Geist and F. Walther (eds) *The behaviour of ungulates and its relation to management.* IUCN Publications. No. 24, Vol *1,* 235–46

Geist, V. (1974b) On the relationship of social evolution and ecology in ungulates. *Am. Zool. 14,* 205–20

Geist, V. (1977) A comparison of social adaptation in relation to ecology in gallinaceous bird and ungulate societies. *Ann. Rev. Ecol. Systematics 8,* 193–207

Geist, V. (1978a) *Life strategies, human evolution, environmental design.* Springer, New York, 495 pp.

Geist, V. (1978b) On weapons, combat and ecology. In L. Krames, P. Pliner and T. Alloway (eds) *Aggression, dominance and individual spacing.* Plenum, New York, pp. 1-30

Geist, V. (1981) On reproductive strategies in ungulates and some problems of adaptation. In G.G.E. Scudder and J.L. Reveal (eds) *Evolution today. (Proc. 2nd Int. Congr. Systematics and Evolutionary Biology.)*

Geist, V. (1982) Adaptive strategies. In J.W. Thomas and D.E. Toweill (eds) *Elk of North America.* Stackpole Books, Harrisburg, PA, pp. 219–77

Geist, V. (1983) On the evolution of Ice Age mammals and its significance to an understanding of speciation. *ASB Bull. 30,* 109–33

Geist, V. (1985a) On Pleistocene bighorn sheep: some problems of adaptation, and relevance to today's megafauna. *Wildl. Soc. Bull. 13,* 351–9

Geist, V. (1985b) On evolutionary patterns in the Caprinae with comments on the punctuated mode of evolution, gradualism and a general model of mammalian evolution. In S. Lovari (ed.) *The biology and management of mountain ungulates.* Croom Helm, London, pp. 15–30

Geist, V. (1986a) The paradox of the Great Irish Stags. *Nat. Hist. 95* (3), 54–64

Geist, V. (1986b) On the evolution and adaptation of *Alces. Viltrevy* (in press)

Geist, V. (1987) On speciation in ice age mammals. *Can. J. Zool. 65,* (in press)

Geist, V. and R. Petocz. (1977) Bighorn sheep in winter: do rams maximise reproductive fitness by spatial and habitat segregation from ewes? *Can. J. Zool. 55,* 1802-10

Goldschmidt, R. (1940) *The material basis of evolution.* Yale University Press, New Haven, CT, 436 pp.

Gottschlich, J.H. (1965) Biotop und Wuchsform: eine Craniometrische Studie an Europaeischen Populationen von *Cervus elaphus. Beitrag zur Jagd und Wildforsch. 4,* 83-101

Gray, D. (1974) The defence formation of the musk-ox. *The Musk-ox. Inst. North. Studies, Univ. Sask. 14,* 25-9

Groves, C.P. and P. Grubb (1985) Reclassification of serow and goral. In S. Lovari (ed.) *The biology and management of mountain ungulates.* Croom Helm, London, pp. 45-50

Guthrie, R.D. (1984) Mosaics, allelochemics and nutrients (an ecological theory of late Pleistocene megafaunal extinctions). In P.S. Martin and S. Klein (eds) *Quaternary Extinctions.* University of Arizona Press, Tucson, Ariz., pp. 259-98

Halternorth, T. (1961) Lebensraum, Lebensweise und Vorkommen des Mesopotamischen Damhirsches, *Cervus mesopotamicus,* Brooke, 1875. *Säugetierkundl. Mitt. 9,* (1), 15-39

Hamilton, W.J. III. (1973) *Life's colour code.* McGraw Hill, New York, 238 pp.

Hamilton, W.J. III, R.L. Tilson and L.G. Frank. (1986) Sexual monomorphism in spotted hyena (*Crocuta crocuta*). *Ethology 71,* 63-73

Hamilton, W.R. (1973) The lower Miocene ruminants of Gebel Zelten, Libya. *Bull. British Museum (Nat. Hist.) Geology. 21,* (3), 150 pp.

Hammond, J. (1960) *Farm animals* (4th edn 1971). Edward Arnold, London 293 pp.

Heptner, V.G., A.A. Nasimovic and A.G. Bannikov. (1961) *Die Saugetiere der Sowjetunion.* V.E.B. Gustav Fischer edition 1966. Jena, 939 pp.

Hibbs, D.L. (1967) Foodhabits of mountain goat in Colorado. *J. Mammal. 48* (2), 242-8

Hight, M.E. and C.F. Nadler. (1976) Relationships between wild sheep and goats and the audad (Caprinae) studied by immunodiffusion. *Comp. Biochem. Physiol. 54B,* 265-9

Hofer, H. (1972) Prolegomena primatologia. In H. Hofer and G. Altner (eds) *Die Sonderstellung des Menschen.* G. Fischer Verlag, Stuttgart, pp. 113-48

Hofmann, R.R. (1973) The ruminant stomach. *East Afr. Monogr. Biol. 2,* 354 pp.

Hofmann, R.S. (submitted) Paper delivered at Theriological Congress, Edmonton, August 1985. To be published

Hornocker, M.G. (1970) An analysis of mountain lion predation upon mule deer and elk in Idaho. *Wildl. Monogr.* No. 21, 39 pp.

Ingebrigtsen, O. (1923) *Das Norwegische Rotwild* (*Cervus elaphus* L.). Mitteilungen Zool. Abteilung, Museum zu Bergen, Bergen 242 pp.

James, F.C. (1983) Environmental components of morphological differentiation in birds. *Science 211,* 184-6

Kishimoto, R. (1981) Behavior and spatial organisation of the Japanese Serow. MSc. thesis, Osaka University, Osaka, 59 pp.

Knaus, W. and W. Schroeder. (1983) *Das Gamswild* (3rd edn). Paul Parey, Berlin, 232 pp.

Korobitsyna, K.V., C.F. Nadler, N.N. Vorontsov and R.S. Hoffman. (1974) Chromosomes of Siberian snow sheep, *Ovis nivicola,* and implications concerning the origin of Amphiberingian wild sheep (subgenus *Pachyceros*). *Quart. Res.* *4*, 235–54

Kramer, A. (1969) Soziale Organisation und Sozialverhalten einer Gemspopulation (*Rupicapra rupicapra* L.) der Alpen. *Z. Tierpsychol. 26*, 889–964

Kuck, L. (1977) The impact of hunting on Idaho's Phasimore mountain goat herd. In W. Samuel and W.G. Macgregor (eds) *Proc. 1st International Mountain Goat Symposium.* Fish and Wildlife Branch, Province of British Columbia, Victoria, pp. 114–25

Kurten, B. (1968) *Pleistocene mammals of Europe.* Weidenfeld & Nicolson, London, 317 pp.

Kurten, B. and E. Anderson. (1980) *Pleistocene mammals of North America.* Columbia University Press, New York, 442 pp.

Lay, D.M., C.F. Nadler and J. D. Hassinger. (1971) The transferrins and hemoglobins of wild Iranian sheep (*Ovis* linnaeus). *Comp. Biochem. Physiol. 40B*, 521–9

Lovari, S. (1985) (ed.) *The biology and management of mountain ungulates.* Croom Helm, London, 271 pp.

Lovari, S. and C. Scala. (1980) Revision of *Rupicapra* genus I. A statistical reevaluation of Coutourie's data on the morphometry of six chamois species. *Bull. Zool. 47*, 113–24

Margalef, R. (1963) On certain unifying principles in ecology. *Am. Natur. 97*, 357–74

Masini, F. (1985) Würmian and Holocene chamois of Italy. In S. Lovari (ed.) *The biology and management of mountain ungulates.* Croom Helm, London, pp. 31–44

Mayr, E. (1966) *Animal species and evolution.* Balknap/Harvard University Press, Cambridge, Mass. 796 pp.

Meunier, K. (1984) Der Spanische Hirsch, ein sehr altes Eiszeitrelikt. *Jagd und Hege 16* (3), 9–12

Nadler, C.F. and T.D. Bunch. (1977) G-band patterns of the Siberian snow sheep (*Ovis nivicola*) and their relationship to chromosomal evolution in sheep. *Cytogenet. Cell Genet. 19*, 108–17

Nadler, C.F., R.S. Hoffmann and A. Woolf. (1973) G-band patterns as chromosomal markers, and the interpretation of chromosomal evolution in wild sheep (*Ovis*). *Experientia 29*, 117–19

Nadler, C.F., R.S. Hoffmann and A. Woolf. (1974) G-band patterns, chromosomal homologies, and evolutionary relationships among wild sheep, goat and audads (Mammalia, Artiodactyla). *Experientia 30*, 744–6

Nascetti, G., S. Lovari, P. Lanfranchi, c. Berducou, S. Mattiocci, L. Rossi and L. Bullini. (1985) Revision of *Rupicapra* genus 3. Electrophoretic studies demonstrating species distinction of chamois population of the Alps from those of the Appennines and Pyrenees. In S. Lovari (ed.) *The biology and management of mountain ungulates.* Croom Helm, London, pp. 56–62

Nievergelt, B. (1981) *Ibex in an African environment.* Springer, Berlin, 189 pp.

Packard, J.M. and L.D. Mech. (1983) Population regulation in wolves. In F.L. Bunnell, D.S. Eastman and J.M. Peek (eds) *Natural regulation of wildlife populations.* Forest, Wildl. and Range Exp. Station. Proj. No. 14. Univ. Idaho, Moscow, Idaho, pp. 151–74

Pedersen, A. (1958) *Der Moschusochs*, Neue Brehm-Bücherei, Nr. 215. Wittenberg, Lutherstadt, 34 pp.

Petzsch, H. (1956) Reflexionen zur Phylogenie der Capridae im allgemeinen und der Hausziege im besonderen. *Wissenschaftliche Zs. Martin Luther Univ., Halle-Wittenberg 6* (6), 995–1019

Pfeffer, P. (1967) Le mouflon de Corse (*Ovis ammon musimon* Schreber, 1982); position systematique, écologie et éthologie comparées. *Mammalia 31* (Supplement) 262 pp.

Sakurai, M. (1981) Socio-ecological study of Japanese serow *Capricornis crispus* (Temminck). (Mammalia: Bovidae) with reference to flexibility of its social structure. *Ecol. Physiol. Japan. 18*, 163–212

Scala, C. and S. Lovari. (1984) Revision of *Rupicapra* genus 2. A skull and horn statistical comparison of *Rupicapra rupicapra ornata* and *R. rupicapra pyrenaica* chamois. *Bull. Zool. 51*, 285–94

Schaller, G. (1977) *Mountain Monarchs*. University of Chicago Press, Chicago, Ill., 425 pp.

Schmitt, J. (1963) *Ammotragus lervia* Pallas, Mahnenschaf oder Mahnenziege? *Z. Saugetierkunde 28*, 7–12

Schroder, W. (1971) Untersuchungen zur Ökologie des Gamswildes (*Rupicapra rupicapra* L.) in einem Vorkommender Alpen. Teil I. *Z. Jagdwissenschaft 17* (3), 113–68; Teil I, *17* (4), 197–235

Scott, W.B. (1937) *A history of land mammals in the western hemisphere*. Hafner, New York, 786 pp.

Shackleton, D.M. (1973) Population quality and bighorn sheep. PhD Thesis., University of Calgary, 227 pp.

Shank, C.C. (1972) Some aspects of social behaviour in a population of feral goats (*Capra hirrus* L.). *Z. Tierpsychol., 30*, 488–528

Shank, C.C. (1979) Sex dimorphism in the ecology of bighorns. PhD Thesis, University of Calgary, 193 pp.

Shank, C.C. (1982) Age–sex differences in the diet of wintering Rocky Mountain Bighorn Sheep. *Ecology 63* (3), 627–33.

Suzuki, K. and S. Takatsuki. (1986) *Winter food habits and sexual monomorphism in Japanese serow*. Biological Institute, Faculty of Science, Tohoku University, Sendai, 980, Japan, 8 pp.

Szunyoghy, J. (1963) *Das Ungarische Rotwild*. Muzeumok Rotauemeben, Budapest, 193 pp.

Thenius, E. (1969) Stammesgeschichte der Saugetiere. *Handb. Zool. 8*, 1–722

Thenius, E. and H. Hofer. (1960) *Stammesgeschichte der Saugetiere*. Springer, Berlin, 322 pp.

Thompson, D'Arcy. (1961) *On growth and form* (abridged, J.T. Bonner). Cambridge University Press, London, 345 pp.

Treisman, M. (1975) Predation and the evolution of gregariousness. *Anim. Behav. 23*, 779–800; 801–25

Valdez, R. (1982) *The wild sheep of the world*. Wild Sheep and Goat International. Mesilla, New Mexico, 186 pp.

Valdez, R., C.F. Nadler and T.D. Bunch. (1978) Evolution of wild sheep in Iran. *Evolution 32*, 56–72

Vogt, F. (1936) *Neue Wege der Hege*. Verlag Neumann, Neudamm, 165 pp.

Vogt, F. (1948) *Das Rotwild*. Oesterreichischer Jagd und Fischerei Verlag, Vienna, 207 pp.

Vogt, F. and F. Schmidt. (1951) *Das Rehwild*. Oesterreichischer Jagd und Fischerei Verlag, Vienna, 111 pp.

Vorreyer, F. (1978) In F. von Raesfeld and F. Vorreyer (eds) *Das Rotwild* (8th edn). Paul Parey Verlag, Berlin, 397 pp.

Vrba, E.S. (1980) Evolution, speciation and fossils: how does life evolve? *S. Afr. J. Sci. 76*, 61–84

Wagenknecht, E. (1981) *Rotwild*. Verlag. J. Neumann-Neudamon, Berlin, 484 pp.

Watson, K. and R. Moss. (1972) A current model of population dynamics in red

grouse. *Proc. 15th Int. Ornithological Congress.* Brill, Leiden pp. 134–49
Wilson, E.O. (1975) *Sociobiology.* Belknap/Harvard University Press, Cambridge, Mass., 697 pp.
Wilson, M. (1980) Morphological dating of Late Quaternary bison on the northern plains. *Can. J. Anthropol. 1,* 81–5
Yamomoto, S. (1966) Breeding Japanese serow in captivity. *Int. Zoo Yrbk 7,* 174-5
Zuh-Ming, D. (1961) The Formosan serow (*Capricornis swinhoi,* Gray). *Quart. J. Taiwan Museum 16,* 97-100.

2

The saiga (*Saiga tatarica*) in captivity, with special reference to the Zoological Society of San Diego

James M. Dolan Jr

Zoological Society of San Diego, PO Box 551, San Diego,
California 92112-0551, USA

The saiga, *Saiga tatarica*, and its close relative the chiru or Tibetan antelope, *Pantholops hodgsoni*, are the only members of the tribe Saigini within
the subfamily Caprinae. These two very distinct genera appear to form a
convenient bridge between the antelopes on the one hand and the rupi-
caprines on the other. Palaeontology does little to help us trace the origins of
these two peculiar genera. Nevertheless, it appears that they have evolved
along separate lines since the latter part of the Miocene or early Pliocene.
The chiru was once distributed throughout suitable habitat on the Tibetan
Plateau. Today it appears to be in decline. George Schaller, in a recent
faunal survey of various parts of Quinghai Province in northeastern Tibet,
where the animal was once widespread, found that it now survives only in
the southwestern corner of the province. Its future may well be a source of
concern.

Reports received in the early 1960s suggested that the chiru may have
been kept in one or more of the Chinese zoos. This information, however,
was erroneous. According to recent information received from Chinese
colleagues, attempts to bring the chiru into captivity have all proven unsuccessful.

Whereas overexploitation and habitat degradation may just now be
affecting chiru populations, recent saiga populations began a downward
crash about 1850. Unlike the chiru, the saiga once had an enormous range.
Vereschchagin and Baryshnikov (1984) write that

> During the Pleistocene the saiga ranged from England in the west to
> Alaska in the east and from the New Siberian Islands in the north to the
> Central Asiatic deserts in the south. It is possible, however, that a separ-
> ate species (*Saiga borealis* Tscherskyi) inhabited the arctic steppes of
> Siberia.

This enormous range can be extended further to the east to include the

41

Northwest Territory of Canada. A southern spread of this genus in North America was prevented by ecological factors. The saiga entered North America during the Illinoian and survived on this continent until the last Wisconsinian. During the 4-Würm in Europe, corresponding to the Wisconsinian in North America, it occurred in England, Belgium and France. Like the musk-ox and reindeer, the saiga is a good example of fluctuation in range size and regional extinction. According to Vereschchagin and Baryshnikov (1984):

> Progressive reduction of a species range under natural influences often retraces the route of its original spread. Some Central Asiatic steppe species — the yellow lemming, the corsac fox, the kulan and the saiga — provide an example. Over the last few centuries their ranges have shrunk rapidly from west to east, first as a result of ecological factors (as yet poorly known) and secondly as a result of human activity in the nineteenth and twentieth centuries.

As recently as 300 years ago, saigas ranged as far west as the foothills of the Carpathians and were common on the steppes of the Dnestr and Dnepr basins. Information is sparse as to its abundance in the early years of the nineteenth century, though earlier texts indicate that it was a very abundant animal. In 1770, Pallas wrote that it was common where the land had not yet been settled by the Kirghiz and particularly along the Ural River. It was so numerous in 1773 that his cossacks subsisted on saiga meat. However, beginning approximately in 1850, saiga populations began to plummet due to overexploitation by nomadic herdsmen and farmers. The animal was particularly vulnerable around water sources. At the onset of the 1920s only about a thousand saigas remained in the Soviet Union and but a few dozen in Mongolia.

Pohle (Chapter 16, this volume) points out that in the Soviet Union, under stringent protection, the animal has made a remarkable recovery after a period of very serious decline between 1850 and the early 1920s. Nevertheless, the contemporary range from the west bank of the Volga to Kazakhstan is considerably less than it was 200 years ago, and the fact that it is now confined to a much constricted range is not without inherent problems.

There are two extant subspecies, the Russian saiga, *Saiga tatarica tatarica* from the western portion of the range, which is the larger of the two forms. In this subspecies adult males stand approximately 75 to 80 cm at the withers, and the females are somewhat smaller. They range in weight from 36 to 69 kg. In their general appearance, saigas resemble sheep. Only the males carry horns, which are from 33 to 36 cm; they are ringed and waxy amber in colour. The latter are much sought after by the Chinese pharmaceutical trade for use in folk medicine. In winter the coat is thick and nearly white without conspicuous markings, whereas in summer it changes to a dull yellowish–tan with a whitish throat and indistinct dark facial markings. The short tail is coloured like the body as are the short, thickly haired ears.

The isolated Mongolian subspecies, *Saiga tatarica mongolica*, is of more recent origin and is separated from the Russian animal by the Mongolian Altai. It occurs in the Dzungarian Gobi and the western portions of the Mongolian People's Republic, where it is considered endangered. Smaller in size, it is sandy–grey in colour with a distinct brown spot on the croup. The horns are shorter and weaker than those of the Russian subspecies. The saiga's most obvious feature is the inflated, mobile muzzle, with downward-projecting nostrils. The muzzle contains convoluted choanae and mucous glands for warming and moistening dry steppe air. During the rut, the nose becomes even more inflated, producing a ridge on its dorsal surface. The nose is less bulbous in the Mongolian subspecies. The latter is unknown in captivity.

The first saiga exhibited outside of Russia was a male received at the London Zoo on 21 November 1864. At the same time saigas were also to be seen in the Moscow Zoo. There is no record of longevity for the first London saiga, but a specimen received on 10 November 1866 lived until 22 October 1869. Other European collections to exhibit saigas prior to the turn of the present century were Berlin 1872, Cologne 1874, Hamburg 1877, Antwerp 1878, and Bremen 1889. The saiga was a great rarity at the time and must have been a prized exhibit. Professor Alex Pagenstecher on a visit to the Cologne Zoo from Moscow in 1874 was astounded to see saigas in the Cologne collection as attempts to keep them in Moscow had been unsuccessful. Unfortunately, two of the Cologne saigas did not survive the year 1874, dying of head injuries after crashing into their enclosure. Longevities for the majority of the early saigas have been lost in time, though data do exist for the Berlin pair of 1872. I do not have the day or month of arrival for these animals, but the female lived until 15 October 1873 and the male survived until 12 October 1875.

Two shipments of saigas, one in 1902 the other in 1906, were received at Woburn Abbey by the eleventh Duke of Bedford from the firm Hagenbeck. Both importations proved disappointing with only a single animal surviving longer than two years. In his autobiography, Hastings Russell, twelfth Duke of Bedford (1949) presents a pessimistic picture of the attempts at Woburn:

We had two importations of the extraordinary saiga antelope and both had the same curious and ultimately disappointing history. They recovered quickly from the effects of their journey, bred and throve amazingly for nearly two years, and then all died off with the exception of a solitary individual, from what cause I do not know ... It is possible that the tantalizing third-year epidemic might be avoided if saigas were moved annually on to fresh pasture, the old pasture being heavily salted and limed.

As I pointed out in 1977, a greater portion of the Woburn ungulate collection proved unsuccessful due to the high incidence of parasitic infection. It was the major cause for the lack of success with the Przewalski's horse (*Equus przewalskii*) herd.

Their rarity in captivity prior to the Second World War and their general lack of success has been underlined by Mohr (1943) who writes:

> The saigas which have been kept in captivity generally have not survived long. In German zoos during the past decade I have only seen six to eight animals of both sexes in the gardens of Berlin, Stellingen and Hannover, where they only survived for a short time. Perhaps this lack of success has to do with the fact that their social behaviour is not satisfied, as the Russians prefer to send two animals together in a crate, rather than a single animal.

Although Mohr only mentions six to eight animals having been seen by her in German zoos, it is obvious from the captive history that many more animals passed through that country as part of the international animal trade. The firm Ruhe, which operated a facility at Alfeld as well as the Hannover Zoo, was responsible for the exportation of saigas to the United States prior to 1940. The first US saiga was a male received at the St. Louis Zoological Park on 28 May 1934, which died in August 1939. The National Zoological Park, Washington, DC, obtained a pair on 5 December 1934 and an additional male on 14 May 1935. Two males were obtained from Germany by the New York Zoological Society, one in 1936 and the other in 1937. The St. Louis Zoological Park purchased a pair in May of 1936 and an additional female in July of 1937. St. Louis produced the first saiga born in North America on 15 May 1939. This female lived until 19 June 1945. Her mother, received in July of 1937, died on 7 January 1945. The Second World War brought an end to saiga groups, both in Western Europe and in the United States.

In the post-war years, saigas first appeared in captivity outside of the Soviet Union with the arrival of a pair at the Prague Zoo in 1950. From that point on, large numbers of saigas were exported from the Soviet Union through the transit points at Prague and most importantly at Tierpark Berlin, German Democratic Republic. From November of 1950 until October of 1972, a total of 148 saigas (52 males, 84 females, 12 unsexed) passed through the Prague Zoo, the majority destined for other collections. Of these 76 (29 males, 43 females, 4 unsexed) died before they could be trans-shipped. In 1958 the first saigas arrived at Tierpark Berlin, and shipments continued to arrive from the Soviet Union so that a total of 143 animals were received until 1972 (Pohle 1974). By 1986, this number had risen to 332 individuals. Pohle (1974 and Chapter 16, this volume) has discussed the keeping of saigas at Tierpark Berlin in detail.

With Prague as a transit point, the St. Louis Zoological Park obtained a trio, and the National Zoological Park, Washington, DC, two males and one female in 1955. From the years 1955 until 1969, saigas were imported by the following United States zoological gardens: New York Zoological Park 1956; Lincoln Park Zoo, Chicago, 1958; Dallas Zoo 1959 and 1962; Philadelphia Zoological Garden 1960, 1961, 1962; San Diego, San Francisco and Toledo zoos 1962; Rio Grande Zoological Park, Albuquerque, Oklahoma City and Omaha zoos 1969. All of these saiga groups are now extinct. In Canada, the groups formerly at the Alberta Game Farm, Edmonton, and the Assiniboine Park Zoo, Winnipeg, are also a thing of the past.

Saigas were reestablished at the Oklahoma City Zoo with the importation of two males and eight females on 21 June 1977, followed by two females on 20 October 1977 and an additional male on 30 January 1978. With the exception of two females all of the imported animals were dead by the conclusion of 1981. One female imported on 21 June 1977 lived in the collection for 8 years and 7 months, and a female of the 20 October 1977 importation for 6 years, 6 months and 24 days. When considered as a group, the longevities for the imported animals average less than four years. The present Oklahoma City herd consists of 4 males and 13 females, the oldest animals being 2 females born in 1982.

The Zoological Society of San Diego received its first saigas, a pair, on 17 March 1972. These animals were housed in a relatively small enclosure at the San Diego Zoo with the consequence that the male sustained a compound fracture of the right foreleg and succumbed on 18 March 1972. The female died on 2 January 1973 as the result of shock syndrome during the removal of a pin from a fractured femur. This experience only underlines Pohle's directive (Chapter 16, this volume) that saigas should not be housed in quarters with insufficient flight space as the animals will dash themselves against barriers.

With the establishment of the San Diego Wild Animal Park, the keeping of saigas was again attempted by the Zoological Society of San Diego. On 25 November 1970 a sibling pair was received from the Rio Grande Zoological Park where they had been born in May of the same year. Since this was a sibling pair, the male was sold to the Alberta Fame Farm on 14 July 1971 and replaced by a male from the San Francisco Zoo which arrived on 22 January 1971. This animal lived in the collection until 25 November 1973 without reproducing. A new male, born in the Oklahoma City Zoo on 1 May 1973, was purchased from a dealer on 16 August 1973. This animal and the female received on 25 November 1970 formed the foundation for the present San Diego herd. The male died on 1 November 1976 and the female on 8 August 1978. On 13 December 1979 two males and one female were received on breeding loan from the Dallas Zoo. An additional pair, from the same source, was received on 5 March 1981. Unfortunately, none of these animals reproduced and they were all dead by November of 1981.

The first saigas born at the San Diego Wild Animal Park were a pair of twins produced on 13 May 1975. The male twin died on 15 May 1975 and the female on 22 May 1977. A second pair of twins was born on 8 May 1976. This pair produced its first offspring, a single male, on 6 May 1978. Between 13 May 1975 and 24 May 1985, 32 saigas had been born in the collection. Three of these were stillborn. Of the total number born, 31 occurred in May and a single birth in June. Only one saiga has ever been shipped out of the collection, a female which was sent to the Dallas Zoo on 19 December 1980. Lambs were born in all years except 1982 when there was no adult male in the herd.

The female born on 8 May 1976 produced a single lamb on 6 May 1978, twins on 15 May 1979, a single lamb on 16 May 1980, and no lambs in 1981, 1982 or 1983. Although this animal achieved the greatest longevity recorded at the San Diego Wild Animal Park, 8 years, 10 months and 27 days (8 May 1985−4 April 1985), she was an unproductive animal with only four lambs to her credit. The present herd consists of five males and nine females, the oldest animals being two females born in 1979. Saigas have been exhibited since 1970 and bred here since 1975 without interruption and without the introduction of new animals into the herd. As far as I have been able to ascertain, this group holds the captive record for continuity.

Saigas at the San Diego Wild Animal Park are maintained in an enclosure of approximately 20 ha in association with other Asiatic mammals such as Indian gaur, blackbuck, Indian rhinoceros, goitred gazelles, and various deer species. Separate enclosures are only provided for the Indian rhinoceros but are only used when the rhino cows are about to calf. The saigas have never been enclosed and are allowed to roam the entire area throughout the year. The saiga group, however, has chosen to occupy a relatively barren portion of the enclosure where grass does not grow. I have never seen them in the green areas of this exhibit. Interspecific conflict has occurred on only one occasion, when a male saiga was injured by a male blackbuck. Unlike the animals in Cologne (Zimmermann 1980) and Tierpark Berlin (Pohle, Chapter 16, this volume) the males have not had their horns covered with hose, nor have they been separated from the females. However, the size of the exhibit does allow an animal to escape the aggressive advances of another individual. In smaller areas I would have to agree with Zimmermann and Pohle that it would be most wise to separate the males from the females after the rut and to cap his horns.

During the rut, saiga bucks are extremely aggressive not only towards other animals but also towards human beings. A male in Dallas Zoo suffered a broken leg inflicted by a keeper who was trying to protect himself from attack. When attempting to keep these animals, it would be most advantageous to provide as much space as possible and to be very cautious in the placement of barriers. When frightened, the animals seem unable to see barriers and crash into them, resulting in traumatic injuries. As Zimmermann (1980) has pointed out, we are still a long way from being successful with the saiga in captivity. It is 122 years since the first saiga entered a zoological garden, and although great strides have been made in the establishment of a broad spectrum of ungulate species in captivity, the saiga still eludes us.

Of the many zoological gardens which have kept this animal, few have published data that might help us to understand better the problems associated with keeping this genus in captivity. Even Crandall (1964) in his classic work is very brief in his coverage. He does mention that four animals received at the New York Zoological Park were rather unsatisfactory, with the greatest longevity being 2 years, eight months and 18 days. Crandall felt that the major problem with this animal was dietary, and drew upon Mohr (1943) for support. As we know today, the feeding of saigas is a very uncomplicated process. Pohle (Chapter 16, this volume) has given a successful diet used at Tierpark Berlin, and the animals at the San Diego Wild Animal park are fed on a prepared pelleted diet used for ungulates with great success. Pohle (see Chapter 16) does caution that the food should be cut into small pieces as two saigas died at Tierpark Berlin as the result of potatoes lodging in their throats. a similar incident occurred at New York Zoological Park.

In 1965, Fontaine published a short article on the breeding of saigas in the Dallas Zoo. He reported that:

> The Dallas Zoo imported a male and two female Saiga antelopes on 25th June 1959. One female was injured and died two days after arrival, the second died on 10th August 1959. The male was given injections against enterotoxaemia. Two new females were imported on 18th March 1962, and were approximately one year old. The animals were treated against enterotoxaemia as the male had been. The Dallas Zoo veterinarian recommends the use of enterotoxaemia vaccine for these animals as they are closely related to sheep.

One of the Dallas females imported in 1962 produced triplets in May of 1964. The Dallas herd suffered an outbreak of Johne's disease which brought about its end.

Four years later Voss (1969) described his experiences at the Assiniboine Park Zoo, Winnipeg, Manitoba, Canada. More recently there are the papers of Pohle (1974), Dolan (1977) and Zimmermann (1980). On the whole, however, captive experiences are poorly documented.

My remarks of 1977 still hold true today:

> In reviewing the pathology of the saiga under captive conditions one sees a recurrence of similar problems. In the USA, the major causes of death can be attributed to trauma and associated stress, parasitism and bacterial infection, in that order. The same is true of Prague, where 30 animals died from traumatic causes. In East Berlin the picture has been somewhat different. Pohle (1974) reports that of the 53 post-mortems performed on imported saiga, 15 animals died of bacterial infection and 15 of infections of the stomach and intestinal tract, but he gives no further details. In only ten cases were parasites found, principally *Cysticercus tenuicollis* and *Taenia hydatigena.*

Zimmermann (1980) reports similar problems with saigas kept in the Cologne Zoo. One of the major killers is *Sphaerophorus necrophorus* for which there is no effective cure. However, this necrobacillis appears to be a secondary invader and may be controlled by creating a better captive environment. In the wild one of the major causes of death, which results in mass die-offs, is a foot and mouth disease. It is particularly prevalent where saigas come into contact with domestic livestock (Heptner *et al.* 1966). Saigas are known to be parasitised by 35 species of worms, and in some areas parasites account for the majority of deaths. Until the turn of the present century, saigas were heavily parasitised by the fly, *Pallasiomyia antilopum.* The larvae of this fly developed in large numbers under the skin and caused the animals great discomfort and sapped their energy. When saiga populations reached their low point in the early 1920s this fly became extinct in Russia. It does, however, still occur in the Mongolian population.

The following pathology can be reported for saigas kept by the Zoological Society of San Diego:

♂ — received 17 March 1962 — died 18 March 1962, compound fracture of right foreleg

♀ — received 17 March 1962 — died 2 January 1963, shock syndrome during the removal of a pin from a fractured femur

♀ — received 25 November 1970 — died 8 August 1978, cancer of the neck

♂ — received 22 January 1971 — died 26 November 1973, inflamed lymph glands

♂ — received 16 August 1973 — died 21 November 1976, abscess of gastric tract

♂ — born 13 May 1975 — died 15 May 1975, maternal neglect

♂ — born 8 May 1976 — died 30 December 1979, trauma

♀ — born 8 May 1976 — died 4 April 1985, anaesthetic shock
♂ — born 6 May 1978 — died 25 May 1978, fracture of left tibia
♂ — born 8 May 1978 — died 15 November 1979, osteomyelitis
♀ — born 8 May 1978 — died 4 March 1981, osteomyelitis
♀ — born 13 May 1979 — died 14 November 1985, traumatic hip injury
♂ — received 13 December 1979 — died 5 January 1980, trauma
♂ — received 13 December 1979 — died 1 March 1980, fracture of femur
♀ — received 13 December 1979 — died 10 August 1981, possible enteritis
♂ — born 16 May 1980 — died 1 January 1984, toxaemia
♂ — received 5 March 1981 — died 6 July 1982, acute pulmonary oedema and pleural effusion
♂ — born 19 May 1984 — died 10 July 1985, osteoporotic femur fracture
♂ — born 20 May 1984 — died 5 September 1984, darting injury
♂ — born 15 May 1985 — died 16 May 1985, atelectasis
♀ — born 15 May 1985 — died 16 May 1985, trauma
♀ — born 13 May 1985 — died 16 May 1985, birth injury
3♂♂ — stillborn

Captive longevities for saigas are, for the most part, discouraging. The average ranges between 3 and 5 years. Heptner *et al.* (1966) report that the longevity in the wild is unknown, although males of 5 to 6 years of age have been recorded and females twice that age. Bannikov (1963) gives the maximum age, not mentioning sex, to be 12 years under optimal conditions. Pohle (Chapter 16, this volume) provides more interesting information by stating that the majority of the animals in the wild live only 4 to 5 years. It may well be that we have placed our expectations too high for captive saigas, and what we have seen in captive populations is the norm. Certainly in looking over the record we find that males die at a much younger age than do females. Nevertheless, the average captive longevities seem to parallel those of wild living populations.

Perhaps as we look towards the future we should consider the following factors. Obviously, it is to great advantage to provide these nervous animals with as much room as we can and to remove all sight barriers whenever possible. When available space is at a premium, the herds should be managed, as pointed out by both Zimmerman (1980) and Pohle (Chapter 16, this volume), with the horns of the bucks covered to prevent injury to the ewes and the animals separated after the rut. The utmost care must be exercised to prevent undue stress, and periodic checks must be made to ensure the elimination of parasitic infection. It is my opinion that the animals should never be kept in pairs, but that two or more females should be provided for each buck in order to ensure aggressive dissipation, particularly during the breeding season. In addition, if the reality is that we are dealing with a naturally short-lived species, herds should be started with a substantial number of females and care must be taken to ensure the survival of the lambs, if captive saiga herds are not to die out as they have done in the past.

REFERENCES

Bannikov, A.G. (1963) *Die Saiga-Antilope.* Neue Brehm-Bücherei, No. 320, 1–143, Wittenberg, Lutherstadt

Crandall, L.S. (1964) *The Management of wild mammals in captivity.* University of Chicago Press, Chicago, Ill.

Dolan, J. (1977) The saiga (*Saiga tatarica*): a review as a model for the management of endangered species. *Int. Zoo. Yrbk. 17*, 25–32

Fontaine, P. (1965) Breeding saiga antelope, *Saiga tatarica,* at Dallas Zoo. *Int. Zoo Yrbk. 5*, 57–8

Hastings, Duke of Bedford (1949) *The years of transition.* Andrew Dakers, London

Heptner, V.G., A.A. Nasimovic and A.G. Bannikov (eds) (1966) *Die Säugetiere Sowjetunion.* Jena, 1.

Mohr, E. (1943) Einiges über die Saiga, *Saiga tatarica* L. *Der Zool. Garten Leipzig (NF) 15*, 175–85

Pohle, C. (1974) Haltung und Zucht der Saiga-Antilope (*Saiga tatarica*) im Tierpark Berlin. *Der Zoologische Garten Jena 44*, 387–409

Vereschchagin, N.K. and G.F. Baryshnikov. (1984) Quaternary mammalian extinction in northern Eurasia, in P.F. Martin and R.G. Klein (eds) *Quaternary extinctions.* University of Arizona Press, Tucson, Ariz., pp. 483–516

Voss, G. (1969) Breeding the pronghorn antelope, *Antilocapra americana,* and the saiga antelope, *Saiga tatarica,* at Winnipeg Zoo. *Int. Zoo Yrbk. 9*, 116–18

Zimmermann, W. (1980) Zur Haltung und Zucht von Saiga-Antilopen (*Saiga tatarica tatarica*) im Kölner Zoo. *Z. Kölner Zoo 4* (23), 120–7

3

Evolutionary aspects of the biology of chamois, *Rupicapra* spp. (Bovidae, Caprinae)

Sandro Lovari

Institute of Zoology, University of Parma, Strada dell'Università No. 12, I-43100 Parma, Italy

INTRODUCTION

Kurtén (1968) wrote that the origin of *Rupicapra* 'is a mystery'. Today we know a little more on its evolution, although we are still far from having a clear picture of how chamois evolved. This is somewhat unusual when compared with the relatively extensive information available on other tribes of the subfamily Caprinae. On the other hand, it seems to be a common pattern in the Rupicaprini tribe. Why are palaeontological remains of Rupicaprini so rare? Most likely, such rarity depends on the nature of the terrain on which *Rupicapra* and the other members of its tribe seem to have always been dependent: a rocky, rugged, steep ground in which bones are easily crushed and eroded, thus preventing fossilisation (Masini and Lovari, in prep.). This fact makes it difficult to reconstruct the evolutionary biology of the genus *Rupicapra*, as we are unable to observe the development of body features and structures which suddenly appear in their 'modern' form in the late Mindel–early Riss (Mid-Pleistocene).

In this chapter I shall outline and summarise some aspects of the evolution and of biologically adaptive features of *Rupicapra*.

PALAEONTOLOGY AND SYSTEMATICS

Rupicapra's morphology tends to isolate it from the other Rupicaprini (F. Masini and S. Lovari, in prep.), as (1) its nasals are relatively small, (2) its orbits jut outwards considerably, (3) its basicranium is strongly flexed: a feature shared only with *Nemorhaedus*. Furthermore, the morphology of the frontal bones at the implant site of the horn cores would also separate *Rupicapra* from the other Rupicaprini.

The only form which shows a close similarity to *Rupicapra* is the European *Procamptoceras*, which survived up to the Mid-Pleistocene. However, some derived characteristics of the horn bases and of the teeth of the latter would not suggest a direct ancestry, but only that *Procamptoceras* is the closest known form to *Rupicapra*'s direct ancestor, whichever it is. The few palaeontological data available would suggest that *Rupicapra* belongs, together with *Procamptoceras* and perhaps the North African *Numidocapra*, to a phyletic branch already distinct in the Villafranchian (2–1 million years ago) from the ancestral rupicaprid stock (e.g. Masini and Lovari, in prep.). The sudden occurrence of *Rupicapra* fossils in the European Mid-Pleistocene, after a time in which apparently no rupicaprid had been present in the area, might mean an immigration from the East during a cold period, as suggested by the associated fauna: *Rangifer*, *Praeovibos*, *Hemitragus* and *Ovis* (Masini and Lovari, in prep.). If so, a migration from some mountain region west of the Himalayas, along the ranges of the Alpine system, seems the most probable hypothesis. Alternatively, its arrival via steppe is unlikely because of *Rupicapra*'s dependency on — and adaptations to — a rocky, steep terrain. The lack of any *Rupicapra* form in the East and the fact that its phylogenetically closest fossils have all been found in Europe and North Africa would confirm its origin in eastern Europe or south-west Asia.

Two species of *Rupicapra* are recognised: *R. pyrenaica* and *R. rupicapra* (Lovari 1984a; Lovari 1985, for a short review of relevant papers; Masini and Lovari, in prep.). *R. pyrenaica* appears to be the oldest form, showing some 'primitive' features such as the little divergent horn cores, inclined backwards at their base and slightly jutting forwards in their upper section (*cf. R. p. pyrenaica*): all features which are present in *Procamptoceras* and *Numidocapra* as well. Furthermore, its distribution and some biological features, as I shall point out later, would support such a statement.

Chamois appeared in the Riss glaciations (250000–150000 years ago) and the only specimen so far taxonomically identified appears to belong to *R. pyrenaica* (Masini and Lovari, in prep.). However, during the Würm glaciations (80000–12000 years ago), chamois fossils became relatively numerous, and, while *R. pyrenaica* was present in Spain and peninsular Italy, *R. rupicapra* seems to have colonised Europe westwards (Masini 1985).

The present natural distribution of *Rupicapra* includes *R. pyrenaica* in south-west Europe and *R. rupicapra* in the northeastern part of the genus range (Figure 3.1). Subspecies of the latter have been introduced for hunting purposes to areas to which they did not belong originally (Figure 3.2).

The two species look different (Figure 3.3) (e.g. Lovari 1985), have significantly different cranial features (Table 3.1) (Lovari and Scala 1980, R. Fondi S. Lovari and C. Scala, unpublished data), have different behavioural repertoires (Table 3.2) — especially in the patterns related to reproductive behaviour (Lovari 1985) — and they show a genetic divergence of 0.108 (Nascetti *et al.* 1985), with two loci discriminating at a level of at least 99 per cent and a third locus showing significantly different allele frequencies between the two species.

Figure 3.1: Natural distribution range of chamois species and subspecies (updated to 1984)

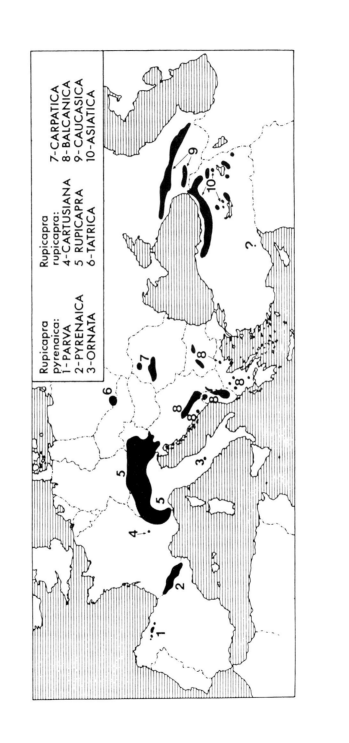

Rupicapra
pyrenaica:
1 - PARVA
2 - PYRENAICA
3 - ORNATA

Rupicapra
rupicapra:
4 - CARTUSIANA
5 - RUPICAPRA
6 - TATRICA

7 - CARPATICA
8 - BALCANICA
9 - CAUCASICA
10 - ASIATICA

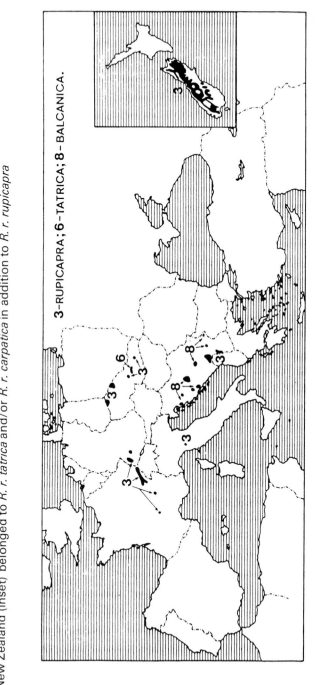

Figure 3.2: Introduced populations of chamois (*Rupicapra rupicapra* sspp.). There are rumours that the chamois brought to New Zealand (inset) belonged to *R. r. tatrica* and/or *R. r. carpatica* in addition to *R. r. rupicapra*

3–RUPICAPRA; **6**–TATRICA; **8**–BALCANICA.

Figure 3.3: The strikingly different winter-coat pattern between *Rupicapra rupicapra* and *R. pyrenaica* is maintained unaltered in all their respective subspecies. (A), *Rupicapra rupicapra rupicapra*; (B), *R. pyrenaica ornata*; 1–3, winter coat; 2–4, summer coat (*R. pyrenaica* sspp. have a reddish coat, and *R. rupicapra* sspp. are brownish)

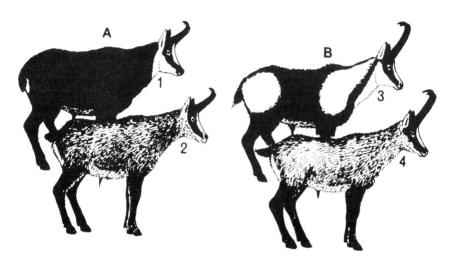

Table 3.1: Main cranial differences between the two species of chamois. The angle between the horn core and the frontal bone has been quantitatively assessed only for *R. r. rupicapra* and *R. pyrenaica ornata*

Main cranial differences	*Rupicapra pyrenaica*	*Rupicapra rupicapra*
Ethmoidal vacuity	absent	present
	┌──── significantly ────┐	
Horn distance at base	smaller	greater
Skull length	smaller	greater
Skull width at parietal level	smaller	greater
Angle horn core/frontal bone	greater	smaller

Source: Lovari and Scala (1980), R. Fondi, S. Lovari and C. Scala, unpublished data

Genetic divergence (Table 3.3) is much lower (0.009) between *R. p. pyrenaica* and *R. p. ornata* (Nascetti *et al.* 1985), being just nil between *R. p. parva* and *R. p. pyrenaica* (L. Bullini, J.M. Fernandez, G. Nascetti and S. Lovari, unpublished data), indicating that gene flow took place between these taxa till very recently. Finally, the two species bear Hippoboscidae parasitic flies belonging to different genera (Kock, *in litteris*): *Lipoptena* in *R. pyrenaica*, *Melophagus* in *R. rupicapra*.

Table 3.2: Main behavioural differences between chamois species. All patterns, except for the 'conflict posture', are related to reproductive behaviour (see also Lovari 1985). + = present; − = absent; (+) = rare

Social behaviour patterns	Rupicapra pyrenaica	Rupicapra rupicapra
Conflict posture	+	−
Dynamic head-down	+	−
Flank stroke	+	−
Herding	−	+
Kick	−	(+)
Mock suck	+	−

Table 3.3: Average values of Nei's (1972) genetic identity (above the diagonal) and genetic distance (below the diagonal) between *rupicapra, pyrenaica* and *ornata*

	R. r. rupicapra	R. p. pyrenaica	R. p. ornata
R. rupicapra rupicapra	−	0.909	0.886
R. pyrenaica pyrenaica	0.096	−	0.991
R. pyrenaica ornata	0.121	0.009	−

Source: Nascetti *et al.* (1985)

It has been suggested (Lovari and Scala 1980) that the first, Rissian chamois belonged to the *R. pyrenaica* type, while *R. rupicapra* may have evolved in the East or North-east some hundred thousand years later. The latter, upon its immigration to the west during the early Würm, could have competed with and eliminated the remnant populations of Rissian chamois which had survived the gap between the major glaciations by retreating to 'islands' on top of mountains.

Actually, palaeontological evidence (Masini 1985) suggests that a wedge of *R. rupicapra* did arrive down to the northern Apennines in Würm II and the Holocene, thus separating *R. p. pyrenaica* from *R. p. ornata*. Behavioural barriers related to courtship could have prevented a significant interbreeding between the two species (Lovari 1985), as their present genetic distance proves (Nascetti *et al.* 1985, Nascetti, *ex verbis*). In the light of the wealth of recent data, lumping *R. rupicapra* with *R. pyrenaica* 'would simply negate the purpose of taxonomy by obscuring rather than clarifying relationships' (Schaller 1977) a point made also by Schaller (1977, p. 37) about urials (*Ovis orientalis*) and argalis (*Ovis ammon*).

Is *R. rupicapra* a 'colonising, advanced form' (*sensu* Geist 1978)? Some comparative data would suggest that:

(1) it does show combat adaptations such as stout short horns (longer and/

or thinner break more easily), a stocky muscular body, and a drab coat (Lovari and Scala 1980);

(2) it is a most successful coloniser (e.g. Caughley 1963, Knaus and Schröder 1975).

ANATOMICAL ADAPTATIONS

I shall now provide an outline of the main anatomical features that make the chamois an animal particularly well adapted to the mountain terrain. Couturier (1938) showed that chamois have much thicker heart walls than most mammals of the same size: this organ can resist over 200 beats/ minute for a relatively long time. Furthermore, their large lungs and unusually great number of red blood cells (12–13 million/ml) make its respiratory and circulatory systems well adapted to life on the mountains. Its hoof structure (Couturier 1938, Blahout 1976) also enables this goat-antelope to move on difficult terrain with greater ease than the majority of the other ungulate species:

(1) the external edges of its hooves are sharp and hard so that the animal can make use of even small projections from the rocks it climbs,
(2) the frontal part of the hooves points slightly downwards, which helps the chamois to move on ice;
(3) the sole of the foot is rubbery, sturdy and slightly concave to prevent skidding on steep rock surfaces;
(4) the front hoot-halves can spread for over 90°, which facilitates braking while running down hill.

It can jump almost 2 m in height and at least 6 m in length, and it can run to a speed of about 50 km/h on uneven ground (Blahout 1976). It is thus an animal that is physically able to cope with most conditions which life on rocky, high mountains may provide. One could say that its ability to climb, to jump and to run well, together with the relevant body adapt-ations, make it *specialised because of its lack of specialisations*, which may not be said for most or all of the other mountain ungulates. Its growing numbers, well over 550000 chamois altogether (Lovari 1984a), are strong evidence of its biological success.

SOCIAL ORGANISATION AND ECOLOGY

Chamois biology has been moulded in the mountain environment. How do chamois cope with it? I shall now summarise some aspects of its social behaviour and behavioural ecology, drawing from data available in the literature and chiefly from my own research on *R. p. ornata*, the Apennine chamois, which survives with several hundred head only in the Abruzzo National Park, Central Italy.

Chamois have two major living areas throughout the year (e.g. Knaus and Schröder 1975, Lovari and Cosentino 1986): in the warm season they inhabit the Alpine meadows never further than several hundred metres from the cliffs where these animals seek refuge if attacked by a predator (e.g. the wolf); the winter ranges must include very steep (or windswept) areas where snow cannot accumulate, thus allowing the chamois to have access to suitable food resources (Berducou 1985, Lovari and Cosentino 1986). It is in these limited areas that the major impact on vegetation may occur, particularly in snowy winters when the concentration of chamois in such areas becomes the greatest. Then, plants smaller than 1.50 m will be the most damaged, often permanently (Berducou 1985). The chamois is a selective herbivore: it shows a low rate of bites of grass per minute (about 25 per minute) (Lovari and Rosto 1985) and it does not feed on plant species in relation to their availability (Ferrari and Rossi 1985), which indicates that food selection is exerted. Furthermore, it has been shown that in late spring, summer and autumn, the Apennine chamois depends mainly on the *Festucetum–Trifolietum thalii* plant community, which grows mainly in areas with a long-lasting snow cover (Ferrari and Rossi 1985; Ferrari and Lovari 1986; Cavani, Ferrari and Rossi, in prep.): the local summer flock size of females and young chamois would seem to be directly correlated with the size of areas with such a plant community. Moreover, Berducou (1985) has shown that the selection operated by chamois on their plant food can be used to detect the signs of excessive population density. In fact, there is a hierarchy of plants which the chamois normally eat. At high densities the most preferred plants become rare by over grazing (or overbrowsing). Then, the less preferred ones are eaten and even those that are mildly toxic, e.g. *Veratrum* spp. The simple detection of frequent eating signs on such 'less preferred' plants is an indicator of excessive grazing/browsing.

Altitudinal movements of chamois are seasonally synchronised (e.g. Hamr 1984, Lovari and Cosentino 1986). In the warm season they stay at altitudes over 1500 m, but they descend to less than 1100 m in the winter. Such a pattern is the same for all age and sex classes except for the fully grown-up males, which tend to stay at much lower altitudes than those used by the other chamois (e.g. Hamr 1984, Shank 1985). Only at rut time, in November, do they move to the Alpine meadows where there are the flocks of females.

Large flocks form in the summer and autumn, whereas normally these tend to scatter when chamois live in woodland in winter (Lovari and Cosentino 1986). The scarce and scattered food resources available in the cold season, and perhaps anti-predator strategies, would underlie such a pattern (Elsner-Schack 1985, Lovari and Cosentino 1986). Births occur in May–early June, and where predators are present in the area pregnant females isolate themselves on cliffs to deliver.

Females are usually fertilised in late November. Earlier in that month fully grown-up males form 'harems'. Only the older males succeed in holding females, which they herd with physically demanding techniques, behaving just as sheepdogs do. Techniques directed towards the central section of the flock are used significantly more often than centrifugal ones by the experienced males holding large groups of sexually mature females. In comparison, less successful males tend to use a significantly higher number of centrifugal techniques (Lovari 1984b, Lovari and Locati, in prep.). Thus, they run more — and obtain less — than the former, which instead maximise the effects of their herding efforts. Males threaten each other vocally in the first part of a dispute — very much like red deer do (cf. Clutton-Brock and Albon 1979). If the intruder comes forward, a series of tense visual displays may follow, ending up usually with a chase of variable duration. The chase is probably the most demanding dominance display which chamois have: the opponents chase each other up and down hill from a few metres to several hundred metres. If the pursuer catches up with the chased male, he will try to hook it, especially if the latter is a younger, thus subordinate, chamois. In some cases, turns will be switched and the opponents may alternatively be pursuer and pursued. Once more, chamois balance their efforts: the length of such energetically demanding runs up and down rocky cliffs is correlated significantly with the potential dangerousness of the opponent (Lovari 1984b, Lovari and Locati, in prep.): young intruders are chased significantly less far from the harem than prime males. Furthermore, alternate chases, i.e. where opponents take turns in chasing each other, occur significantly more often when these belong to the same age class, i.e. full expenditure of energy in running up and down is restricted to chases between equally matched rivals (Lovari and Locati, in prep.).

Therefore, the 'law of least effort' (see, for example, Geist 1978), which dictates that individuals ought to minimise expenditures on maintenance in order to maximise reproductive fitness, thereby sparing maximum resources for reproduction, is quite well known in all its facets to the chamois. This can be expected in a medium-sized ungulate which has so successfully thrived in the harsh mountain environment for over 80000 years.

ACKNOWLEDGEMENTS

W. Schröder read an earlier draft of this paper. F. Masini provided useful discussions and information on the section dealing with palaeontology. The author's research on the Apennine chamois has been financially supported by the Abruzzo National Park Administration and by the Italian Ministry of Education (MPI).

REFERENCES

Berducou, C. (1985) Analyse de quelques relations entre une population‚ de chamois des Pyrénées françaises et son environnement. In T. Balbo, P. Lanfranchi, P.G. Meneguz and L. Rossi. (eds) *Atti Simposio Internazionale Cheratocongiuntivite Infettiva Camoscio*, Centro Stampa Univerità, Turin, pp. 125–46

Blahout, M. (1976) *Kamzičia zver* Priroda, Bratislava

Caughley, G. (1963) Dispersal rates of several ungulates introduced into New Zealand. *Nature 200*, 280–1

Clutton-Brock, T.H. and S.D. Albon (1979) The roaring of red deer and the evolution of honest advertisement. *Behaviour 69*, 145–70

Couturier, M. (1938) *Le chamois*. Arthaud, Grenoble

Elsner-Schack, I. von (1985) Seasonal changes in the sizes of chamois groups in the Ammergauer Mountains, Bavaria. In S. Lovari (ed.) *The biology and management of mountain ungulates*, Croom Helm, London, pp. 148–53

Ferrari, C. and S. Lovari. (1986) Premesse ambientali per una reintroduzione del camoscio appenninico (*Rupicapra pyrenaica ornata*). In *Atti Conv. Prog. Faunistico Appenninico*. Fed. Ital. Caccia, Pescara, pp. 86–9

Ferrari, C. and G. Rossi. (1985) Preliminary observations on the summer diet of the Abruzzo chamois (*Rupicapra pyrenaica ornata* Neum.). In S. Lovari (ed.) *The biology and management of mountain ungulates*. Croom Helm, London, pp. 85–92

Geist, V. (1978) *Life strategies, human evolution, environmental design*. Springer, New York and Berlin

Hamr, J. (1984) Home range size and determinant factors in habitat use and activity of the chamois (*Rupicapra rupicapra* L.) in Northern Tyrol, Austria. PhD thesis, Leopold Franzens Universität, Innsbruck

Knaus, W. and W. Schröder (1975) *Das Gamswild*. Paul Parey, Hamburg

Kurtén B. (1968) *Pleistocene mammals of Europe*. Weidenfeld & Nicolson, London

Lovari, S. (1984a) *Il popolo delle rocce*. Rizzoli, Milan

Lovari, S. (1984b) Herding strategies of male Abruzzo chamois on the rut. *Acta Zool. Fenn. 172*, 91–2

Lovari, S. (1985) Behavioural repertoire of the Abruzzo chamois, *Rupicapra pyrenaica* Neumann, 1899 (Artiodactyla: Bovidae). *Säugetierkundl. Mitt. 32*, 113–36

Lovari, S. and R.F. Cosentino. (1986) Seasonal habitat selection and group size of the Abruzzo chamois (*Rupicapra pyrenaica ornata*). *Boll. Zool. 53*, 73–8

Lovari, S. and G. Rosto. (1985) Feeding rate and social stress of female chamois foraging in a group. In S. Lovari (ed.) *The biology and management of mountain ungulates*. Croom Helm, London, pp. 102–5

Lovari, S. and C. Scala. (1980) Revision of *Rupicapra* genus. 1. A statistical re-evaluation of Couturier's data on the morphometry of six chamois subspecies. *Boll. Zool. 47*, 113–24

Masini, F. (1985) Würmian and Holocene chamois of Italy. In S. Lovari (ed.) *The biology and management of mountain ungulates*, Croom Helm, London, pp. 31–44

Nascetti G., S. Lovari, P. Lanfranchi, C. Berducou, S. Mattiucci, L. Rossi and L. Bullini. (1985) Revision of *Rupicapra* genus. 3. Electrophoretic studies demonstrating species distinction of chamois populations of the Alps from those of the Apennines and Pyrenées. In S. Lovari (ed.) *The biology and management of mountain ungulates.* Croom Helm, London, pp. 56–62

Nei, M. (1972) Genetic distance between populations. *Am. Natur. 106*, 283–92

Schaller, G.B. (1977) *Mountain monarchs. Wild sheep and goats of the Himalaya.* University of Chicago Press, Chicago, Ill.

Shank, C.C. (1985) Inter- and intra-sexual segregation of chamois (*Rupicapra rupicapra*) by altitude and habitat during summer. *Z. Säugetierkunde 50*, 117–25

4

Evolutionary pathways of karyotypes of the tribe Rupicaprini

Hiroaki Soma, Hidemi Kada and Kunio Matayoshi
Department of Obstetrics & Gynaecology, Tokyo Medical College, Shinjuku, Tokyo
160, Japan

INTRODUCTION

Serows (*Capricornis*), chamois (*Rupicapra*), gorals (*Nemorhaedus*) and Rocky Mountain goats (*Oreamnos*) are all members of the tribe Rupicaprini within the family Bovidae. Ancestral bovines prospered on grassland during the Pliocene epoch 7 million years ago. Rupicaprini appeared in the Pleistocene and were distributed throughout the Eurasian continent. However, the evolutionary paths of today's species are difficult to trace because fossil records in the mountains are poor. From taxonomic studies of the tribe Rupicaprini on the basis of the skull and skin, *Rupicapra*, *Oreamnos*, *Capricornis* and *Nemorhaedus* are sufficiently similar to be combined into one tribe. In addition to morphological affinities, genetic comparisons of *Capricornis* species also confirm the ancestral stocks of the serow lineage.

THE KARYOTYPES OF *RUPICAPRA* SPECIES

Until now, cytogenetic studies on Rupicaprini species such as Rocky Mountain goat and chamois have been reported by Wurster and Benirschke (1968) and Gropp *et al.* (1970). According to chromosomal studies of Rocky Mountain goats, first reported by Wurster and Benirschke, this species has a chromosome number of 42 with nine pairs of metacentrics and a Nombre Fondamentale (NF) of 60. Karyotypes of the chamois investigated by Gropp *et al.* showed 58 acrocentrics with one pair of large metacentrics and an NF of 60. According to chromosome studies of the tribe Caprines by Nadler *et al.* (1973), there are two main lineages among surviving caprines. Starting from a hypothetical Rupicaprini ancestor with a primitive $2n = 60$, $NF = 60$ karyotype, one lineage evolved through an intermediate aoudad-like form to the true sheep with reduction in diploid number.

62

THE KARYOTYPES OF *CAPRICORNIS* SPECIES

The karyotype of *Capricornis crispus* (Japanese serow) obtained by skin biopsies at the Japan Serow Center was first investigated by Benirschke *et al.* (1972), and later was affirmed by Soma *et al.* (1981). The chromosomes of the Japanese serow showed a diploid number of $2n = 50$ with five pairs of metacentrics and submetacentrics, the remainder being acrocentric, including the sex chromosomes. The NF is therefore 60. The unusually large first element was particularly revealed by the Giemsa banding pattern (Figure 4.1).

Although Formosan serow (*Capricornis crispus swinhoei*) has lived in Taiwan as a subspecies of the Japanese serow, no cytogenetic investigation had hitherto been undertaken. We examined karyotyping of Formosan serows obtained by skin biopsies at the Taipei Zoo, Taiwan. The karyotypes of the Formosan serow showed a diploid number of $2n = 50$ with five pairs of metacentrics and submetacentrics and 20 pairs of acrocentrics including the sex chromosomes (Figure 4.2). A large No. 1 metacentric element was also seen. The NF of this serow showed the same number of 60. Thus, the karyotype of the Formosan serow is virtually identical to that of the Japanese serow.

Figure 4.1: G-banding karyotype of a female Japanese serow. $2n = 50$

Figure 4.2: G-banding karyotype of a female Formosan serow. $2n = 50$

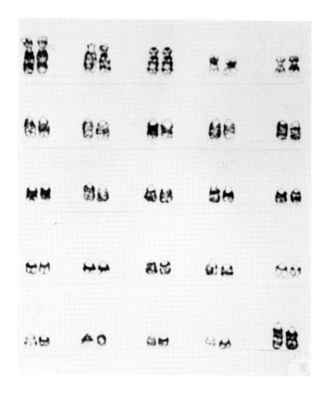

The Sumatran serow (*Capricornis sumatrensis*) consisting of 11 sub-species is the largest *Capricornis* species (Dolan 1963). The mainland serows are found in the forest mountain areas of Malaya, Sumatra, Thailand, Burma and South China. Although they are taxonomically closely related to Japanese and Formosan serows, they are as large as a calf in body size when compared with Japanese serows. The first cytogenetic study on a male Sumatran serow using blood samples obtained at Dusit Zoo, Bangkok, was reported by Fischer and Höhn (1972). The karyotype showed a diploid number of $2n = 46$ including five pairs of metacentrics and the unusual metacentric sex chromosomes giving a NF of 58.

In order to clarify these unusual karyotypic findings reported by Fischer and Höhn, we undertook a chromosomal investigation on cultures of skin fibroblasts obtained from a pair of Sumatran serows at the Dusit Zoo, Bangkok, in 1982 (Soma *et al.* 1982). The Giemsa banding patterns of the Sumatran serows had a diploid number of $2n = 48$ with six pairs of meta-centrics and submetacentrics and 18 pairs of acrocentrics including the sex chromosomes (Figure 4.3). Accordingly, the NF of the Sumatran serow is

Figure 4.3: G-banding karyotype of a male Sumatran serow. $2n = 48$

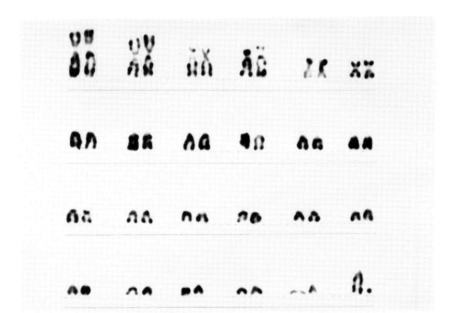

60, the same number as those of Japanese and Formosan serows. From a comparison of the Giemsa banding pattern of each chromosome of the Japanese serow, Formosan serow and Sumatran serow, similar Giemsa banding patterns in all three species were recognised (Soma and Kada 1984). However, in the Sumatran serow, an additional large metacentric pair (No. 2) is found. That is, pair No. 2 of the metacentrics seem to be the result of a Robertsonian fusion of pairs No. 9 and No. 13. Thus, the Sumatran serow has the same NF number as that of other serows (Soma 1984).

THE KARYOTYPES OF *NEMORHAEDUS GORAL*

Nemorhaedus goral is widely distributed from the Himalayas via China to North Korea and is taxonomically classified into nine or ten subspecies (Dolan 1963). One of the subspecies is commonly called a Korean serow, and resembles the Japanese serow in gross appearance (Figure 4.4).

The karyotypes of a male *Nemorhaedus cranbrooki* (red goral) and a female *Nemorhaedus griseus* (West Chinese goral) were shown by Wurster (1972) to have an unusual diploid number of chromosomes of 55. The question of whether the goral has a diploid number of chromosomes of 54 or 56 has still not been determined. On the other hand, it still remains a

65

Figure 4.4: North Korean gorals in Tierpark Berlin

matter of opinion as to whether *Nemorhaedus goral* could have evolved from a species of serow, or whether the two groups of species could have evolved from a common ancestral taxon.

Recently Groves and Grubb (1985) suggested that goral might have derived from a species of serow because of a suggestive resemblance between the small serows and gorals. In order to compare the chromosomal variations of *Nemorhaedus goral* in each district, we investigated the karyotypes of gorals from Nepal, China and North Korea, respectively. Karyotypes of a pair of *Nemorhaedus g. caudatis* sent from China to the Japan Serow Center showed a diploid number of $2n = 56$, which consisted of 28 pairs of acrocentrics including the sex chromosomes (Soma *et al.* 1980). The karyotypes of *Nemorhaedus g. goral* examined at Kathmandu Central Zoo in Nepal showed the same diploid number of $2n = 56$, consisting of acrocentrics including sex chromosomes. The C-banding pattern of the goral chromosomes displayed heterochromatin staining of the centromeres only. Accordingly, the NF number of goral is 56. We attempted to clarify the differences in a chromosome study of so-called Korean serow and Japanese serow. Recently four *Nemorhaedus g. raddeanus* were sent from Pyongyang, North Korea, to Tierpark Berlin, East Germany (Pohle and Rodloff 1985). Fortunately we were able to examine the karyotypes of five gorals including two pairs of North Korean gorals (*Nemorhaedus g. raddeanus*) and one female Chinese goral (*Nemorhaedus g. arnouxianus*) by courtesy of Tierpark Berlin in December 1985. The karyotypes of Mid-Chinese goral and North Korean gorals showed the diploid number of $2n = 56$, consisting of 28 pairs of acrocentrics including sex chromosomes (Figure 4.5). A comparison of the Giemsa banding patterns of the chromosomes of four subspecies among gorals in different areas revealed the identical structure of all the chromosomes (Figure 4.6). The C-banding patterns of North Korean gorals displayed heterochromatin staining of the centromeres.

Figure 4.5: G-banding karyotype of a male Korean goral. $2n = 56$

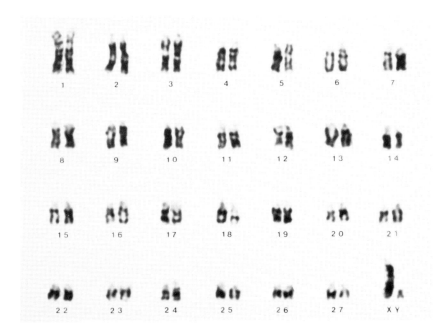

In order to know the possible site of rRNA genes, the nucleolus organiser regions (NORs) of the Korean gorals were stained by a silver staining technique and compared with those of the Sumatran serows. The NORs in the Sumatran serow were found to occur only in the long arm of No. 4 metacentric (Figure 4.7), whereas the NORs in the Korean goral were present on chromosome pairs No. 1, No. 4 and No. 10 (Figure 4.8). The NORs in Mid-Chinese goral were located in the long arm of Nos. 1, 3 and 4 metacentrics. This seems to be a common finding between the karyotypes of *Capricornis* and *Nemorhaedus*. The result obtained from NOR

Figure 4.6: A comparison of G-banding patterns among *Nemorhaedus goral*. J, *N. g. caudatis* (Japan Serow Center). N, *N. g. goral* (Central Zoo, Kathmandu, Nepal). MC, Mid-Chinese goral (*N. g. griseus*) (Tierpark Berlin). NK, North Korean goral (*N. g. raddeanus*) (Tierpark Berlin)

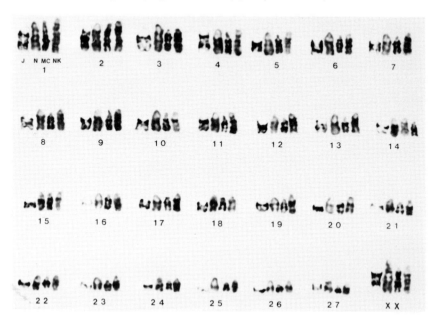

staining may hint at a close genetic relation between *Nemorhaedus goral* and *Capricornis*.

DISCUSSION

It seems difficult to explain the reason why the chromosome number of these gorals was 56, being different from other Rupicaprini, even though both animals are the most primitive living species. In addition, their exact evolutionary divergence from their hypothetical ancestor still remains unclear, and therefore further verification may be required. However, judging from the Giemsa banding analysis of chromosomes of the Korean goral, it seems probable that the long acrocentric chromosomes of two pairs, No. 1 and No. 2, might have been constructed by a tandem fusion, thereby leading to a reduction of the diploid number to 56. The NF number of goral was originally 60. A future study of the highly repetitive DNA in these species may help to elucidate gene expression.

Volf (1983) reported successful cross-breeding at Prague Zoo between *Nemorhaedus g. griseus* which originated from West China and *N. g.*

Figure 4.7: A silver-stained karyotype of a male Sumatran serow. Arrow indicates localisation of NORs

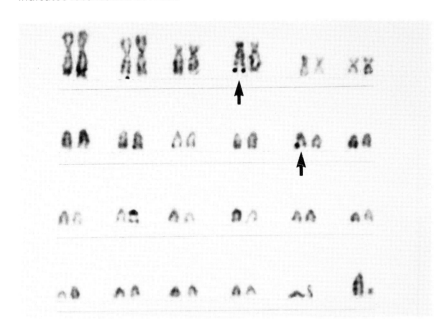

raddeanus which came from North China but their hybrids were sterile. He suggested that North Chinese goral may constitute an independent species, so chromosomal research is necessary. However, we were unable to find significant differences of chromosomal constitution between Mid-Chinese goral and North Korean goral using a Giemsa banding technique, even though the presence of the NORs between both gorals was different in this study. Furthermore, chromosomal analysis of the hybrid between both gorals is required.

We have already reported that the karyotypes of Mongolian gazelles and saigas exhibit a similar chromosome pattern with a diploid number of $2n = 60$ showing a fundamental type of all the acrocentrics (Soma *et al.* 1979). Accordingly, from the results obtained, we suggest that three chromosomal lineages consisting of centric fusion, tandem fusion and fundamental number type could have been evolved from the Rupicaprini.

According to our conjecture about the genetic evolutional paths of the tribe Rupicaprini, precursors of the Japanese serow and Formosan serow migrated from the major continent to each island when the islands were still connected to the mainland during the Pleistocene. These serows have settled in the highlands of both islands. Accordingly, both serows remained small in size, retaining primitive lineage ancestral characters, and they may have the old primitive chromosome number. The Sumatran serow is the

Figure 4.8: A silver-stained karyotype of a male North Korean goral. Arrow indicates localisation of NORs

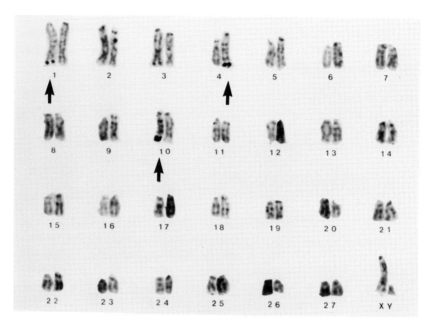

tropical giant among the living forms of the rupicaprids (Geist 1983), and the mainland serow that derived from the same lineage as the other serows has become larger and more advanced than other serows and spread from south-east Asia to China (Soma and Kada 1984). As the mainland serows radiated, they enlarged and reduced chromosome number (V. Geist, pers. comm.). The current mainland serow may well vary in chromosome numbers.

On the other hand, according to Schaller (1977), gorals separated from the ancestor of the Rupicaprini during the Miocene and Lower Pliocene, and spread from the Himalayas via China to Korea and Siberia over a long period. their chromosome constitutions showed the identical karyotypic situations at each district, giving rise to a reduction of chromosome number due to probable tandem fusion. However, such differences of cytogenetic stock between the serows and the gorals will await future study.

REFERENCES

Benirschke, K., H. Soma and T. Ito. (1972) The chromosomes of the Japanese serow, *Capricornis crispus* (Temminck). *Proc. Japan Acad. 48*, 608–12

Dolan, J.M. (1963) Beitrag zur Systematischen Gliederung des Tribes Rupicaprini Simpson. *Z. Zool. Syst. Evolutionsforsch. 1*, 311–407

Fischer, H. and H. Höhn. (1972) Der Karyotyp des Serau (*Capricornis sumatrensis*, Bechstein, 1799). *Giess Beitr. Erbpath. Zuchthyg. 4*, 8–15

Geist, V. (1983) On the evolution of ice age mammals and its significance to an understanding of speciations. *ASB Bulletin 30*, 109–33

Gropp, A., D. Giers, R. Fernandez-Donoso, L. Tiepolo and M. Fraccaro. (1970) The chromosomes of the chamois (tribe Rupicaprini Simpson). *Cytogenetics 9*, 1–8

Groves, C.P. and P. Grubb (1985) Reclassification of the serow and gorals (*Nemorhaedus*: Bovidae). In S. Lovari (ed.) *The biology and management of mountain ungulates*. Croom Helm, London, pp. 45–50

Nadler, C.F., R.S. Hoffmann and A. Woolf (1973) G-band patterns, chromosomal homologies, and evolutionary relationships among wild sheep, goats, and aoudads (Mammalia, Artiodactyla). *Experientia 30*, 744–6

Pohle, C. and K. Rodloff. (1985) Gorale im Tierpark Berlin. *Milu, Berlin 6*, Band 6, Heft 1/2 161–78

Schaller, G.B. (1977) *Mountain monarchs*. University of Chicago Press, Chicago, Ill.

Soma, H. and H. Kada. (1984) Evolutionary pathway of chromosomes of the *Capricornis*. In O.A. Ryder and M.L. Byrd (eds) *One Medicine*. Springer, Berlin, pp. 109–18

Soma, H., H. Kada, K. Matayoshi, T. Kiyokawa, T. Ito, M. Miyashita and K. Nagase. (1980) Some chromosomal aspects of *Naemorhedus goral* (goral) and *Procapra gutturosa* (Mongolian gazelle). *Proc. Japan Acad. 56B*, 273–7

Soma, H., H. Kada, K. Matayoshi, M.T. Tsai, T. Kiyokawa, T. Ito, K-P. Wong, B.P.C. Chen and S-C. Tseng. (1981) Cytogenetic similarities between the Formosan serow (*Capricornis swinhoei*) and the Japanese serow (*Capricornis crispus*). *Proc. Japan Acad. 57*, 254–9

Soma, H., H. Kada, K. Matayoshi, K. Suzuki, C. Meckvichai, A. Mahannop and B. Vatanaromya. (1982) The chromosomes of the Sumatran serow (*Capricornis sumatrensis*). *Proc. Japan Acad. 58B*, 265–9

Soma, H., T. Kiyokawa, K. Matayoshi, I. Tarumoto, M. Miyashita and K. Nagase (1979) The chromosomes of the Mongolian gazelle (*Procapra gutturosa*), a rare species of antelope. *Proc. Japan Acad. 55B*, 6–9

Volf, J. (1983) Die Zucht von Goralen (genus *Nemorhaedus* Smith, 1827) im Zoologischen Garten in Prag. *Zool. Garten N.F., Jena 53*, 354–8

Wurster, D.H. (1972) Sex-chromosome translocation and karyotypes in bovid tribes. *Cytogenetics 11*: 197, cited in T.C. Hsu and K. Benirschke (1973) *An atlas of mammalian chromosomes*, Vol. 7. Springer, New York, p. 342

Wurster, D.H. and K. Benirschke. (1968) The chromosomes of the Rocky Mountain goat (*Oreamnos americanus*). *Mamm. Chromo. Newsletter 9*, 80–1

Part Two:
Ecological Distribution
and Behaviour of *Capricornis*

5

Survey of 217 Japanese serows, *Capricornis crispus*, bred in captivity

Atsushi Komori

Japanese Association of Zoological Gardens and Aquariums, 38 Ueno Park,
Taito-ku, Tokyo 110, Japan

INTRODUCTION

The breeding of Japanese serow, *Capricornis crispus*, in captivity began in 1965. Before that, since 1950 there had been three cases of dams giving birth soon after their capture, that is the dams were caught when pregnant. The first case of the export of Japanese serow from Japan abroad was the sending of a pair of them to Peking Zoo, China, in 1973. Meanwhile work on an international register of the species was started in 1971 by myself at Tama Zoological Park. I have therefore been ideally placed to collect information concerning the breeding of Japanese serow in captivity. Here I will attempt to survey this information.

FLUCTUATIONS IN ANNUAL BREEDING

From 1950 to 1984, 217 Japanese serows were born in 24 institutions, including seven cases of dams giving birth soon after their capture from the wild (hereafter referred to as semi-wild breeding). A complete list of them and their founder group is given in Tables 5.1 to 5.3. Excluding the three early cases of semi-wild breeding, the first offspring of Japanese serow bred in captivity was the female bred at Japan Serow Center, Mt. Gozaisho, Mie Prefecture, on 26 August 1965 and marked as 'B004, JSC-2b'.

The mark (number and name) means as follows: the capital letter at the beginning , 'B', means 'bred in captivity'; 'C' means 'caught from the wild'. The number ('004' in this example) is the series number according to the date of birth or capture. The group of three capital letters ('JSC' in this case) are the code name of the site of birth or capture. The following number ('2' in this case) is the series number of capture in each site of capture. The following small letter ('b' in this case) is the alphabetical series of offspring of a dam's line. As a result, in the example, 'JSC-2b' means 'the second offspring of the dam JSC-2', which is the first generation in captivity.

Table 5.1: Japanese serows, *Capricornis crispus*, bred in captivity

No.	Sex	Studbook name	Date of birth	Dam	Sire
B001	♂	UEN-7a	1950. 5.27	C007 UEN-7	Wild
B002	♀	KYT-2a	1955. 5. 8	C026 KYT-2	Wild
B003	♀	JSC-2a	1964. 6.17	C039 JSC-2	Wild
B004	♀	JSC-2b	1965. 8.12	C039 JSC-2	C038 JSC-1
B005	♀	KOB-1a	1965. 8.26	C047 KOB-1	C048 KOB-2
B006	♂	KOB-1b	1966. 6.10	C047 KOB-1	C048 KOB-2
B007	♂	JSC-2c	1966. 8. 1	C039 JSC-2	C038 JSC-1
B008	♀	KOB-1c	1967. 5.24	C047 KOB-1	C048 KOB-2
B009	♂	KYT-7b	1967. 7. 7	C044 KYT-7	C036 KYT-4
B010	♀	JSC-2d	1967. 7.22	C039 JSC-2	C038 JSC-1
B011	♀	KOB-1d	1968 5 1	C047 KOB-1	C048 KOB-2
B012	♂	JSC-2e	1968. 5.13	C039 JSC-2	C038 JSC-1
B013	♂	TYM-1a	1968. 5.17	C051 TYM-1	C054 TYM-4
B014	♀	JSC-2ba	1968. 5.24	B004 JSC-2b	C059 JSC-3
B015	♀	KYT-7c	1968. 5.27	C044 KYT-7	C036 KYT-4
B016	♂	KOB-1aa	1968. 6.18	B005 KOB-1a	C048 KOB-2
B017	♂	JSC-2aa	1968. 7. 3	B003 JSC-2a	C059 JSC-3
B018	♂	KOB-1e	1969. 5. 3	C047 KOB-1	C048 KOB-2
B019	♀	TYM-1b	1969. 5.10	C051 TYM-1	C054 TYM-4
B020	♀	KYT-7d	1969. 5.14	C044 KYT-7	C036 KYT-4
B021	♂	JSC-2bb	1969. 6.14	B004 JSC-2b	C059 JSC-3
B022	♂	JSC-2ab	1969. 6.20	B003 JSC-2a	C059 JSC-3
B023	♂	KOB-1ab	1969. 6.24	B005 KOB-1a	C048 KOB-2
B024	♂	KOB-1ca	1969. 7.15	B008 KOB-1c	C048 KOB-2
B025	♀	JSC-2f	1969. 7.22	C039 JSC-2	C038 JSC-1
B026	♂	JSC-2g	1970. 5.13	C039 JSC-2	C072 JSC-5
B027	♂	TYM-1c	1970. 5.15	C051 TYM-1	C054 TYM-4
B028	♂	OMC-12a	1970. 5.29	C076 OMC-12	C055 OMC-4
B029	♀	JSC-2ac	1970. 6.11	B003 JSC-2a	B007 JSC-2c
B030	♂	KOB-1cb	1970. 6.12	B008 KOB-1c	C048 KOB-2
B031	♂	JSC-2bc	1970. 6.16	B004 JSC-2b	C059 JSC-3
B032	♂	KYT-7e	1970. 7. 8	C044 KYT-7	C036 KYT-4
B033	♂	KOB-1f	1970. 9. 2	C047 KOB-1	C048 KOB-2
B034	♀	KOB-1ac	1970.10.19	B005 KOB-1a	B006 KOB-1b
B035	♂	JSC-2ad	1971. 5.18	B003 JSC-2a	B007 JSC-2c
B036	♀	JSC-2da	1971. 5.18	B010 JSC-2d	C059 JSC-3
B037	♀	OMC-12b	1971. 5.20	C076 OMC-12	C055 OMC-4
B038	♂	JSC-2bd	1971. 5.24	B004 JSC-2b	C059 JSC-3
B039	♀	TYM-1d	1971. 5.25	C051 TYM-1	C054 TYM-4
B040	♂	JSC-2baa	1971. 5.30	B014 JSC-2ba	B012 JSC-2e
B041	♀	KYT-7f	1971. 6.20	C044 KYT-7	C036 KYT-4
B042	♀	KOB-1cc	1971. 6.30	B008 KOB-1c	C048 KOB-2
B043	♂	WKY-3a	1972. 4.20	C092 WKY-3	C091 WKY-2
B044	♂	JSC-2h	1972. 4.24	C039 JSC-2	C072 JSC-5
B045	♀	KOB-1ad	1972. 4.29	B005 KOB-1a	B006 KOB-1b
B046	♂	KOB-1cd	1972. 5.19	B008 KOB-1c	C048 KOB-2
B047	♂	JSC-2ae	1972. 5.20	B003 JSC-2a	B007 JSC-2c
B048	♀	TYM-1ba	1972. 5.21	B019 TYM-1b	C054 TYM-4
B049	♀	OMC-12c	1972. 5.22	C076 OMC-12	C055 OMC-4
B050	♀	OMC-11a	1972. 6. 3	C074 OMC-11	B028 OMC-12a
B051	♂	TYM-1e	1972. 6. 7	C051 TYM-1	C054 TYM-4
B052	♂	JSC-2fa	1972. 6.17	B025 JSC-2f	B022 JSC-2ab
B053	♀	JSC-2be	1972. 6.18	B004 JSC-2b	C059 JSC-3
B054	♀	JSC-2bab	1972. 7. 2	B014 JSC-2ba	B012 JSC-2e

Site of birth	Owner	Since	Date of death	Note
Ueno Zool. Gardens	Ueno Zool. Gardens	Since Birth	1950. 8. 7	
Kyoto Zoo	Kyoto Zoo	Since Birth	1955. 7.28	
Japan Serow Center	Japan Serow Center	Since Birth	1976. 7.27	
Japan Serow Center	Japan Serow Center	Since Birth	1975. 2. 8	
Kobe Muni. Arboretum	Kobe Oji Zoo	Since Birth	1983. 5.29	
Kobe Muni. Arboretum	Kobe Oji Zoo	Since Birth	1981. 1.20	
Japan Serow Center	Japan Serow Center	Since Birth	1981. 2. 1	
Kobe Muni. Arboretum	Kobe Oji Zoo	Since Birth	1974. 9. 9	
Kyoto Zoo	Kyoto Zoo	Since Birth	1967. 9.15	
Japan Serow Center	Japan Serow Center	Since Birth	1976.11 4	
Kobe Muni. Arboretum	Kobe Oji Zoo	Since Birth	1968. 9.24	
Japan Serow Center	Tama Zool. Park	1972.10.21	1976. 2.29	
Tateyama Fudokigaoka	Toyama Prefecture	Since Birth	1968. 7.30	
Japan Serow Center	Japan Serow Center	Since Birth	1978. 7.30	
Kyoto Zoo	Kyoto Zoo	Since Birth	1968. 9.18	
Kobe Muni. Arboretum	Kobe Oji Zoo	Since Birth	1968. 6.21	
Japan Serow Center	Japan Serow Center	Since Birth	1968. 7.18	
Kobe Muni. Arboretum	Kobe Oji Zoo	Since Birth	1969. 6.21	
Tateyama Fudokigaoka	Maruyama Zoo	1975.10. 8	1984. 1.30	
Kyoto Zoo	Kyoto Zoo	Since Birth	1969. 8.17	
Japan Serow Center	Tama Zool. Park	1977.11. 7	1980.10.13	
Japan Serow Center	Japan Serow Center	Since Birth	1978.10. 1	
Kobe Muni. Arboretum	Kobe Oji Zoo	Since Birth	1973. 9. 3	
Kobe Muni. Arboretum	Kobe Oji Zoo	Since Birth	1971.10.17	
Japan Serow Center	Japan Serow Center	Since Birth	1971. 1. 8	
Japan Serow Center	Japan Serow Center	Since Birth	1971. 1.12	
Tateyama Fudokigaoka	Toyama Prefecture	Since Birth		Alive
Omachi Alpine Museum	Peking Zoo, China	1973. 4. 3		Alive
Japan Serow Center	Tama Zool. Park	1972.10.21		Alive
Kobe Muni. Arboretum	Kobe Oji Zoo	Since Birth	1970. 6.22	
Japan Serow Center	Japan Serow Center	Since Birth	1970. 6.23	
Kyoto Zoo	Kyoto Zoo	Since Birth		Alive
Kobe Muni. Arboretum	Kobe Oji Zoo	Since Birth	1971. 4. 5	
Kobe Muni. Arboretum	Kobe Oji Zoo	Since Birth	1983. 7.24	
Japan Serow Center	Japan Serow Center	Since Birth	1971.10.22	
Japan Serow Center	Japan Serow Center	Since Birth	1971. 5.31	
Omachi Alpine Museum	Omachi Alpine Museum	Since Birth	1975. 8. 8	
Japan Serow Center	Los Angeles Zoo	1976.10.14	1980.11. 3	
Tateyama Fudokigaoka	Higashiyama Zoo	1977. 5.25		Alive
Japan Serow Center	Japan Serow Center	Since Birth	1972. 2. 4	
Kyoto Zoo	Kyoto Zoo	Since Birth		Alive
Kobe Muni. Arboretum	Kobe Oji Zoo	Since Birth	1978.10. 9	
Oto-mura	Oto-mura	Since Birth	1979. 5. 3	
Japan Serow Center	Japan Serow Center	Since Birth	1972.11.17	
Kobe Muni. Arboretum	Kobe Oji Zoo	Since Birth	1972. 5. 1	
Kobe Muni. Arboretum	Oto-mura	1974. 6. 8	1974.12.12	
Japan Serow Center	Japan Serow Center	Since Birth	1972. 6.11	
Tateyama Fudokigaoka	Ishikawa Arboretum	1979. 9.20		Alive
Omachi Alpine Museum	Omachi Alpine Museum	Since Birth	1973.12.10	
Omachi Alpine Museum	Omachi Alpine Museum	Since Birth	1973. 2.13	
Tateyama Fudokigaoka	Toyama Prefecture	Since Birth	1974. 4.24	
Japan Serow Center	Japan Serow Center	Since Birth	1981. 2.25	
Japan Serow Center	Japan Serow Center	Since Birth	1972.11.11	
Japan Serow Center	Japan Serow Center	Since Birth	1972.12.25	

77

Table 5.1 continued

No.	Sex	Studbook name	Date of birth	Dam	Sire
B055	♂	KYT-7g	1972. 9.19	C044 KYT-7	C036 KYT-4
B056	♂	KOB-1ce	1973. 4.28	B008 KOB-1c	C048 KOB-2
B057	♀	JSC-2i	1973. 5. 3	C039 JSC-2	C072 JSC-5
B058	♂	JSC-2af	1973. 5. 7	B003 JSC-2a	B007 JSC-2c
B059	♂	TYM-1f	1973. 5.12	C051 TYM-1	B027 TYM-1c
B060	♀	JSC-2bf	1973. 5.24	B004 JSC-2b	C059 JSC-3
B061	♀	OMC-12d	1973. 5.24	C076 OMC-12	C055 OMC-4
B062	♂	TAM-J-2a	1973. 5.27	B029 JSC-2ac	B012 JSC-2e
B063	♂	KYT-7fa	1973. 6. 4	B041 KYT-7f	B032 KYT-7e
B064	♀	JSC-2db	1973. 6. 4	B010 JSC-2d	B021 JSC-2bb
B065	♀	TYM-1bb	1973. 6. 6	B019 TYM-1b	B027 TYM-1c
B066	♀	JSC-2fb	1973. 6. 8	B025 JSC-2f	B022 JSC-2ab
B067	♀	WKY-5a	1973. 6.10	C094 WKY-5	C093 WKY-4
B068	♂	KYT-7h	1973. 7.30	C044 KYT-7	C036 KYT-4
B069	♂	TYM-1da	1973. 9. 5	B039 TYM-1d	C054 TYM-4
B070	♂	KOB-1cd	1974. 5. 3	B008 KOB-1c	C048 KOB-2
B071	♂	TYM-1g	1974. 5.10	C051 TYM-1	B051 TYM-1e
B072	♂	JSC-2ag	1974. 5.10	B003 JSC-2a	B007 JSC-2c
B073	♂	JSC-2bg	1974. 5.15	B004 JSC-2b	C059 JSC-3
B074	♀	JSC-2bac	1974. 5.16	B014 JSC-2ba	B022 JSC-2ab
B075	♂	TYM-1bc	1974. 5.24	B019 TYM-1b	B027 TYM-1c
B076	♀	KYT-2fb	1974. 5.29	B041 KYT-7f	B032 KYT-7e
B077	♀	OMC-12e	1974. 5.30	C076 OMC-12	C055 OMC-4
B078	♀	TYM-1db	1974. 6. 9	B039 TYM-1d	C054 TYM-4
B079	♂	OMC-12ba	1974. 6.22	B037 OMC-12b	B028 OMC-12a
B080	♀	PKN-0-2a	1974. 8.25	C106 OMC-20	B028 OMC-12a
B081	♂	KYT-7i	1974. 9.18	C044 KYT-7	B032 KYT-7e
B082	♀	TAM-J-2b	1975. 4.13	B029 JSC-2ac	B012 JSC-2e
B083	♂	TYM-1h	1975. 5. 5	C051 TYM-1	C054 TYM-4
B084	♂	WKY-5b	1975. 5. 6	C094 WKY-5	B043 WKY-3a
B085	♂	KYT-7fc	1975. 5.16	B041 KYT-7f	B032 KYT-7e
B086	♂	TYM-1baa	1975. 5.20	B048 TYM-1ba	B027 TYM-1c
B087	♂	TYM-1bd	1975. 5.22	B019 TYM-1b	B059 TYM-1f
B088	♀	JSC-2bad	1975. 5.31	B014 JSC-2ba	B022 JSC2ab
B089	♀	KOB-1cca	1975. 6.19	B042 KOB-1cc	C093 WYK-4
B090	♂	PKN-0-2b	1975. 6.27	C106 OMC-20	B028 OMC-12a
B091	♀	WKY-5aa	1975. 7.10	B067 WKY-5a	B043 WKY-3a
B092	♂	OMC-12bb	1975. 7.18	B037 OMC-12b	C055 OMC-4
B093	♀	TAM-J-2c	1976. 5. 2	B029 JSC-2ac	B012 JSC-2e
B094	♂	TYM-1i	1976. 5. 3	C051 TYM-1	C054 TYM-4
B095	♀	TYM-1dc	1976. 5. 3	B039 TYM-1d	B069 TYM-1da
B096	♂	WKY-5c	1976. 5. 3	C094 WKY-5	B043 WKY-3a
B097	♂	KOB-1ccb	1976. 5. 7	B042 KOB-1cc	C093 WKY-4
B098	♂	KYT-7fd	1976. 5. 8	B041 KYT-7f	B032 KYT-7e
B099	♀	TYM-1bab	1976. 5. 9	B048 TYM-1ba	B027 TYM-1c
B100	♀	NRA-5a	1976. 5.22	C142 NRA-5	C114 NRA-3
B101	♀	PKN-0-2c	1976. 5.24	C106 OMC-20	B028 OMC-12a
B102	♀	JSC-2bae	1976. 6. 1	B014 JSC-2ba	B022 JSC-2ab
B103	♀	WKY-5ab	1976. 6. 1	B067 WKY-5a	B043 WKY-3a
B104	♂	OMC-12f	1976. 6. 7	C076 OMC-12	C055 OMC-4
B105	♀	JSC-2dc	1976. 6.20	B010 JSC-2d	B021 JSC-2bb
B106	♂	OMC-23a	1976. 7. 7	C116 OMC-23	C140 OMC-27
B107	♂	TYM-1dba	1976. 7.25	B078 TYM-1db	B069 TYM-1da
B108	♂	WKY-5d	1977. 4.23	C094 WKY-5	B043 WKY-3a

Site of birth	Owner	Since	Date of death	Note
Kyoto Zoo	Ayameike Park Zoo	1978. 6.16	1983. 7.18	
Kobe Muni. Arboretum	Kobe Oji Zoo	Since Birth	1974.12. 1	
Japan Serow Center	Japan Serow Center	Since Birth	1973. 5. 8	
Japan Serow Center	Japan Serow Center	Since Birth	1973. 5.16	
Tateyama Fudokigaoka	Maruyama Zoo	1975.10. 87	1975.10.08	
Japan Serow Center	Japan Serow Center	Since Birth	1974.10.16	
Omachi Alpine Museum	Omachi Alpine Museum	Since Birth	1973. 8.28	
Tama Zoological Park	Tama Zoological Park	Since Birth	1976. 7.17	
Kyoto Zoo	Kyoto Zoo	Since Birth	1973. 8.12	
Japan Serow Center	Japan Serow Center	Since Birth	1973.10.22	
Tateyama Fudokigaoka	Toyama Prefecture	Since Birth	1974. 2.21	
Japan Serow Center	Japan Serow Center	Since Birth	1976.10.14	
Oto-Mura	Oto-Mura	Since Birth	1979 8.30	
Kyoto Zoo	Kyoto Zoo	Since Birth	1975. 7.11	
Tateyama Fudokigaoka	Higshiyama Zoo	1977. 5.25		Alive
Kobe Muni. Arboretum	Kobe Oji Zoo	Since Birth	1974. 7.15	
Tateyama Fudokigaoka	Nihondaira Zoo	1975. 9. 7	1975. 9. 7	
Japan Serow Center	Japan Serow Center	Since Birth	1974. 6. 2	
Japan Serow Center	Japan Serow Center	Since Birth	1974. 9.11	
Japan Serow Center	Japan Serow Center	Since Birth	1974.12.20	
Tateyama Fudokigaoka	Toyama Prefecture	Since Birth	1974. 6.27	
Kyoto Zoo	Kyoto Zoo	Since Birth	1974. 9. 5	
Omachi Alpine Museum	Omachi Alpine Museum	Since Birth	1974.11.28	
Tateyama Fudokigaoka	San Diego Zoo	1977.11.30	1978. 8.24	
Omachi Alpine Museum	Omachi Alpine Museum	Since Birth	1974.10. 2	
Peking Zoo	Peking Zoo	Since Birth	1975. 9. 9	
Kyoto Zoo	Kyoto Zoo	Since Birth	1974.10. 2	
Tama Zoological Park	Tama Zoological Park	Since Birth	1975. 5. 4	
Tateyama Fudokigaoka	San Diego Zoo	1977.11.30	1978. 9. 5	
Oto-Mura	Oto-Mura	Since Birth	1978. 5. 6	
Kyoto Zoo	Kyoto Zoo	Since Birth	1975. 6.17	
Tateyama Fudokigaoka	Toyama Prefecture	Since Birth	1975. 5.20	
Tateyama Fudokigaoka	Maruyama Zoo	1975.10. 8	1975.10.26	
Japan Serow Center	Japan Serow Center	Since Birth	1975.10.30	
Kobe Muni. Arboretum	Kobe Oji Zoo	Since Birth	1976.11.19	
Peking Zoo	Peking Zoo	Since Birth	1975. 7. 1	
Oto-mura	Oto-mura	Since Birth	1979. 9.29	
Omachi Alpine Museum	Omachi Alpine Museum	Since Birth	1975. 8. 9	
Tama Zoological Park	Tama Zoological Park	Since Birth	1976. 9. 1	
Tateyama Fudokigaoka	San Diego Zoo	1979.12. 6		Alive
Tateyama Fudokigaoka	San Diego Zoo	1979.12. 6	1980. 5. 9	
Oto-mura	Oto-mura	Since Birth	1976. 6.10	
Kobe Muni. Arboretum	Kobe Oji Zoo	Since Birth		Alive
Kyoto Zoo	Kyoto Zoo	Since Birth	1976. 5.11	
Tateyama Fudokigaoka	Peking Zoo	1978.11. 5		Alive
Ayameike Park Zoo	Ayameike Park Zoo	Since Birth	1979. 8.28	
Peking Zoo	Peking Zoo	Since Birth	1978. 6. 1	
Japan Serow Center	Japan Serow Center	Since Birth	1976.11. 1	
Oto-mura	Oto-mura	Since Birth	1984. 7. 7	
Omachi Alpine Museum	Schoenbrun Zoo	1984.11. 5		Alive
Japan Serow Center	Japan Serow Center	Since Birth	1976. 6.21	
Omachi Alpine Museum	Omachi Alpine Museum	Since Birth	1980. 6.30	
Tateyama Fudokigaoka	Peking Zoo	1978.11. 5		Alive
Oto-mura	Oto-mura	Since Birth	1977. 4.29	

Table 5.1 continued

No.	Sex	Studbook name	Date of birth	Dam	Sire
B109	♀	TYM-1j	1977. 5. 1	C051 TYM-1	C054 TYM-4
B110	♀	KYT-7fe	1977. 5. 7	B041 KYT-7f	B032 KYT-7e
B111	♂	NGY-T-1d	1977. 5.13	B039 TYM-1d	B069 TYM-1da
B112	♂	TYM-1bac	1977. 5.14	B048 TYM-1ba	B027 TYM-1c
B113	♀	PKN-0-2d	1977. 5.18	C106 OMC-20	B028 OMC-12a
B114	♂	WKY-5ac	1977. 5.19	B067 WKY-5a	B043 WKY-3a
B115	♂	OMC-23b	1977. 5.31	C116 OMC-23	C140 OMC-27
B116	♂	TYM-1dbb	1977. 6.21	B078 TYM-1db	B083 TYM-1h
B117	♂	TYM-5a	1977. 6.21	C159 TYM-5	C054 TYM-4
B118	♂	KOB-1ccc	1977. 6.23	B042 KOB-1cc	B006 KOB-1b
B119	♀	WKY-3b	1978. 4.16	C092 WKY-3	C091 WKY-2
B120	♀	SAN-N-2b	1978. 4.20	C142 NRA-5	C114 NRA-3
B121	♂	JSC-2baf	1978. 5. 4	B014 JSC-2ba	B022 JSC-2ab
B122	♀	TYM-1k	1978. 5.11	C051 TYM-1	C054 TYM-4
B123	♂	NGY-T-1e	1978. 5.13	B039 TYM-1d	B069 TYM-1da
B124	♂	PKN-0-2e	1978. 5.16	C106 OMC-20	B028 OMC-12a
B125	♂	WKY-5e	1978. 5.19	C094 WKY-5	B043 WKY-3a
B126	♂	WKY-5ad	1978. 5.20	B043 WKY-3a	B068 WKY-5a
B127	♀	WKY-5aaa	1978. 5.20	B091 WKY-5aa	B084 WKY-5b
B128	♂	KYT-7ff	1978. 5.23	B041 KYT-7f	B032 KYT-7e
B129	♀	TYM-1dca	1978. 6. 7	B095 TYM-1dc	C094 TYM-1i
B130	♂	KOB-1ccd	1978. 6. 9	B042 KOB-1cc	B006 KOB-1b
B131	♂	TYM-1baba	1978. 6.18	B099 TYM-1bab	B107 TYM-1dba
B132	♀	OMC-23c	1978. 6.26	C116 OMC-23	B104 OMC-12f
B133	♀	TAM-J-2d	1978. 7.24	B029 JSC-2ac	B021 JSC-2bb
B134	♂	WKY-5aba	1978. 9.14	B103 WKY-5ab	B043 WKY-3a
B135	♀	OMC-12g	1978.11. 9	C076 OMC-12	B104 OMC-12f
B136	♀	TYM-1bad	1979. 4.24	B048 TYM-1ba	B027 TYM-1c
B137	♂	NGY-T-1f	1979. 5. 3	B039 TYM-1d	B069 TYM-1da
B138	♂	TAM-J-2e	1979. 5. 6	B029 JSC-2ac	B021 JSC-2bb
B139	♀	TYM-11	1979. 5.18	C051 TYM-1	C054 TYM-4
B140	♀	TYM-1dcb	1979. 5.28	B095 TYM-1dc	C094 TYM-1i
B141	♂	GER-2a	1979. 5.29	C233 GER-2	Wild
B142	♀	PKN-0-2f	1979. 5.30	C106 OMC-20	B028 OMC-12a
B143	♀	TYM-5b	1979. 6.11	C159 TYM-5	B116 TYM-1dbb
B144	♂	OMC-23d	1979. 6.12	C116 OMC-23	C140 OMC-27
B145	♂	HRS-3a	1979. 6. 0	C133 HRS-3	C131 HRS-1
B146	♀	SPR-T-1e	1979. 7. 5	B019 TYM-1b	B112 TYM-1bac
B147	♂	KYT-7fg	1979. 9.29	B041 KYT-7f	B032 KYT-7e
B148	♀	GER-18a	1980. 4. 2	C304 GER-18	Wild
B149	♂	TAM-J-2f	1980. 4.22	B029 JSC-2ac	B021 JSC-2bb
B150	♀	NGY-T-1g	1980. 5. 8	B039 TYM-1d	B069 TYM-1da
B151	♀	PKN-0-2g	1980. 5. 4	C106 OMC-20	B028 OMC-12a
B152	♂	WKY-5abb	1980. 5.12	B103 WKY-5ab	C091 WKY-2
B153	♂	LAG-J-2a	1980. 6.18	B066 JSC-2fb	B014 JSC-2ba
B154	♀	TYM-5c	1980. 6.19	C159 TYM-5	B116 TYM-1dbb
B155	♂	TYM-1ka	1980. 7. 9	B122 TYM-1k	B117 TYM-5a
B156	♂	KYT-T-1a	1980. 7.22	B109 TYM-1j	B032 KYT-7e
B157	♀	OMC-23e	1980. 8. 6	C116 OMC-23	C140 OMC-27
B158	♂	PKN-T-1b	1980. 8.26	B099 TYM-1bab	B107 TYM-1dba
B159	♀	ISK-T-1e	1981. 4.29	B048 TYM-1ba	B131 TYM-1baba
B160	♂	PKN-0-2h	1981. 5. 9	C106 OMC-20	B028 OMC-12a
B161	♀	KYT-T-1b	1981. 5.13	B109 TYM-1j	B032 KYT-7e
B162	♀	NGY-T-1h	1981. 5.20	B039 TYM-1d	B069 TYM-1da

Site of birth	Owner	Since	Date of birth	Note
Tateyama Fudokigaoka	Kyoto Zoo	1978.10. 6		Alive
Kyoto Zoo	Kyoto Zoo	Since Birth	1977.10.16	
Higashiyama Zoo	Los Angeles Zoo	1979.12.10		Alive
Tateyama Fudokigaoka	Maruyama Zoo	1978.10. 8	1979.10.17	
Peking Zoo	Peking Zoo	Since Birth	1977. 9.10	
Oto-mura	Oto-mura	Since Birth	1983. 4.30	
Omachi Alpine Museum	Tama Zool. Park	1982.11.20		Alive
Tateyama Fudokigaoka	Toyama Prefecture	Since Birth	1980.12. 5	
Tateyama Fudokigaoka	Toyama Prefecture	Since Birth		Alive
Kobe Muni. Arboretum	Toyama Family Park	1984.10.20		Alive
Oto-mura	Oto-mura	Since Birth	1982. 5.26	
Hsian Zoo, China	Hsian Zoo, China	Since Birth	1979. 6.25	
Japan Serow Center	Japan Serow Center	Since Birth	1978. 5. 4	
Tateyama Fudokigaoka	Toyama Prefecture	Since Birth		Alive
Higashiyama Zoo	Maruyama Zoo	1980.10.21	1980.12.16	
Peking Zoo, China	Shihchiachuang Zoo	1981. 4. 4	1984.10. 5	
Oto-mura	Oto-mura	Since Birth	1984.10.13	
Oto-mura	Oto-mura	Since Birth	1979. 5. 5	
Oto-mura	Oto-mura	Since Birth	1980. 4. 3	
Kyoto Zoo	Kyoto Zoo	Since Birth	1978. 7. 2	
Tateyama Fudokigaoka	Los Angeles Zoo	1980.12.10		Alive
Kobe Muni. Arboretum	Kobe Oji Zoo	Since Birth	1978.11.24	
Tateyama Fudokigaoka	Ishikawa Arboretum	1979. 9.21		Alive
Omachi Alpine Museum	Omachi Alpine Museum	Since Birth	1978. 8.19	
Tama Zoological Park	Tama Zoological Park	Since Birth	1978.10.27	
Oto-mura	Oto-mura	Since Birth	1978. 9.14	
Omachi Alpine Museum	Omachi Alpine Museum	Since Birth	1979. 1. 4	
Tateyama Fudokigaoka	Toyama Prefecture	Since Birth	1979.11.12	
Higashiyama Zoo	Maruyama Zoo	1982. 7. 7	1984. 1.13	
Tama Zoological Park	Tama Zoological Park	Since Birth	1979. 5. 6	
Tateyama Fudokigaoka	Kobe Oji Zoo	1980.11.11		Alive
Tateyama Fudokigaoka	Los Angeles Zoo	1981. 5.12		Alive
Gero-machi	Gero-machi	Since Birth		Alive
Peking Zoo, China	Peking Zoo, China	Since Birth	1979. 7.17	
Tateyama Fudokigaoka	Kobe Oji Zoo	1980.11.11		Alive
Omachi Alpine Museum	Omachi Alpine Museum	Since Birth		Alive
Hirosaki Park	Hirosaki Park	Since Birth	1979. 6. 0	
Maruyama Zoo	Maruyama Zoo	Since Birth	1979. 8.22	
Kyoto Zoo	Kyoto Zoo	Since Birth		Alive
Gero-machi	Gero-machi	Since Birth		Alive
Tama Zoological Park	Tama Zoological Park	Since Birth	1980. 5. 9	
Higashiyama Zoo	Higashiyama Zoo	Since Birth	1980. 5. 9	
Peking Zoo, China	Peking Zoo, China	Since Birth	1980. 6.29	
Oto-mura	Oto-mura	Since Birth	1980. 7.19	
Los Angeles Zoo	Los Angeles Zoo	Since Birth	1980. 7.15	
Tateyama Fudokigaoka	San Diego Zoo	1982. 1.15		Alive
Tateyama Fudokigaoka	San Diego Zoo	1982. 1.15	1982. 6.16	
Kyoto Zoo	Kyoto Zoo	Since Birth	1980. 9.17	
Omachi Alpine Museum	(Schoenbrun Zoo)	1984.11. 5	1984.11. 5	
Peking Zoo, China	Tienchin Zoo, China	1981. 6.30	1984. 0. 0	
Ishikawa Arboretum	Ishikawa Arboretum	Since Birth		Alive
Peking Zoo, China	Kuangchou Zoo, China	1982. 8. 6		Alive
Kyoto Zoo	Kyoto Zoo	Since Birth	1981. 7. 5	
Higashiyama Zoo	Higashiyama Zoo	Since Birth		Alive

81

Table 5.1 continued

No.	Sex	Studbook name	Date of birth	Dam	Sire
B163	♂	KZK-17a	1981. 6.10	C292 KZK-17	C288 KZK-15
B164	♂	TYM-5d	1981. 6.18	C159 TYM-5	B116 TYM-1dbb
B165	♀	TYM-1kb	1981. 6.19	B122 TYM-1k	B117 TYM-5a
B166	♀	KZK-14a	1981. 6.20	C285 KZK-14	C232 KZK-6
B167	♂	KYT-7fh	1981. 6.23	B041 KYT-7f	B032 KYT-7e
B168	♀	KOB-T-2a	1981. 7.24	B143 TYM-5b	B079 KOB-1ccb
B169	♀	PKN-0-2i	1982. 4.23	C106 OMC-20	B028 OMC-12a
B170	♂	NGY-T-1i	1982. 5.19	B039 TYM-1d	B069 TYM-1da
B171	♀	KOB-T-1a	1982. 5.19	B139 TYM-11	B079 KOB-1ccb
B172	♂	TYM-1kc	1982. 5.31	B122 TYM-1k	B117 TYM-5a
B173	♀	KYT-7fi	1982. 5.31	B041 KYT-7f	B032 KYT-7e
B174	♀	OMC-23f	1982. 6. 1	C116 OMC-23	C140 OMC-27
B175	♂	KYT-T-1c	1982. 6. 3	B109 TYM-1j	B147 KYT-7fg
B176	♂	KOB-T-2b	1982. 6. 6	B143 TYM-5b	B079 KOB-1ccb
B177	♀	KZK-17b	1982. 6.25	C293 KZK-17	C288 KZK-15
B178	♀	NTG-1a	1982. 7.14	C321 NTG-1	C346 NTG-2
B179	♀	GER-18aa	1982. 7.27	B148 GER-18a	B141 GER-2a
B180	♂	TYM-5e	1982. 8. 3	C159 TYM-5	C054 TYM-4
B181	♂	PKN-T-1c	1982. 8.15	B099 TYM-1bab	B107 TYM-1dba
B182	♀	PKN-0-2j	1983. 4.20	C106 OMC-20	B028 OMC-12a
B183	♀	TYM-1kd	1983. 5. 5	B122 TYM-1k	B117 TYM-5a
B184	♀	NGY-T-1j	1983. 5.11	B039 TYM-1d	B069 TYM-1da
B185	♀	SDG-T-3a	1983. 5.12	B154 TYM-5c	B094 TYM-1i
B186	♀	KYT-T-1d	1983. 5.14	B109 TYM-1j	B147 KYT-7fg
B187	♀	OMC-23g	1983. 5.18	C116 OMC-23	C140 OMC-27
B188	♀	WKY-6a	1983. 5.25	C308 WKY-6	C091 WKY-2
B189	♀	NTG-4a	1983. 5.30	C348 NTG-4	C346 NTG-2
B190	♀	KOB-T-1b	1983. 6.10	B139 TYM-11	B079 KOB-1ccb
B191	♀	SPR-T-1f	1983. 7. 8	B019 TYM-1b	B137 NGY-T-1f
B192	♂	TYM-1m	1983. 7.11	C051 TYM-1	B027 TYM-1c
B193	♀	PKN-T-1d	1983. 7.19	B099 TYM-1bab	B107 TYM-1dba
B194	♂	TYM-5f	1983. 8.16	C159 TYM-5	C335 TYM-8
B195	♀	NTG-1b	1983.10. 3	C321 NTG-1	C347 NTG-3
B196	♂	TAM-J-2g	1984. 4.25	B029 JCS-2ac	B115 OMC-23b
B197	♀	TYM-1ke	1984. 4.28	B122 TYM-1k	B117 TYM-5a
B198	♂	KYT-T-1e	1984. 5.24	B109 TYM-1i	B147 KYT-7fg
B199	♀	OMC-23h	1984. 5.26	C116 OMC-23	C140 OMC-27
B200	♂	KOB-T-1c	1984. 5.26	B136 TYM-11	B097 KOB-1ccb

The number of offspring in each year and the juvenile death among them are shown in Table 5.4. 'Juvenile death' means death within a year after birth.

It is notable that there was a rather high rate of the death within a year after birth, that is juvenile death, and there is no change for the better over the period studied. The worst rate of juvenile death, 66.6 per cent, is found in 1974; the total juvenile death reaches 43.8 per cent. The most frequent cause of death in the fawn is acute enteritis at the beginning of winter, when the fawn reaches an age of about 6 months or so.

Site of birth	Owner	Since	Date of death	Note
Kozaka-machi	Kozaka-machi	Since Birth		Alive
Tateyama Fudokigaoka	Shenyang Zoo, China	1982. 9.30		Alive
Tateyama Fudokigaoka	Shenyang Zoo, China	1982. 9.30		Alive
Kozaka-machi	Kozaka-machi	Since Birth	1984. 9.11	
Kyoto Zoo	Kyoto Zoo	Since Birth	1981. 9.25	
Kobe Muni. Arboretum	Kobe Oji Zoo	Since Birth	1981. 9.26	
Peking Zoo, China	Peking Zoo, China	Since Birth	1982. 6.10	
Higashiyama Zoo	Higashiyama Zoo	Since Birth	1982. 5.19	
Kobe Muni. Arboretum	Kobe Oji Zoo	Since Birth	1982.10.24	
Tateyama Fudokigaoka	Japan Serow Center	1983.10.12	1984. 8.22	
Kyoto Zoo	Kyoto Zoo	Since Birth	1982. 8.11	
Omachi Alpine Museum	Omachi Alpine Museum	Since Birth	1982. 6.22	
Kyoto Zoo	Kyoto Zoo	Since Birth	1982. 9.28	
Kobe Muni. Arboretum	Kobe Oji Zoo	Since Birth	1982. 8.14	
Kozaka-machi	Kozaka-machi	Since Birth	1982. 7.14	
Nakatsugawa-shi	Nakatsugawa-shi	Since Birth		Alive
Gero-machi	Gero-machi	Since Birth	1983.10.15	
Tateyama Fudokigaoka	Yagiyama Zoo	1983.11.15		Alive
Peking Zoo, China	Peking Zoo, China	Since Birth		Alive
Peking Zoo, China	Peking Zoo, China	Since Birth	1983. 5.12	
Tateyama Fudokigaoka	Toyama Family Park	1984.10.20		Alive
Higashiyama Zoo	Higashiyama Zoo	Since Birth		Alive
San Diego Zoo	San Diego Zoo	Since Birth	1983. 5.12	
Kyoto Zoo	Kyoto Zoo	Since Birth	1983. 8. 3	
Omachi Alpine Museum	Omachi Alpine Museum	Since Birth		Alive
Oto-mura	Oto-mura	Since Birth		Alive
Nakatsugawa-shi	Nakatsugawa-shi	Since Birth		Alive
Kobe Muni. Arboretum	Kobe Oji Zoo	Since Birth		Alive
Maruyama Zoo	Maruyama Zoo	Since Birth	1983. 9. 9	
Tateyama Fudokigaoka	Toyama Prefecture	Since Birth		Alive
Peking Zoo, China	Peking Zoo, China	Since Birth		Alive
Tateyama Fudokigaoka	Toyama Prefecture	Since Birth		Alive
Nakatsugawa-shi	Nakatsugawa-shi	Since Birth		Alive
Tama Zoological Park	Tama Zoological Park	Since Birth		Alive
Tateyama Fudokigaoka	Toyama Prefecture	Since Birth		Alive
Kyoto Zoo	Kyoto Zoo	Since Birth	1984. 7.22	
Omachi Alpine Museum	Omachi Alpine Museum	Since Birth		Alive
Kobe Muni. Arboretum	Kobe Oji Zoo	Since Birth		Alive

BREEDING SEASON OF JAPANESE SEROW

The number of births of Japanese serow in every ten days of each month and also the number of juvenile deaths among them are shown in Table 5.5.

In total, over 45 per cent were born in May, and over 25 per cent in June. There is no case of birth in December, February, and March. The last ten days in May are the best season for breeding of Japanese serow.

Table 5.2: Founder group of captive bred Japanese serows, *Capricornis crispus*

DAM

#	R-No.	Sex	Studbook name	Date of caught	Site of caught
1	C007	♀	UEN-7	1950. 3.13	Kitamuro-gun, Mie
2	C026	♀	KYT-2	1955. 3. 1	Koga-gun, Shiga
3	C039	♀	JSC-2	1962. 4.16	Mie-gun Mie
4	C044	♀	KYT-7	1963. 4.15	Tsuruga-shi, Fukui
5	C047	♀	KOB-1	1964. 2.18	Koga-gun, Shiga
6	C051	♀	TYM-1	1965. 3. 3	Mt. Yakushi, Toyama
7	C074	♀	OMC-11	1968. 9. 1	Higashimuro, Wakayama
8	C076	♀	OMC-12	1968. 9.28	Minamiazumi, Nagano
9	C092	♀	WKY-3	1970.12.20	Nishimuro-gun, Wakayama
10	C094	♀	WKY-5	1971. 1.17	Nishimuro-gun, Wakayama
11	C106	♀	OMC-20	1972. 3.15	Kaniina-gun, Nagano
12	C116	♀	OMC-23	1972.11.15	Omachi-shi, Nagano
13	C133	♀	HRS-3	1973. 6.17	Shimokita-gun, Aomori
14	C142	♀	NRA-5	1973.12.19	Yoshino-gun, Nara
15	C159	♀	TYM-5	1974. 5.27	Uozu-shi, Toyama
16	C186	♀	AKZ-2	1976. 5. 5	Kitaakita-gun, Akita
17	C233	♀	GER-2	1978.11. 7	Nakatsugawa-shi, Gifu
18	C238	♀	JSC-16	1978.11.30	Kumano-gun, Mie
19	C283	♀	AKZ-4	1979. 7.13	Akita-shi, Akita
20	C285	♀	KZK-14	1979.12. 8	Mashita-gun, Gifu
21	C292	♀	KZK-17	1980. 2.21	Mashita-gun, Gifu
22	C304	♀	GER-18	1980. 4. 2	Mashita-gun, Gifu
23	C308	♀	WKY-6	1980. 5.19	Higashimuro, Wakayama
24	C321	♀	NTG-1	1981. 2.16	Nakatsugawa-shi, Gifu
25	C348	♀	NTG-4	1981.10. 7	Ena-gun, Gifu
26	C358	♀	KMR-1	1982. 1.25	Kiso-gun, Nagano
27	C359	♀	NTG-6	1982. 2. 6	Nakatsugawa-shi, Gifu
28	C419	♀	NGN-1	1983.11.18	Kiso-gun, Nagano
29	C442	♀	TKC-8	1984. 3. 2	Ena-gun, Gifu

SIRE

#	R-No.	Sex	Studbook name	Date of caught	Site of caught
1	C036	♂	KYT-4	1962. 1.10	Koga-gun, Mie
2	C038	♂	JSC-1	1962. 4.11	Mie-gun, Mie
3	C048	♂	KOB-2	1964. 2.20	Koga-gun, Shiga
4	C054	♂	TYM-4	1965. 5.19	Shimohei-gun, Iwate
5	C055	♂	OMC-4	1965. 6. 9	Ogata-gun, Nagano
6	C059	♂	JSC-3	1966. 3.18	Mie-gun, Mie
7	C072	♂	JSC-5	1968. 3. 3	Mie-gun, Mie
8	C091	♂	WKY-2	1970.12.17	Nishimuro-gun, Wakayama
9	C093	♂	WKY-4	1971. 1.13	Nishimuro-gun, Wakayama
10	C114	♂	NRA-3	1972. 9.18	Yoshino-gun, Nara
11	C131	♂	HRS-1	1973. 6. 7	Minamitsugaru, Aomori
12	C140	♂	OMC-27	1973.10.24	Omachi-shi, Nagano
13	C232	♂	KZK-6	1978.11. 7	Mashita-gun, Gifu
14	C288	♂	KZK-15	1980. 1.27	Takayama-shi, Gifu
15	C335	♂	TYM-8	1981. 5.28	Uozu-shi, Toyama
16	C346	♂	NTG-2	1981.10. 7	Nakatsugawa-shi, Gifu
17	C347	♂	NTG-3	1981.10. 7	Ena-gun, Gifu
18	C349	♂	NTG-5	1981.10. 7	Ena-gun, Gifu
19	C365	♂	AKZ-5	1982. 6.11	Senpoku-gun, Akita
20	C383	♂	KMR-2	1983. 1.19	Kiso-gun, Nagano

Owner (first)	Owner (further)	Since	Date of death	Note
Ueno Zool. Gardens	Ueno Zool. Gardens		1950. 9.25	
Kyoto Zoo	Kyoto Zoo		1955. 5.20	
Japan Serow Center	Japan Serow Center		1975. 2.17	
(Site of caught)	Kyoto Zoo	1963. 9.10	1975. 6.15	
(Site of caught)	Kobe Oji Zoo	1964. 2.21	1971. 5.10	
Toyama Prefecture	Toyama Prefecture			Alive
Omachi Alpine Mus.	Omachi Alpine Mus.		1984.11.13	
Omachi Alpine Mus.	Omachi Alpine Mus.		1979.11.12	
Oto-mura	Oto-mura		1979. 5.31	
Oto-mura	Oto-mura		1979. 6. 3	
Omachi Alpine Mus.	Peking Zoo	1973. 4. 3		Alive
Omachi Alpine Mus.	Omachi Alpine Mus.			Alive
Hirosaki Park	Hirosaki Park		1981. 1.25	
Ayameike Park Zoo	Hsian Zoo	1977.10. 3	1979. 7.29	
Toyama Prefecture	Toyama Prefecture			Alive
Omoriyama Zoo	Omoriyama Zoo			Alive
Gero-machi	Gero-machi		1980	
Japan Serow Center	Japan Serow Center			Alive
Omoriyama Zoo	Omoriyama Zoo		1985.11.19	
Kozaka-machi	Kozaka-machi		1982. 9.11	
Kozaka-machi	Kozaka-machi		1983.11.21	
Itadori-mura	Itadori-mura	1980. 8.27	1980. 8.27	
Oto-mura	Oto-mura			Alive
Nakatsugawa-shi	Nakatsugawa-shi			Alive
Nakatsugawa-shi	Nakatsugawa-shi			Alive
Komoro Zoo	Komoro Zoo			Alive
Nakatsugawa-shi	Nogeyama Zoo	1983. 5.25		Alive
Chausuyama Zoo	Chausuyama Zoo			Alive
Tsukechi-machi	Tsukechi-machi			Alive

Owner (first)	Owner (further)	Since	Date of death	Note
Kyoto Zoo	Kyoto Zoo		1973.10.15	
Japan Serow Center	Japan Serow Center		1969. 3.11	
(Site of Caught)	Kobe Oji Zoo	1964. 2.21	1974.11.28	
(Site of Caught)	Toyama Prefecture	1965.11. 2		Alive
(Site of Caught)	Omachi Alpine Mus.	1965. 6.24	1980. 4.25	
(Site of Caught)	Japan Serow Center	1966. 4. 13	1980. 3. 4	
Japan Serow Center	Japan Serow Center		1974. 1.31	
Oto-mura	Oto-mura			Alive
Oto-mura	Kobe Oji Zoo	1974. 6. 8	1976.10.22	
Ayameike Park Zoo	Hsian Zoo	1977.10. 3	1979. 6.26	
Hirosaki Park	Hiroski Yayoi P.	1984. 5.18	1985.12.19	
Omachi Alpine Mus.	Omachi Alpine Mus.		1985. 9.30	
Kozaka-machi	Kozaka-machi		1984. 7.25	
Kozaka-machi	Kozaka-machi		1982. 1.26	
Toyama Prefecture	Toyama Prefecture			Alive
Nakatsugawa-shi	Nakatsugawa-shi		1983.10.11	
Nakatsugawa-shi	Nakatsugawa-shi			Alive
Nakatsugawa-shi	Nogeyama Zoo	1983. 5.25		Alive
Omoriyama Zoo	Omoriyama Zoo			Alive
Kmoro Zoo	Kmoro Zoo			Alive

Table 5.3: Japanese serows, *Capricornis crispus*, bred in captivity 1950-1985

No.	Sex	Code	House name	Date of birth	Sire		Dam	
B201	♂	TYM-5g	Yuu	1984. 5.27	C335	TYM-8	C159	TYM-5
B202	♂	JSC-16a		1984. 5.27	B172	TYM-1kc	C238	JSC-16
B203	♂	WKY-6b	Yutaro	1984. 5.27	C091	WKY-2	C308	WKY-6
B204	♂	NGN-1a	Nozomi	1984. 5.28		Wild	C419	NGN-1
B205	♀	TKC-8a		1984. 6. 4		Wild	C442	TKC-8
B206	♀	YKH-N-2a	Azusa	1984. 6.15	C349	NTG-5	C359	NTG-6
B207	♂	KMR-1a		1984. 6.23	C383	KMR-2	C358	KMR-1
B208	♀	KOB-T-2c	Junko-III	1984. 6.24	B097	KOB-1ccb	B143	TYM-5b
B209	♀	PKN-T-1e	84-2	1984. 6.25	B107	TYM-1dba	B099	TYM-1bab
B210	♀	SDG-T-3b	sowap3	1984. 7. 7	B094	TYM-1i	B154	TYM-5e
B211	♀	OMC-23ea		1984. 7.20	B104	OMC-12f	B157	OIMC-23e
B212	♀	NTG-4b	Goku	1984. 7.28	C347	NTG-3	C348	NTG-4
B213	♀	AKZ-4a		1984. 7.28	C365	AKZ-5	C283	AKZ-4
B214	♂	AKZ-2a		1984. 8.27	C365	AKZ-5	C186	AKZ-2
B215	♀	SNY-T-2a		1984. 9.13	B164	TYM-5d	B165	TYM-1kb
B216	♀	NGY-T-1k	Akiko	1984. 9.23	B069	TYM-1da	B039	TYM-1d
B217	♀	NTG-1c	Rise	1985. 1.14	C347	NTG-3	C321	NTG-1
B218	♂	KYT-T-1f	Akira	1985. 5.14	B147	KYT-7fg	B109	TYM-1i
B219	♂	WKY-6aa		1985. 5.14	C091	WKY-2	B118	WKY-6a
B220	♀	NTG-1d	Muu	1985. 5.25	C374	NTG-3	C348	NTG-4
B221	♂	TYM-1kf		1985. 5.28	B117	TYM-5a	B122	TYM-1k
B222	♂	KMR-1b		1985. 6. 3	C383	KMR-2	C358	KMR-1
B223	♀	TYM-5h		1985. 6. 5	C335	TYM-8	C159	TYM-5
B224	♂	NGN-1b		1985. 6. 6	C420	NGN-2	C419	NGN-1
B225	♀	OMC-23i		1985. 6. 8	C140	OMC-27	C116	OMC-23
B226	♂	TAM-J-2h		1985. 6.20	B115	OMC-23b	B029	JSC-2ac
B227	♂	AKZ-4b	Tsurutaro	1985. 7.22	C365	AKZ-5	C283	AKZ-4
B228	♀	AKZ-2b		1985. 7.23	C365	AKZ-5	C186	AKZ-2
B229	♀	KOB-T-1d	Akiko-4	1985. 8.22	B097	KOB-1ccb	B136	TYM-11

The earliest date of birth of Japanese serow is 2 April and the latest 14 January. However, the last case seems to be somewhat abnormal, and excluding this abnormal case the latest date of birth is 9 November. Concerning the scattering of month of birth in an individual dam, the cases of some multiparous dams are shown in Table 5.6.

Among the above-mentioned dams, the individual 'C051, TYM-1' has produced 13 offspring over a continuous breeding period of 12 years during which there is only one case of juvenile death. Furthermore the scattering of the month of birth is very narrow.

(O in Note means being of Breeding Next Generation)

Site of birth	Present owner	Date of remove	Date of death	Longevity yrs. days	Note
Tateyama Fudokigaoka	Toyama Prefecture	Since Birth			Alive
Japan Serow Center	Japan Serow Center	Since Birth	1984.11.10	0 163	
Oto-mura	Oto-mura	Since Birth			Alive
Chausuyama Zoo	Chausuyama Zoo	Since Birth			Alive
Tsukechi-cho, Gifu	Tsukechi-cho, Gifu	Since Birth			Alive
Nogeyama Zoo (Kanazwa)	Nogeyama Zoo	Since Birth			Alive
Komoro Zoo	Komoro Zoo	Since Birth	1984. 8. 6	0 43	
Kobe Muni. Arboretum	Kobe Oji Zoo	Since Birth			Alive
Peking Zoo	Peking Zoo	Since Birth			Alive
San Diego Zoo	San Diego Zoo	Since Birth	1984. 7. 7	0 0	
Omachi Alpine Museum	Omachi Alpine Museum	Since Birth	1984.10.11	0 81	
Nakatsugawa-shi	Nakatsugawa-shi	Since Birth			Alive
Omoriyama Zoo	Omoriyama Zoo	Since Birth	1984. 8.30	0 32	
Omoriyama Zoo	Omoriyama Zoo		1984. 8.28	0 1	
Shijiazhuang Zoo	Shijiazhuang Zoo	Since Birth	1984.10. 7	0 24	
Higashiyama Zoo	Higashiyama Zoo	Since Birth			Alive
Nakatsugawa-shi	Nakatsugawa-shi		1985. 2. 6	0 22	
Kyoto Zoo	Kyoto Zoo	Since Birth			
Oto-mura	Oto-mura	Since Birth	1985.10.16	0 152	
Nakatsugawa-shi	Nakatsugawa-shi	Since Birth			Alive
Tateyama Fudokigaoka	Toyama Prefecture	Since Birth			Alive
Komoro Zoo	Komoro Zoo	Since Birth			Alive
Tateyama Fudokigaoka	Toyama Prefecture	Since Birth			Alive
Chausuyama Zoo	Chausuyama Zoo	Since Birth			Alive
Omachi Alpine Museum	Omachi Alpine Museum	Since Birth	1985. 6.28	0 20	
Tama Zool. Park	Tama Zool. Park		1985. 6.20	0 0	
Omoriyama Zoo	Omoriyama Zoo	Since Birth			Alive
Omoriyama Zoo	Omoriyama Zoo	Since Birth	1985. 7.23	0 0	
Kobe Muni. Arboretum	Kobe Oji Zoo	Since Birth			Alive

SEXUAL MATURATION

The age of the first reproduction in captive-bred dams is shown in Table 5.7.

About the age of first reproduction, there are 11 cases of around 2 years, 12 cases of around 3 years and 4 cases of around 4 years. The youngest age of first reproduction is 1 year 11 months and 4 days. The dam in this case reached sexual maturity at $1^1/_2$ years.

Table 5.4: Number of births and juvenile deaths in each year

Year	Total births[a]	(Semi-wild)	Juvenile deaths	(Semi-wild)
1950	1 (1/0)	(1/0)	1 (1/0)	(1/0)
1955	1 (0/1)	(0/1)	1 (0/1)	(0/1)
1964	1 (0/1)	(0/1)	0	(0)
1965	2 (0/2)		0	
1966	2 (2/0)		0	
1967	3 (1/2)		1 (1/0)	
1968	7 (4/3)		5 (3/2)	
1969	8 (5/3)		2 (1/1)	
1970	9 (7/2)		3 (3/0)	
1971	8 (3/5)		3 (2/1)	
1972	13 (7/6)		7 (3/4)	
1973	14 (7/7)		6 (2/4)	
1974	12 (7/5)		8 (5/3)	
1975	11 (7/4)		7 (5/2)	
1976	15 (7/8)		5 (2/3)	
1977	11 (8/3)		3 (1/2)	
1978	17 (9/8)		8 (5/3)	
1979	12 (6/6)	(1/0)	5 (2/3)	(0)
1980	11 (6/5)	(0/1)	6 (4/2)	(0)
1981	10 (4/6)		3 (1/2)	
1982	13 (6/7)		8 (3/5)	
1983	14 (2/12)		4 (0/4)	
1984	22 (9/13)	(1/1)	9 (4/5)	
Total	217 (108/109)	(3/4)	95 (48/47)	(1/1)

Note: a. The number in parentheses means males/females

JUVENILE DEATHS

As mentioned before, in captive-bred Japanese serow, there is a rather high ratio of juvenile death. It reaches 43.8 per cent in total. It seems to be related to breeding between one generation and the next, or to a high degree of inbreeding, as well as to the vitality of the species itself or to some technical problem in the breeding of the species. Table 5.8 shows the number of births in each crossing generation. Here, generation means only those in the maternal lines, generations in the paternal lines being neglected in this case.

In Table 5.8 it can be seen that the worst result is in the crossing of the second by second generation, in which the ratio of juvenile death reaches 88.9 per cent. The second is in the crossing of the first by first generation, in which the ratio of juvenile death reaches 63.9 per cent. These facts suggest that the high degree of inbreeding through successive generations has some influence upon the degree of success in breeding of Japanese serow in captivity.

Table 5.5: The births and juvenile deaths in each month

Month		Birth	Juvenile death		Birth (M/F)	Juvenile death (M/F)
April	I	1	0	}	16 (6/10)	8 (3/5)
	II	5	2			
	III	10	6			
May	I	27	15	}	98 (52/46)	45 (23/22)
	II	35	16			
	III	36	14			
June	I	21	9	}	55 (27/28)	22 (13/9)
	II	21	7			
	III	13	6			
July	I	10	6	}	26 (11/15)	12 (4/8)
	II	6	2			
	III	10	4			
August	I	3	0	}	10 (6/4)	1 (1/0)
	II	3	0			
	III	4	1			
September	I	2	1	}	8 (6/2)	5 (4/1)
	II	4	4			
	III	2	0			
October	I	1	0	}	2 (0/2)	0
	II	1	0			
	III	0				
November	I	1	1	}	1 (0/1)	1 (0/1)
	II	0	0			
	III	0				
December	I	0		}	0	
	II	0				
	III	0				
January	I	0		}	1 (0/1)	1 (0/1)
	II	1	1			
	III	0				

The data in Table 5.9 examine the correlation between juvenile death and the coefficient of inbreeding.

In Table 5.9, it is not so clear that the coefficient of inbreeding has an effect on juvenile death, but it is recognised that there is some tendency towards a worse result with a high degree of inbreeding. For example, at a coefficient of inbreeding of 0.25 there is the worst juvenile death result.

The founder group of captive-bred Japanese serows consists of 20 sires and 29 dams which were caught from the wild. It seems that this group is rather small for 217 offspring in captivity. This fact caused the coefficient of inbreeding to increase, and as a result it seems to be the reason for the high ratio of juvenile death. The enlargement of the founder group of this species is a very important priority for the future.

Table 5.6: Some cases of scattering of the month of birth (asterisk indicates juvenile death)

CO51, TYM-1 (Founder)

1968	1969	1970	1971	1972	1973	1974	1975	1976	1977	1978	1979	1980-2	1983
May*	May	May	May		May	May	May	May	May	May	May	—	
				June									
													July

C106, OMC-20 (Founder)

1974	1975	1976	1977	1978	1979	1980	1981	1982	1983
								April*	April*
		May	May*	May	May*	May*	May		
	June*								
August									

CO39, JSC-2 (Founder)

1964	1965	1966	1967	1968	1969	1970	1971	1972	1973
							—	April*	
				May		May*			May*
June									
			July		July				
	August	August							

C116, OMC-23 (Founder)

1976	1977	1978	1979	1980	1981	1982	1983	1984
	May				—		May	May
		June*	June			June*		
July								
				August				

CO44, KYT-7, (Founder)

1967	1968	1969	1970	1971	1972	1973	1974
	May*	May*					
				June			
July*			July			July	
					September		September*

BO39, TYM-1d (Captive bred)

1973	1974	1975	1976	1977	1978	1979	1980	1981	1982	1983	1984
		—	May	May	May	May	May*	May	May*	May	
	June										
September											September

BO41, KYT-7f (Captive bred)

1973	1974	1975	1976	1977	1978	1979	1980	1981	1982
	May*	May*	May*	May*					May*
June*								June*	
				September					

Table 5.7: Age of the first reproduction

No. and name	Date of birth	First reproduction	Age of first reproduction
B041, KYT-7f	20 June 1971	4 June 1973	1 y 11 m 14 d
B067, WKY-5a	10 June 1973	10 July 1975	2 y 1 m 0 d
B095, TYM-1dc	3 May 1976	7 June 1978	2 y 1 m 4 d
B099, TYM-1bab	9 May 1976	18 June 1978	2 y 1 m 9 d
B143, TYM-5b	11 June 1979	24 July 1981	2 y 1 m 13 d
B078, TYM-1db	9 June 1974	25 July 1976	2 y 1 m 16 d
B008, KOB-1c	24 May 1967	15 July 1969	2 y 1 m 22 d
B122, TYM-1k	11 May 1978	9 July 1980	2 y 1 m 29 d
B039, TYM-1d	25 May 1971	5 September 1973	2 y 3 m 11 d
B103, WKY-5ab	1 June 1976	14 September 1978	2 y 3 m 13 d
B148, GER-18a	2 April 1980	27 July 1982	2 y 3 m 25 d
B004, JSC-2b	12 August 1965	24 May 1968	2 y 9 m 12 d
B005, KOB-1a	26 August 1965	18 June 1968	2 y 9 m 23 d
B091, WKY-5aa	10 July 1975	20 May 1978	2 y 10 m 10 d
B154, TYM-5c	19 June 1980	12 May 1983	2 y 10 m 23 d
B025, JSC-2f	22 July 1969	17 June 1972	2 y 10 m 26 d
B029, JSC-2ac	11 June 1970	27 May 1973	2 y 11 m 16 d
B048, TYM-1ba	21 May 1972	20 May 1975	3 y 0 m 0 d
B139, TYM-11	18 May 1979	19 May 1982	3 y 0 m 1 d
B014, JSC-2ba	24 May 1968	30 May 1971	3 y 0 m 6 d
B019, TYM-1b	10 May 1969	21 May 1972	3 y 0 m 11 d
B037, OMC-12b	20 May 1971	22 June 1974	3 y 1 m 2 d
B109, TYM-1j	1 May 1977	22 July 1980	3 y 2 m 21 d
B165, TYM-1kb	19 June 1981	19 September 1984	3 y 2 m 24 d
B010, JSC-2d	22 July 1967	18 May 1971	3 y 9 m 27 d
B157, OMC-23e	6 August 1980	20 July 1984	3 y 11 m 14 d
B042, KOB-1cc	30 June 1971	19 June 1975	3 y 11 m 19 d
B003, JSC-2a	17 June 1964	3 July 1968	4 y 0 m 16 d
B066, JSC-2fb	8 June 19673	18 June 1980	7 y 0 m 10 d

Table 5.8: The number of births in each crossing generation. The figures in parentheses are the numbers of juvenile deaths (M/F)

Dam \ Sire	Founder 20	I 16	II 6	III 3	IV 1	Total
Founder 29	34/41 (10/11)	11/10 (4/8)	1/0 (1/0)	1/2 (0/0)	— —	47/53 (15/19)
I 18	14/7 (6/2)	19/17 (14/9)	7/10 (3/5)	2/5 (1/3)	— —	42/39 (24/19)
II 10	2/1 (1/0)	9/9 (4/5)	5/4 (4/4)	3/2 (0/0)	0/1 (0/0)	16/15 (9/9)
III 1	— —	— —	— —	3/2 (0/0)	— —	3/2 (0/0)
Total	50/49 (17/13)	39/36 (22/22)	13/4 (9/9)	6/9 (1/3)	0/1 (0/0)	108/109 (48/47)

Table 5.9: Coefficient of inbreeding and juvenile death

Coefficient of inbreeding	Number M/F	Juvenile death M/F	Ratio %
0.375	4/5	1/2	33.3
0.3-0.35	1/0	0/0	0.
0.25	35/18	23/10	62.3
0.15-0.249	6/6	2/4	50.0

CONCLUSION

In addition to this report, a further 13 Japanese serows were born in 10 institutions. They comprise seven males and six females and two males and three females died within a year.

As regards the status of Japanese serows in captivity (31 December 1986), 122 in total (57 males and 65 females) are kept in 35 institutions. Of these 32 males and 36 females were born in captivity. The ratio of captive-born serows to all Japanese serows in captivity is 55.7 per cent.

Outside Japan, Japanese serows are now kept in the US, the People's Republic of China, and Austria. In the autumn of 1986, a pair of serows were sent by the Government of Japan to Berlin Zoo, West Germany.

6

Distribution of Japanese serow in its southern range, Kyushu

Teruo Doi[1], Yuiti Ono[1], Toshitaka Iwamoto[2] and Toshiyuki Nakazono[3]

[1]Department of Biology, Faculty of Science, Kyushu University 33, Fukuoka 812 Japan;
[2]Faculty of Education, Miyazaki University, Miyazaki 880, Japan; and
[3]Soyoh High School, Soyoh-cho, Kumamoto 861-35, Japan

INTRODUCTION

The Japanese serow (*Capricornis c. crispus* Temminck) is an endemic ungulate in Japan and is to be found in the montane regions of Honshu, Shikoku and Kyushu. In the last decade, investigations of the distribution and abundance of the serow have been conducted by the Japan Ministry of Environment (1980) and the Ministry of Culture (Kiuchi *et al.* 1978). In Kyushu, the distribution of the serow was limited to the main mountainous districts of the central part of the island that covered Ohita, Miyazaki and Kumamoto prefectures. This chapter presents the results of a ten-year survey of the distribution and abundance of the serow. The characteristics of the distributional patterns and present situation of the serow in Kyushu are discussed, and the role of secondary forests as the habitat of serows is assessed.

METHOD

We employed two methods to estimate the population densities of the Japanese serow. One is the faecal pellet-group count technique (Morisita *et al.* 1977). Serows deposit faecal pellets on the ground in a group. The numbers of serow (N) in an area (S) can be estimated by counting the numbers of faecal pellet groups (F) in the area as

$$N = \beta \, F / \alpha \, H$$

where β is the instantaneous rate of loss of pellet groups under natural environmental conditions; α is mean discovery rate of pellet groups, and H is the number of pellet groups deposited per unit time per individual, say *per capita* day. In this study, the values of variables α, β and H are measured and fixed as 0.39, 0.0428 and 90 respectively. These values were examined and confirmed as suitable for the environmental conditions in Kyushu (Morisita and Murakami 1979, Ono *et al.* 1976).

The second method is the direct count method. Use of this is limited to only winter or early spring because the individuals are hidden in the thick growth of bushes and trees at other seasons of the year.

A total of 85 sites in Kyushu were studied from 1975 to 1984 (Figure 6.1). Plural quadrats $1\,000-3\,000$ m^2 in area were established in each study site. Each quadrat was surveyed by 10 or more persons at a time, and the number of faecal pellet groups was counted.

RESULTS

The faecal pellet groups of the serow were recorded at 71 sites. The calculated density of the serows and the estimated number of individuals in a habitat are summarised in Table 6.1. The total number of individuals in Kyushu island was estimated as 254 to 322, and the average density was $1.7-1.9$ individuals/km^2. The range of population densities is quite similar to that in the Honshu area (Ministry of Environment 1980).

Figure 6.2 shows the distribution and abundance of the serow in Kyushu. The area showing relatively high population densities (cores) are Sobo-Katamuki mountain ranges, Natsuki-Ohkue mountain ranges, Mt. Osuzu, the Kunimi-Naidaijin mountain range, and Mt. Ichifusa. A remarkable feature of this distribution is that it is fragmented in a patchy pattern that depends on the distribution pattern of the endemic primary forests. In addition, the population of the Tsukigi-Aya mountain ranges is the southernmost limit of the distribution of the serow in Japan.

Table 6.2 shows the yearly changes of the density of faecal pellet groups at a site in the endemic primary forest near Mt. Sobo. The density was fairly stable throughout the year, which may indicate the long-term stability of the population density of the serow in this area. Another example of this stability was obtained at Mt. Katamuki. A number of serows were identified and were observed on the mountain slope of an artificial plantation bordering the endemic primary forest. The observers were arranged on the top of each branch ridge of the mountain. The serows could be easily observed from the ridge top. The total area covered in every observation period was 3.85 km^2 between 650 and $1\,100$ m in altitude on the northwest side of Mt. Katamuki. The observation was made monthly, mainly in winter and spring, from June 1979 to December 1983. A total of 194 observation points were obtained from the 16 observations over 222 person-days. The number of individuals observed in the study area 'Mo' is shown in Figure 6.3. Six adult females and two or three adult males stayed in the same home ranges for the whole study period. These individuals were regarded as residents. The number of females was surprisingly stable

Figure 6.1: Location of the study sites

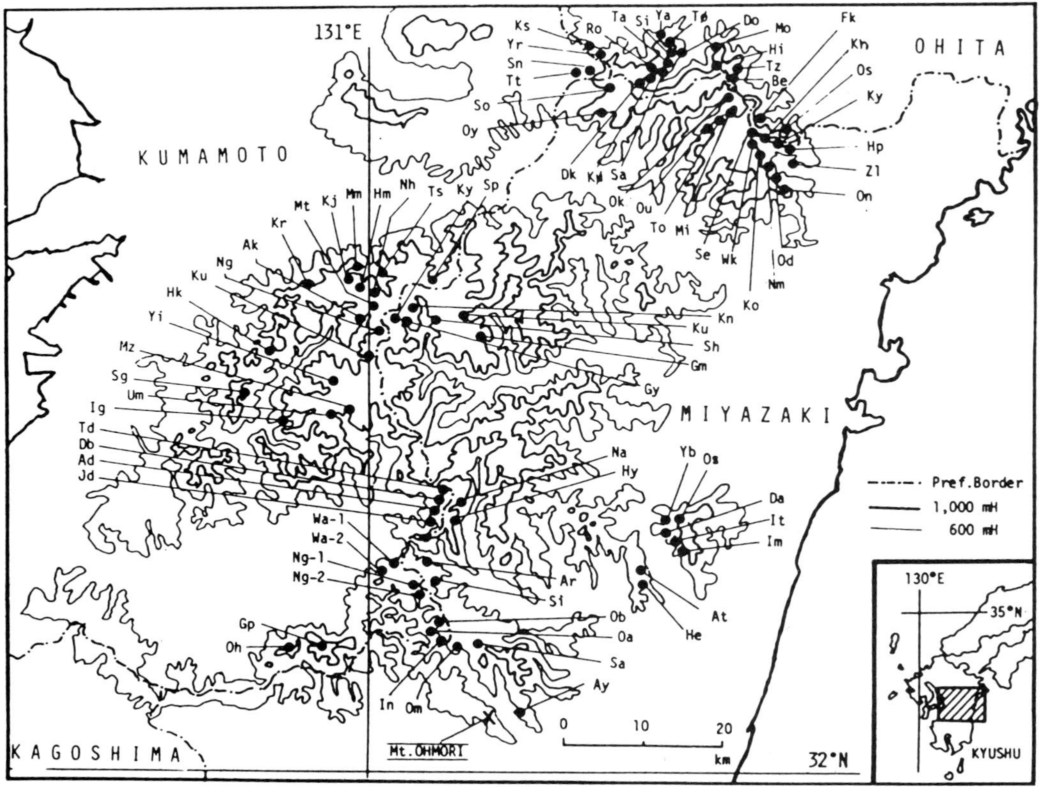

Table 6.1: Density and abundance of the Japanese serow on Kyushu

District	Study plot	Serow Density (n/km^2)	Habitat size (km^2)	No. of serow	
				Min.	Max.
A. Sobo-Katamuki	Ks, Yr	1.7	0.3-0.7	1	1
	Sn	1.5	0.6	1	1
	Tt	13.7	0.7-1.6	10	22
	Ku, Dk, Ya, Tø, Si, Ta, Sa, Do, Ro, Mo, Hi	2-3	37.4	75	113
	Be, Tz	1.9	4.5-5.5	9	11
	Oy	1.6	1.2	2	2
	Ka, Mi, To	3.2	4.2	14	14
	Ok	5.2	1.6	9	9
			50.6-52.2	121	173
B. Ohkue	Se, Wk, Ko, Nm, Od	0.7	12.9	10	10
	Kh, Ky, Hp, Zl	0.2	3.3	1	1
	On	2.5	3.4-4.8	9	12
	Fk	1.0	4.0-5.0	4	5
	Os	0.5	0.5-1.0	1	1
			24.1-27.0	25	29
C. Kunimi-Naidaijin	Gm, Ku, Sh	1.0	11.6	12	12
	Kn, Gy, Ky	0.7	8.1	6	6
	Sp	0.1	2.0-3.3	1	1
	Ts	0.1	1.1-2.6	1	1
	Nh, Hm, Kj, Mm	2.1	14.4-17.4	31	37
	Mt, Ng, Ku	0.3	9.7-13.7	3	4
	Yi	3.3	7.5-7.9	25	27
			54.4-64.6	79	88
D. Ichifusa	Db, Ad, Jd, Td	1.6	3.9-5.5	7	9
	Na	4.2	0.3	2	2
	Hy	1.1	3.1	4	4
			7.3-8.9	13	15
E. Tsukigi-u Aya	Oa	0.5	0.1	1	1
	Ob	1.3	0.3-1.9	1	3
	In	0.3	0.1-0.3	1	1
	Si	1.3	0.1	1	1
	Ar	0.2	0.4	1	1
	Om	0.1	1.7	1	1
	Sa	0.1	0.2	1	1
	Ay	0.9	0.7	1	1
			3.6-5.4	8	10

Table 6.1 continued

District	Study plot	Serow density (n/km^2)	Habitat size (km^2)	No. of serow Min.	Max
F. Osuzu	Oz	0.2	3.2	1	1
	It, Im	0.3	6.2	2	2
	Da	0.5	3.4	1	1
	Yb	0.3	0.9	1	1
	He	0.1	1.0	1	1
	At	0.1	0.3	1	1
			15.0	8	8
			155.0-173.1	254	323

Table 6.2: Yearly changes in the density of faecal pellet groups at a site in the endemic primary forest at the foot of Mt Sobo

Year	Month	Size of quadrat (m^2)	No of faecal pellet group	Density of faecal pellet group (No./m^2)
1973	May	2800	29	0.010
1975	July	2800	40	0.014
1976	January	1400	15	0.011
1977	December	2400	29	0.012
1983	December	2800	35	0.013
1985	March	2800	33	0.012
1986	April	2800	31	0.011

throughout the period. The fluctuation of the number of individuals in this area occurred only by the change of number of fawns and wanderers ('?' in Figure 6.3). Newly born offspring had to disperse to the unoccupied areas when they attained adulthood. Thus, the stability of the population density of the serow might be closely related to its social system.

The number of individuals in the whole observation area was 22 (4.4 individuals/km^2) at maximum, including ones temporarily resident and 17 individuals at minimum. On the other hand, the density estimated for the same area (sites Hi and Mo) with the pellet-group counting method was 4.5 individuals/km^2 in October 1975.

Figure 6.2: The distribution of the Japanese serow in Kyushu. The cores of the distribution are shown by the darker shading

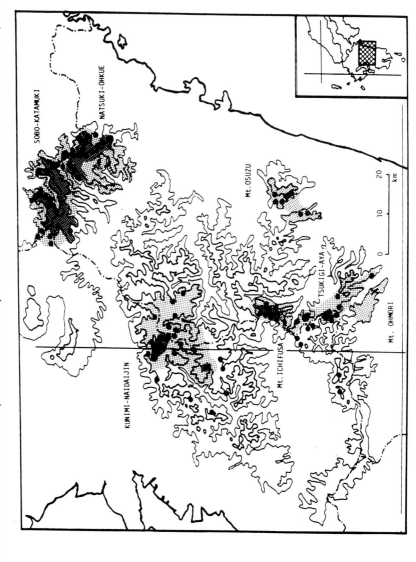

Figure 6.3: The number of individuals counted around site 'Mo' by direct observation. O: Newly born offspring, 1: 1 year old, ?: wanderer

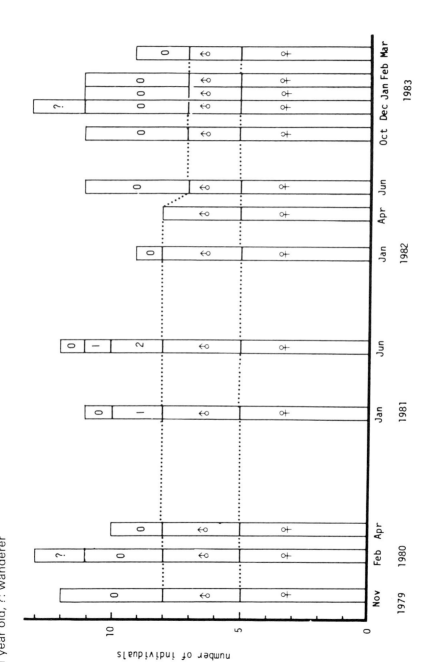

DISCUSSION

Recently, as areas of logging and afforestation spread, the animals started to feed on the leaves of the tree *Chamaecyparis*, which is one of the main plantation species in Japan. The damage is becoming an economical problem in forestry. From another viewpoint, however, it can be said that the utilisation of afforested areas by serows is an inevitable outcome of the drastic change that has occurred in the animals' natural habitat.

We have evaluated the utility of secondary forest after afforestation as the feeding area for serows in the mountainous area of the Sobo-Katamuki range (Ono *et al.* 1978). The population density of the serow in this area has been estimated as 2–4 individuals/km^2 by the faecal pellet-group count method. The abundance of faecal pellet groups in an area can be considered as indicating its utility for serows because they deposit their faecal pellet group at several limited places in their home ranges. Figure 6.4 shows the change in the density of faecal pellet groups in various ages of forest. Afforested areas several years old which border the primary forest were selected for sampling of the faecal pellet groups. The ages of afforestation selected were *Chamaecyparis* and *Cryptomeria* trees $1^1/_2$ years old, 4 years old, 6 years old and 13 years old. The mean density of faecal pellet group in the primary forest which was the original habitat of the serow was 40 ± 12 (95% confidence limit) per hectare. This density decreased drastically after logging, caused by the escape of serows from the area and cessation of additional deposition of pellet groups. However, shortly after afforestation, regeneration of sprouts from stumps and thick growth of herbs occurred at the logging area and supplied plenty of food for the serow. Then the density of faecal pellet group again increased, and reached a higher level, c. 60 per hectare, than in the primary forest, which meant the return of the animals to their original habitat. This extremely high density of faecal pellet groups is seen until six or seven years after logging and afforestation, and the density decreased steeply by only 5 per hectare in the 13-year-old plantation. The growth of trees and the development of the canopy layer inhibit the growth of herbs or grasses on the forest floor and results in a poor supply of food for the animals. This may be the explanation for the decrease in utility in the afforestation area with ageing of the plantation. The changing pattern of density was very different in the secondary forest which was left in its natural state after logging. The density of the pellet groups kept to the initial level after declining from the peak value six or seven years after logging and afforestation.

The serow in Kyushu has a patchy distribution with several cores in the mountainous regions. These locally isolated populations are considered to be vulnerable under heavy afforestation. Conservation of endemic primary forest is therefore to be recommended strongly. Furthermore, the secondary forests that connect these cores should be protected since they will act

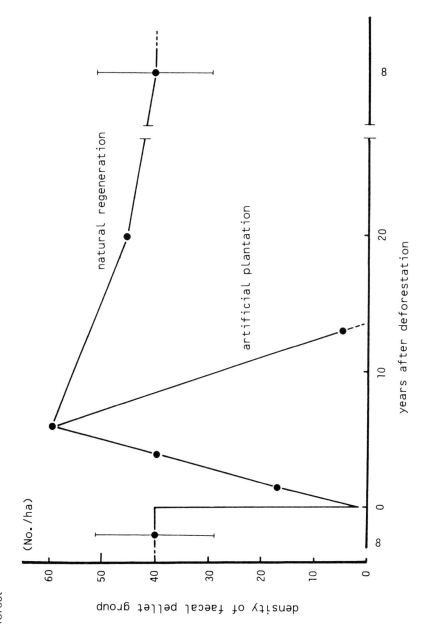

Figure 6.4: Change in the density of faecal pellet groups in various ages of forest. ∞ indicates the endemic primary forest

density of faecal pellet group

(No. /ha)

years after deforestation

natural regeneration

artificial plantation

as corridors for the otherwise isolated populations of the serow in the future.

From an ecological viewpoint, the serow in Kyushu has completely adapted to the specific environmental conditions of the warm-temperate zone, which has a surfeit of food plants throughout the year and has less snow fall in winter compared with the mainland of Japan. Both factors profoundly affect the ecology of the serow. We are evaluating the nutritional value of the herbs and trees as the food of the serow, with a view to assessing the habitat values of the warm-temperate forests in the Sobo-Katamuki range (T. Doi and Y. Ono, unpublished). Results so far indicate that the serow uses a high proportion of the leaves of evergreen trees, particularly in winter. It might therefore be expected that the serow in Kyushu has evolved some specific traits and a social system suitable for life in the warm-temperate forests.

ACKNOWLEDGEMENTS

We would like to thank Mr H. Tatsumoto, Mr S. Akiyoshi, Mr Y. Nakashima, the members of the Miyazaki and Kumamoto Wildlife Research Groups, and the staff of the Ecological Institute of Kyushu University for cooperation in the fieldwork. This study was supported in part by a Grant in Aid for Special Project Research on Biological Aspects of Optimal Strategy and Social Structure from the Japan Ministry of Education, Science and Culture.

REFERENCES

Kiuchi, M., W. Kuramoto and M. Yoshida. (1978) Review of the study of the Japanese serow. Rep. Nature Conservation Soc. Jap. No. 55: 223–46. (In Japanese.)

Ministry of Environment. (1980) Distribution and population density of the Japanese serow. Special Report of Japan Ministry of Environment. (Mimeo., in Japanese.)

Morisita, M., O. Murakami and Y. Ono. (1977) Estimation of population density of the Japanese serow, *Capricornis crispus* Temminck, by the pellet and pellet-group count methods. Studies on methods of estimating population density, biomass and productivity in terrestrial animals. *JIBP Synthesis 17*, 138–60

Morisita, M. and O. Murakami. (1979) Ecological study on the Japanese serow. [*Hakusan no Sizen* (*Wildlife of Hakusan*)] Ishikawa Pref., Ishikawa. (In Japanese with English summary.)

Ono, Y., K. Higashi, T. Doi and T. Yamaguchi. (1976) Population density of the Japanese serow in Sobo-Katamuki mountain range. Rep. Natur. Monument

Study in Ohita Pref. No. 36: 1–10. Education Committee of Ohita Pref. (In Japanese.)

Ono, Y., K. Higashi and T. Doi. (1978) Utility of secondary forest as feeding area for the Japanese serow *Capricornis crispus* Temminck in Sobo-Katamuki Range of Central Kyushu. Rep. Nature Conservation Soc. Jap. No. 55: 189–202. (In Japanese with English summary.)

7

Family break-up in Japanese serow, *Capricornis crispus*

Ryosuke Kishimoto
Laboratory of Animal Sociology, Department of Biology, Faculty of Science, Osaka
City University, Sugimoto, Sumiyoshi-ku, Osaka 558, Japan

INTRODUCTION

Ungulates have a gradational social organisation from a solitary life or monogamous pairs to large gregarious groups. This gradation of social organisation has been pointed out to be correlated with a shift in habitats from closed forest to open grassland or savannah (Estes 1974, Jarman 1974). The dissolution of the mother–young bond, otherwise called family break-up, is presumed to be related to the social organisation of the species, and possibly to the sex of the young. However, family break-up has not been well documented so far, especially in mountain ungulates, because long-term observations of identified individuals are required. The Japanese serow *Capricornis crispus* usually lives alone, and sometimes forms male–female pair units or family groups with up to four individuals (Akasaka and Maruyama 1977, Kishimoto 1981, Sakurai 1981, Ochiai 1983b). In the Japanese serow, the process of reaching independence by offspring has been documented (Ochiai 1983a), but there is not enough information on territory establishment by independent offspring. This chapter aims to describe the process of reaching independence, and that of territory establishment of offspring in the Japanese serow, based on continuous observations in the wild for seven years.

STUDY AREA AND METHODS

The study area, covering about 320 ha (2.5 km long and 2.0 km wide), lay 150–574 m above sea level (39° 48′ N; 140° 15′ E) in Akita Prefecture, Japan. The vegetation was composed primarily of a mixed forest of cedars and deciduous trees, where areas clear cut by lumbering were patchily distributed (one-fifth of the study area in extent). Snow cover (at maximum 1 m deep) lasted from mid-December to early April.

Field observations were made usually from April to December during

1979–1985. Over 110 residents, except kids, were individually identified through natural features, e.g. shape of horns and torn ears. I walked throughout the study area and observed the serows that I encountered. Binoculars (12 × 40) and a spotting scope (25–50 × 60) were used in watching.

RESULTS

Spatial organisation

Adult serows (4 years old and over) were sedentary throughout the year. Figure 7.1 shows a schema of the spatial organisation of adult residents' home ranges. The study area was almost completely covered with the home ranges of the same sex in each year. The mean size of annual home ranges was 13.8 ha ($N = 71$, range = 1.6–33.5 ha) for adult males, and 9.3 ha ($N = 81$, range = 1.2–24.5 ha) for adult females, and the sexual difference was significant ($t = 4.3$, $P<0.001$).

Intrasexual territoriality was strongly suggested in both sexes by the following observations: (1) range overlaps within the same sex were slight in both sexes; (2) close encounters around home-range peripheries

Figure 7.1: Left: schema of spatial organisation of territorial males and females. Right: instances of pairs of and polygyny by intersexual home-range overlaps

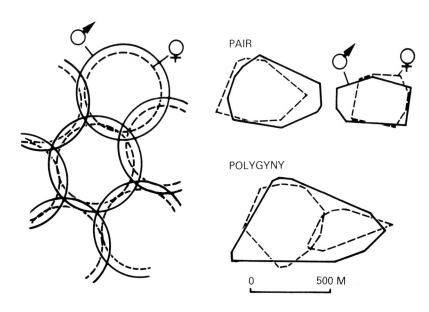

between neighbouring adult residents of the same sex usually resulted into aggressive chases, although there were few incidences of close encounters between adult females because of their almost complete spacing out. Accordingly, it was assumed that the home-range approximately coincided with the territory in both sexes.

Despite solitary ranging, pairs and polygynous males were identified by intersexual home-range overlaps: one adult male range of the pair almost completely overlapped with only one adult female range; one range of the polygynous male similarly overlapped with two adult female ranges (Figure 7.1 shows respective instances). Eight to 15 pair units and one to three polygynous males occurred at the same time of each year (censused in early October). These male–female range units were maintained for one to over seven years.

Dispersal of offspring

Females (3 years old or more) gave birth to single kids usually in May or June. At least 77.3 to 88.3 per cent (mean 80.2 per cent, $N = 5$) of the females gave birth in each year. More than half of the kids disappeared within one year; probably all of them died, because no kids moved alone away from their mothers' ranges, and 58 per cent of 12 corpses, which I collected in the study area, were kids.

The tight mother–young bond usually continued until mothers had their next newborn. The kids followed their mothers in 92.3 per cent of 481 kid observations. Generally yearlings were forced to leave their mothers by the aggressive behaviour of their mothers when they had their next newborn, although some offspring around one year of age appeared to leave their mothers spontaneously. If a mother failed to have the next newborn, the yearling tended to follow her for a further few months. When the newborn was a few months old, mothers appeared to be tolerant to previous off-spring; a mother, her kid and yearling occasionally formed a group. Almost all yearlings remained within their mothers' ranges although they became independent of them.

Figure 7.2 shows a schema of the spatial aspect of dispersal of offspring. Most offspring disappeared from their mothers' ranges when they were 2 or 3 years old and sexually mature. By 5 years of age, all offspring disappeared from the study area or established their own territories in the study area.

Six male offspring who were born in the study area established their territories when they were 3 to 5 years old. After leaving their mothers' ranges, four of them moved into vacant spaces which territorial males had previously occupied (three moved into adjacent spaces, and one moved about 1 km away from its mother's range). The other two took over a half

Figure 7.2: Schema of family breakup and establishment of territory in male (upper) and female (lower) offspring

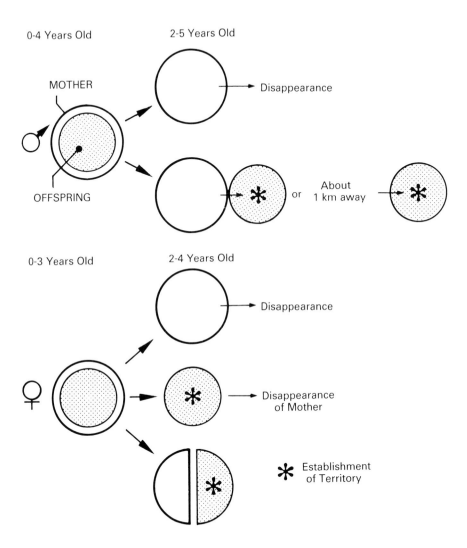

of the home-range of polygynous males (one intruded into an adjacent polygynous range, and one did so about 1 km away from its mother's range), and consequently formed a new pair with one of the two females of the polygyny. Presumably grown-up offspring were exploratory in looking for a vacant space, and repeatedly moved away from and back to their mothers' ranges. Accordingly, it should be quite possible to settle in a vacant space in adjacent areas if there is one available.

Six female offspring (3 to 4 years old) established territories as they remained in their mothers' ranges: three consequently took over their mothers' ranges because of the disappearance of their mothers; the other three took over a half of their mothers' ranges. Some female offspring who disappeared from their mothers' ranges may possibly have established their territories out of the study area. But it was presumed to be rare because only two female newcomers established their territories in the study area during the study period as compared with 18 male newcomers establishing territories in the study area.

DISCUSSION

Japanese serows of both sexes hold intrasexual territories which approximately coincide with their home-ranges, and basically maintain a pair by almost complete range overlaps within the different sexes. This pairing pattern has been regarded as 'solitary-ranging pair' in the common tree shrew *Tupaia glis* by Kawamichi and Kawamichi (1979). In consequence of this territorial society, offspring would have to leave their natal range towards their sexual maturity at 2 to 3 years of age. Actually most offspring disappeared from their mothers' ranges when they were 2 to 3 years old.

In the process of territory establishment of offspring, female offspring tended to remain within their mothers' ranges as compared with male offspring, all of which left their mothers' ranges. Such a sexual difference appeared to be reflected in the different interactions of mothers and their male partners towards offspring (over 1 year old): the male partners usually showed aggressive chases against male offspring whereas they were tolerant to female offspring; mothers were usually tolerant to their offspring of both sexes unless they were accompanied by their newborn kids.

This remaining of the female offspring may have caused father—daughter mating: when mothers disappeared, territorial males formed a pair with their grown-up daughters; when daughters took over half of their mothers' ranges, territorial males became polygynous with the mother and her daughter. However, such male—female bonding if any, would not be continued for a long period, because territorial males were more frequently replaced by a male newcomer or a neighbouring male of the pair than territorial females, and one of two female ranges in the polygyny was often intruded into. Therefore, the pairing pattern of the Japanese serow may have a role in the avoidance of inbreeding.

ACKNOWLEDGEMENTS

I thank Dr T. Kawamichi, Osaka City University, for continuing guidance over the course of the present study and for a critical reading of the manuscript. This work was supported in part by a Grant in Aid for Special Project Research on Biological Aspects of Optimal Strategy and Social Structure from the Japan Ministry of Education, Science and Culture.

REFERENCES

Akasaka, T. and N. Maruyama. (1977) Social organization and habitat use of Japanese serow in Kasabori. *Japan J. Mammal. 7* (2), 87–102
Estes, R.D. (1974) Social organization of the African Bovidae. *IUCN New Series 24*, 166–205. IUCN, Morges
Jarman, P.J. (1974) The social organization of antelope in relation to their ecology. *Behaviour 48*, 215–67
Kawamichi, T. and M. Kawamichi. (1979) Spatial organization and territory of tree shrews (*Tupaia glis*). *Anim. Behav. 27*, 381–93
Kishimoto, R. (1981) Behavior and spatial organization of the Japanese serow (*Capricornis crispus*). Unpubl. MSc. Thesis. Osaka City University, Japan. 59 pp.
Ochiai, K. (1983a) Pair-bond and mother–offspring relationships of Japanese serow in Kusoudomari, Wakinosawa Village. *Japan J. Mammal. 9* (4), 192–203. (In Japanese with English abstract.)
Ochiai, K. (1983b) Territorial behavior of the Japanese serow in Kusoudomari, Wakinosawa Village. *Japan J. Mammal. 9* (5), 253–9. (In Japanese with English abstract.)
Sakurai, M. (1981) Socio-ecological study of the Japanese serow. *Capricornis crispus* (Temminck) (Mammalia: Bovidae) with special reference to the flexibility of its social structure. *Ecol. Physiol. Japan 18*, 163–212

8

Censusing Japanese serow by helicopter in deciduous mountain forests

Manabu T. Abé[1] and Eiji Kitahara[2]

[1]Forestry and Forest Products Research Institute, PO Box, Tsukuba Norin Kenkyu Danchi-Nai, Ibaraki 305, Japan; and [2]Forestry and Forest Products Research Institute Kansai Branch, Nagai Kyutaro, Momoyama, Fushimi 612 Kyoto, Japan

INTRODUCTION

Once the Japanese Serow, *Capricornis crispus crispus* Temminck, endemic to Japan, was an endangered species because of excessive and illegal hunting. In 1934 the serow, which is distributed throughout Japan except Hokkaido and adjacent small islands, was nominated as a Natural Monument of Japan by the Cultural Properties Protection Act of 1925, and in 1955 it was upgraded to a Special Natural Monument by the Ministry of Education. Thereafter it has been protected completely under the Act.

Recently, it has been gradually increasing in number under strict protection, and severely damaging young forest plantations by feeding and horn rubbing. The mountainous areas were originally covered by deciduous broad-leaved trees. The main trees after logging and planting are conifers, such as Hinoki cypress, *Chamaecyparis obtusa*, Japanese red cedar, *Cryptomeria japonica*, and Japanese larch, *Larix leptolepis*. Unfortunately the cypress, which is the most valuable and merchantable tree, is also the most heavily damaged.

The present study is a contribution towards a systematic programme integrating the welfare of both forester and serow through the recording of accurate data on serow distribution and density. Until now, a major stumbling block in the management of serow resident in a forested habitat has been the determination of accurate population densities. Various methods developed elsewhere have been tried. Cahalane (1938) adopted the block count method to ascertain the number of elk, *Cervus canadensis*, in Yellowstone National Park as an estimation technique for big game. The reliability of this method was tested for the Sika deer, *Cervus nippon centralis*, and the Japanese serow (Maruyama and Nakama 1983, Maruyama and Furubayashi 1983). Bennett *et al.* (1940) introduced the pellet-group count method for the first time to estimate the number and distribution of white-tailed deer, *Odocoileus virginianus*. This method was applied to various wild ruminant species by numerous investigators (Eberhardt and

110

Van Etten 1956, Rogers *et al.* 1958, Wallmo *et al.* 1962, Van Etten and Bennett 1965, Smith 1968) to determine population trends and differential habitat use, particularly in woody cover. Neff (1968) reviewed many reports concerning the pellet-group count technique. Morishita and Murakami (1970) and Morishita *et al.* (1977) applied the pellet-group count method to estimate the density of the Japanese serow. Ono *et al.* (1976) and Hirata *et al.* (1976) estimated the number of serow by this method.

The visual count method was developed to estimate the number of the serow by Nakama *et al.* in 1980. The line transect census technique was developed and applied to the Sika deer (Ito 1968, 1976, Ito *et al.* 1973). Though these are all excellent and valuable ways to estimating the number of big game or of determining the trend of population density, they are seldom applicable in Japan because of excessive manpower costs and inaccessible steep terrain. From necessity the block count method has usually been used in Japan, but there has been no way of assessing its accuracy until the work reported here.

Many aerial censusing efforts using light aircraft have been carried out by investigators of many kinds of animals (Buechner *et al.* 1951, Petrides 1953, Evans *et al.* 1966, Goddard 1967, Bergerud and Manuel 1969, Sinclair 1972, Caughley 1974, LeResche and Rausch 1974). However, it has proved impossible to count big game animals living in dense coniferous forest and on rugged terrain from a rapidly moving light airplane. Ohta *et al.* (1972) and Haga *et al.* (1978) tried to count Sika deer, *Cervus nippon yesoensis* from a helicopter in Hokkaido, the northern most island of Japan. Neither research group estimated the whole population in its survey area. The accuracy of this aerial censusing method was tested by Maruyama *et al.* (1976) in Nikko, in the central part of Japan. They counted Sika deer from the air by helicopter and compared this with numbers of deer estimated previously by the block count method (ground count method).

However, a new efficacious aerial serow census technique accurate under Japanese conditions has been developed and validated. The objectives of the present study were:

(1) to conduct an aerial census of the Japanese serow in their autumn and winter range in Mt Iwamatsu (1512 m) of northwestern Gumma Prefecture during late autumn and midwinter of 1981, 1982 and 1983;
(2) to compare the aerial estimate of the serow population with the result of the ground count done in the same area by the block count method;
(3) to test the reproducibility of the aerial count by repeating it over the same area under the same conditions during two consecutive days; and
(4) to determine the practicability of an air-census technique in steep, rugged mountainous topography.

111

This chapter describes this new aerial-censusing technique (aerial driving census method) using helicopter over rugged terrain.

EQUIPMENT AND METHODS

A heliport was established in the study area. The first helicopter used was the large jet-helicopter, Bell 204-B, capable of safe, stable flight in steep mountain valleys. Preliminary flights were carried out on 4–5 February 1982, and in these test flights the helicopter was equipped with a video tape recording system: two cameras with 8.3 mm wide angle lenses were fixed at different angles under the helicopter so that the serow in various types of vegetation could be recorded easily. Counting was carried out from the air with the aid of direct visual count by two observers sitting on both sides of the rear window. At least six persons were required: a pilot, a navigator, two technicians for conducting the VTR system, and two observers. All flights were carried out after leaf fall. The flight courses were parallel to each other with 150 m between, an altitude of 100-150 m, and an air speed of 90 km (50 knots) per hour. Time and voice of observers were recorded simultaneously on the VTR system. The two observers could see both outside of the windows and two colour TV monitoring sets in the cabin. However, no serow were found in the monitoring screen, because serow were too small to be detected in the picture. Also the picture angle was changed frequently because of the rugged terrain: some areas were observed twice and others missed because of such variations. Thus linear flight along parallel courses through mountainous areas without respect to the rugged topography was not suitable for censusing serow. It was revealed that the VTR system was also useless for this purpose.

After several such trials, the linear flight technique was given up and a new flight technique and a direct visual counting method were developed. With this new method the helicopter was successfully used for counting the serow that inhabited the rugged, mountainous area.

The first flight was carried out on 10 March 1982. In this flight, a new aerial-censusing technique, was developed.

In this flight the Aerospacial 350-B type jet-helicopter, slightly smaller than the type used previously, was used to count the serow directly. Four crew members, a pilot, a navigator (recorder) next to the pilot and two observers on the back are sufficient to count the serow by direct observation. The pilot operated the helicopter under instructions from the navigator who was provided with a map on which all streams were marked with blue marker in advance. The map was of sufficient detail to enable the navigator to recognise his location easily and to put a mark on it when observers called out the number of serow. A distribution map was compiled later from these data.

The method is as follows: when a helicopter is slowly ascending a V-shaped mountain valley, the two observers can look round both slopes entirely and see all serow. Two observers, leaning out of their windows, can scan both slopes and the stream bottom, while the helicopter flies as closely as possible to 20–30 m above the ground at a speed of 20–30 km/h.

Efficiency of observation is further improved by removal of the rear helicopter doors so that the observers can easily scan from the bottom of the valley to the top of the ridge. When the helicopter has reached the top of the valley, it flies back at that high level and then descends to fly low up the next stream. Visual counting is made mainly during up-flight with re-affirmation on the way down. To minimise duplicate counting, each valley is flown only once. In the present study, the whole area was covered and scanned entirely. Two experienced survey crew-men, one on each side, classified serow by age (adult or subadult) and called out the results to the recorder (navigator), who spotted each location on a 1:10000 topo-graphical map. As much as 60 to 100 min were spent censusing and buzzing in one flight on an average, providing a census coverage of about 250–300 ha per hour. It proved practicable to fly and observe 5 hours a day even on the shortest days of midwinter.

In order to estimate relative census precision, repeated censuses were carried out under the same conditions, e.g. observer, flight course, hour, helicopter, pilot, navigator, etc. over two consecutive days on a sample area of 1 315 ha, on 30 November and 1 December 1982. The block count (ground count) made prior to the air census was carried out on 9 December 1981 in an area of 316 ha to produce a comparison with the air count. The procedure of the block count was the same as that carried out by Maruyama and Nakama (1983) and by Maruyama and Furubayashi (1983).

In addition, intense aerial-driving censuses were carried out on closed-canopy cypress and cedar forests, aided by the voice of a barking dog on the tape. The forests were checked for serow by two persons on the ground immediately after the air driving count. Although many species of larger birds and mammals could be identified from the air, their enumeration distracted from the accuracy of the serow count. Therefore only two additional species were counted: the hare, *Lepus brachyurus*, and the copper pheasant, *Phasianus colchicus*.

DESCRIPTION OF THE STUDY AREA

Gumma Prefecture is in the temperate forest region of central Japan where predominant tree species were originally two species of oak, *Quercus mongolica* and *Q. serrata*, Siebold's beech, *Fagus crenata*, Yedo hornbean, *Carpinus tschonoskii*, Japanese red pine, *Pinus densiflora*, etc. Several

113

decades ago, the deciduous forest was largely clear cut (except on ridges) for charcoal or pulpwood: old deciduous stands were left as a protection belt along the ridges to protect young planted trees from cold wind and heavy snow. After this, merchantable coniferous trees such as the Japanese red cedar, *Cryptomeria japonica,* the Hinoki cypress, *Chamaecyparis obtusa,* and the Japanese larch, *Larix leptolepis,* were planted.

The study area is located in the steep mountainous National Forest, ranging from 700 to 1512 m in altitude above sea level, in Kamisawatari and Tange, Nakanojo, northwestern Gumma Prefecture, Japan. Approximately two-thirds of the 2300 ha study area consisting of eight compartments is covered by young planted coniferous trees: this provides numerous openings and forage for the serow. Often, planted trees less than 5 years old have been killed by heavy, year-round serow browsing. The whole study area was surveyed by air, but the area of primary concern was the lower-lying winter ranges where 48 per cent of the slope was covered with deciduous forest: mostly mature 11–15 m tall trees. This area held a high serow population during the study periods of February 1982 and 1983.

The topography searched varied from relatively narrow V-shaped valleys with small streams and rills to an easy terrain in the lower parts where the planted young trees and clear-cut logged areas are seen. High, steep cliffs more than 200 m in height are seen here and there, and are used by the serow, in escaping from predators. Altogether, the study area represents a typical year-round habitat for the Japanese serow.

RESULTS

Aerial serow census

Serow were censused in mountainous forests in the northwestern Gumma Prefecture. In February, March and November 1982 and January, March and November 1983, using the new helicopter technique. Many wild species, including serow of course, ran around trying to escape the helicopter. Serows move when disturbed by the helicopter, but do not run far; usually they ran only a short distance to dense cover and remained there until the helicopter had passed on. This makes censusing possible and easy. The total observed population of the serow in 2300 ha was 100 individuals in March 1982 and 134 in March 1983, respectively. We observed 72 solitary serows and 14 pairs in 1982. Each pair probably consisted of an adult female and her young, or an adult male and female. The serow does not aggregate at any season. No group larger than two was observed in either year.

114

Comparison with the ground count

In order to test the accuracy of the much-used block count method (ground count), the aerial serow census of one 316 ha compartment was compared with a block count census made two months prior to it in the same compartment. The block count gave a density of 2.37–2.96 serows per square kilometre on 9 December 1981. In contrast, the aerial census indicated 5.70 head per square kilometre in the same area on 11 March 1982, nearly double the block count.

Verification of reproducibility

To test the precision of the aerial census method described above, identical censuses were conducted on two consecutive days over an area of 1315 ha comprising five compartments. The results of the repeat counts were: 30 November 1982, 48 individuals; 1 December 43 individuals. Numbers of serows in these two counts were not significantly different (Chi-square value is 0.275).

Distribution of the serow in the study area

When a serow was found during the census, a recorder also holding the post of navigator marked its location on a map of 1:10000 scale. Serows were not equally distributed on the study area: most serows were found on the slopes facing in a southwestern direction but a few were observed on the deeper snow of the north-facing slopes. Mature conifer stands had few serow, presumably because of a paucity of available food. No serow was found in the vicinity of humans present in the serow habitat.

Other wildlife detected from the air

It is very easy to find moving wildlife on the ground or in the air during a low-level, slow, helicopter survey because animals move in response to wind and noise. Dried and fallen leaves and fresh snow on the ground are whirled up by the strong wind. The following eight species of birds and mammals other than the serow were identified from the air in the course of this census: the Japanese black bear, *Selenarctos thibetanus japonicus*, the red fox, *Vulpes vulpes japonica*, the hare, the copper pheasant, the rufous turtle dove, *Streptopelia orientalis*, the jay, *Garrulus glandarius*, the Japanese green woodpecker, *Picus awokera*, and the great spotted woodpecker, *Dendrocopos major*.

115

Snow and detectability

Snow cover on the ground made finding wildlife easier in general. But snow was not necessary to discover flying or running individuals. Autumn counts, after leaf shedding but prior to snow fall, gave a satisfactory result, better than summer surveys. After leaf fall, any animal moving on the ground was fairly easy to find even if there was no snow on the ground. Detectability was not noticeably influenced by snow cover. However, coloration of the serow's hair (fur) surface differs individually, some being whitish grey and others nearly black: this would influence detectability slightly under snowless conditions.

Aerial driving count in closed-canopy forest

It was very difficult but not impossible to find serow in dense evergreen conifer forest growth sufficient to obscure ground visibility when the helicopter flew over the forest at routine census speed (i.e. 20−30 km/h). However, it was possible to detect a serow's tracks on the snow in the forest when it flew at the lesser speed of 20 km/h. Two types of dense conifer forests of Japanese red cedar and Hinoki cypress more than 40 to 50 years old, were checked by two persons walking about in the forest to determine whether or not any serow were lurking after the persistent aerial seeking and the recorded dog barking were over. No serow running away from the forest were, however, found by these ground observers in either type of forest.

Copper pheasant censusing

This method was tested for the copper pheasant inhabiting the study area. Censusing the phesant was carried out at the same time as the serow count. The sex of this species can be visually determined by total length: a male has much longer tail feathers than the female. Therefore, locations detected were marked on the map using different marks for each sex.

DISCUSSION

The larger birds and mammals are difficult to see when they are not moving. They are also hidden by the foliage of deciduous trees. Further, steep rocky slopes make observation from the ground slow (and hence expensive) and inefficient. All of these difficulties can be overcome by the helicopter count method if it is conducted after leaf fall and the flight is

116

very low and very slow. If a helicopter moves at tree-top height, very slowly, up steep mountain stream-beds, virtually all larger birds and mammals, startled by the noise and wind, will move away, and so become easily visible. Snow improves visibility, but is not necessary for an adequate count. Where there is strong sexual dimorphism, as in the copper pheasant, sex can easily be distinguished. Although only serow, hare and copper pheasant were tallied in the present study, it would be possible with this method to census virtually any larger birds or mammals.

ACKNOWLEDGEMENTS

Special acknowledgement is made to the following personnel, past or present of the Nakanojo District Forest Office, the Maebashi Regional Forestry Office: Mr Kanjiro Arai, Mr Tsuyoshi Ohkura, Mr Takashi Tamiya and Mr Isamu Shimizu for offering the heliport and for their courtesy in other ways.

We also appreciate very much the help of Akira Kurita of the Japan Agriculture Aviation Association. Special thanks go to Mr Kunihiko Tokita of the Japan Wildlife Research Center who acted as a leader of the block count on the ground. We wish to express our appreciation to the pilot Mr Shigehisa Hirayama of Nihon Norin Helicopter, Inc., whose enthusiastic outlook towards wildlife censusing and confident handling of the helicopter made aerial work in Nakanojo both a pleasure and a more precise technique. We extended grateful thanks to Professor Richard D. Taber of Forest Zoology, College of Forest Resources, University of Washington, who read the manuscript and offered helpful suggestions. Many people other than the above-mentioned participated and assisted in this study. We thank all of them.

REFERENCES

Bennett, L.J., P.E. English and R. McCain. (1940) A study of deer populations by use of pellet-group counts. *J. Wildl. Mgmt 4* (4), 398–403

Bergerud. A.T. and F. Manuel. (1969) Aerial census of moose in central Newfoundland. *J. Wildl. Mgmt 33* (4), 910–16

Buechner, H.K., I.O. Buss and H.F. Bryan. (1951) Censusing elk by airplane in the Blue Mountain of Washington. *J. Wildl. Mgmt 15* (1), 81-7

Cahalane, V.H. (1938) The annual northern Yellowstone elk herd count. *Trans. N. Am. Wildl. Conf. 3*, 388–9

Caughley, G. (1974) Bias in aerial survey. *J. Wildl. Mgmt 38* (4) 921–33

Eberhardt, L. and R.C. Van Etten. (1956) Evaluation of the pellet group count as a deer census method. *J. Wildl. Mgmt 20* (1), 70–4

Evans, C.D., W.A. Troyer and C.J. Lerisink. (1966) Aerial census of moose by quadrat sampling units. *J. Wildl. Mgmt 39* (4), 767–76

Goddard, J. (1967) The validity of censusing black rhinoceros populations from the air. *East Afr. Wildl. J. 5* (1), 18–23

Graves, H.B., E.D. Bellis and W.M. Knuth. (1972) Censusing white-tailed deer by airborne thermal infrared imagery. *J. Wildl. Mgmt 36* (3), 875–84

Haga, R., Y. Fujimaki, S. Takemoto and H. Ogawa. (1978) Mammals in the upper stream of Shizunai River basin (II). Report on the Ecological Survey in Shizuoka River Basin. Mimeo, Obihiro University of Agriculture and Veterinary Medicine (in Japanese)

Hirata, S. *et al.* (1976) Research Report on Japanese Serow. *Res. Rep. Nat. Mon. Aomori Pref.*, 6–59 (in Japanese)

Ito, T. (1968) Ecological studies on the Japanese deer, *Cervus nippon centralis Kishida* on Kinkasan Island. *Bull. Marine Biol. Stn Asamushi Tohoku Univ. 13* (2), 139–49

Ito, T. (1976) Counter line transect census for Sika deer on Kinkasan Island. *Saito Ho-on Kai Museum Res. Bull. 44*, 31–8

Ito, T. *et al.* (1973) The census of the Japanese sika deer, *Cervus nippon nippon* Temminck, on Kinkasan Island carried out by the counter line transect method in 1972. *Ann. Rep.* JIBP-CTS 1972, 197–207

LeResche, R.E. and R.A. Rausch. (1974) Accuracy and precision of aerial moose censusing. *J. Wildl. Mgmt 38* (2), 175–82

Maruyama, N. and K. Furubayashi. (1983) Preliminary examination of block count method for estimating numbers of Sika deer in Fudakake. *J. Mammal. Soc. Japan 9* (6), 274–8

Maruyama, N. and S. Nakama. (1983) Block count method for estimating serow population. *Japan. J. Ecol. 33* (3), 243–51

Maruyama, N., Y. Totake and R. Okabayashi. (1976) Seasonal movements of sika in Omote-Nikko, *Tochigi Pref. J. Mammal Soc. Japan 6* (5, 6), 187–98

Morishita, M. and O. Murakami. (1970) Ecological studies of the Japanese serow, *Capricornis crispus. Sci. Studies Hakusan Nat. Park*, pp. 277–321

Morishita, M., O. Murakami and Y. Ono. (1977) Estimation of population density of the Japanese serow, *Capricornis crispus* Temminck, by the pellet and pellet-group count methods. *JIBP Synthesis 17*, 138–60

Nakama, S., N. Maruyama, S. Hanawa and O. Mori. (1980) On estimating the serow number by new visual methods in Wakinosawa village, Aomori Prefecture. *J. Mammal Soc. Japan 8* (2,3), 59–69 (in Japanese with English abstract)

Neff, D.J. (1968) The pellet-group count technique for big game trend, census, and distribution: a review. *J. Wildl. Mgmt 32* (3), 597–614

Ohta, K., H. Abé, T. Kobayashi, N. Ohtaishi and K. Maeda. (1972) Faunal survey of Oketo area. (I) Report on JIBPCT No. 138, 208–35 (in Japanese with English summary)

Ono, Y. *et al.* (1976) On the population of Japanese serow in the Sobo mountains. *Oh-ita Pref. Cul. Prop. Res. Rep. 36*, 1–12 (in Japanese)

Petrides, G.A. (1953) Aerial deer counts. *J. Wildl. Mgmt 17* (1), 97–9

Rogers, G., O. Julander and W.L. Robinette. (1958) Pellet-group counts for deer census and range-use index. *J. Wildl. Mgmt 22* (2), 193–9

Sinclair, A.R.E. (1972) Long-term monitoring of mammal populations in the Serengeti: census of non-migratory ungulates, 1971. *East. Afr. Wildl. J. 10*, 287–97

Smith, R.H. (1968) A comparison of several sizes of circular plots for estimating deer pellet-group density. *J. Wildl. Mgmt 32* (3), 585–91

Van Etten, R.C. and C.L. Bennett Jr. (1965) Some sources of error in using pellet-group counts for censusing deer. *J. Wildl. Mgmt 29* (4), 723–9

Wallmo, O.C., A.W. Jackson, T.L. Hailey and R.L. Carlisle. (1962) Influence of rain on the count of deer pellet groups. *J. Wildl. Mgmt 26* (1), 50–5

9

Radio tracking of Japanese serow in Akita Prefecture, Japan

Kazuhiko Maita

Section of Natural Conservation, Akita Prefectural Government, 1-1-4 Sanno-cho, Akita City, Akita Prefecture 010, Japan

Seventeen Japanese serow (*Capricornis crispus*) were captured by laying wire traps on Mt. Taihei (1170.6 m), Akita Prefecture, from 2 April 1983 to 31 August 1985, and the activities of fourteen of them were observed in their natural habitat through the use of radio telemetry (Table 9.1).

The Mt. Taihei range has been an upheaved highland since the late Tertiary, and it has gradually been steepened by erosion even though it is not very high above sea level. At an altitude of 200–600 m, the tree species comprise Japanese cedars, ground cypresses, chestnuts and wingnuts, and from 700 to 900 m, beech appears. Above this altitude, the trees consist of stunted beech and oak.

OBSERVATION OF THE MONTHLY ACTIVITIES

Variations in the amount of behavioural patterns of the serows were discernible through recording of the signal strength received from each transmitter, and activity patterns of 14454 h 3 min, which were possible to read out of 16480 h 13 min obtained, were analysed. From these analyses, three patterns of NMP (no modulation pattern), SMP (slight modulation pattern) and IMP (intensive modulation pattern) were obtained (Table 9.2) and (Figures 9.1a–c). IMP included mostly general movements from one place to another to a simple shaking of the head; SMP consisted solely of slow movements without change of position; and NMP was concerned with sleeping and taking a rest without movement. In the classification of each pattern, a certain amount of overlap was involved. Through the observation of monthly variations in the behaviour of a male serow aged 12 years old captured on 14 November 1983, a regular pattern of activity became apparent for eight months until June 1984 despite a heavy snowfall. From this observation, it is suggested that herbivores need to eat food regularly regardless of whether it is day or night.

119

Table 9.1: Tracking duration of Japanese serows at Mt Taihei from 2 April 1983 to 31 August 1985

No.	Sex and age	Weight (kg)	Date radio-marked	Tracking durations in day
1	Male fawn	27.0	2 April 1983	57
2	Male adult	32.0	13 November 1983	181
3	Male adult	35.0	14 November 1983	225
4	Female adult	35.0	15 November 1983	
5	Female adult	44.0	15 November 1983	27
6	Male adult	40.0	16 November 1983	124
7	Male fawn	16.0		
8	Female adult	39.0	5 September 1984	597
9	Female fawn	14.0		
10	Female fawn	18.0		
11	Male adult	48.0	23 September 1984	157
12	Male juvenile	44.5	23 September 1984	584
13	Male adult	48.0	23 September 1984	371
14	Male adult	35.0	29 July 1985	274
15	Male adult	36.0	31 July 1985	272
16	Female adult	42.0	4 August 1985	269
17	Male adult	35.0	8 August 1985	243

Table 9.2: Category of activities among Japanese serows

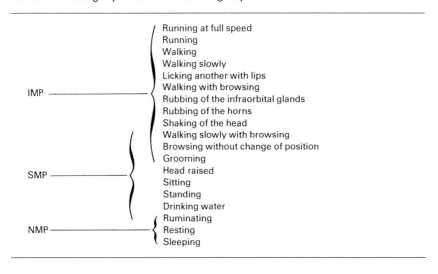

IMP
- Running at full speed
- Running
- Walking
- Walking slowly
- Licking another with lips
- Walking with browsing
- Rubbing of the infraorbital glands
- Rubbing of the horns
- Shaking of the head

SMP
- Walking slowly with browsing
- Browsing without change of position
- Grooming
- Head raised
- Sitting
- Standing
- Drinking water

NMP
- Ruminating
- Resting
- Sleeping

Figure 9.1: Monthly changes of activity of eight radio-marked Japanese serows

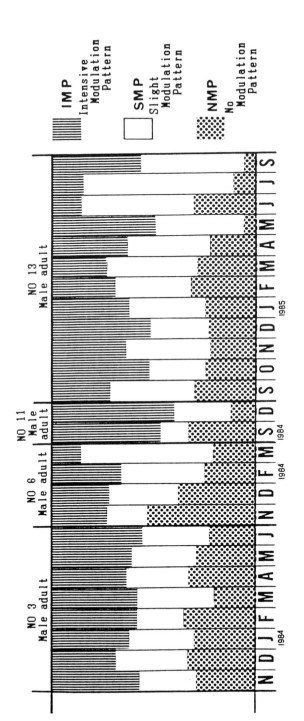

(a)

Figure 9.1 *continued*

(b)

IMP

SMP

NMP

NO 14
Male adult

NO 8
Female adult(with young)

S O N D J F M A M J J A S O N D J F M A A S O N D J F M A
1985   1986   1986

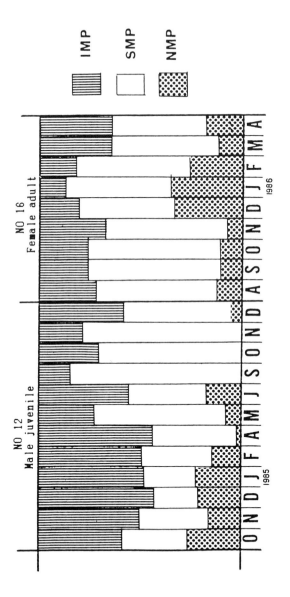

(c)

A male serow aged 18 years old captured on 16 November 1983 was released in a remote place in order to study homing movements. The serow displayed a strong instinct to return to his former habitat and was unsettled. It died in May 1984 after a heavy snowfall which had sharply decreased IMP in that month. The autopsy showed that there was a large quantity of the leaves of Japanese cedars in the stomach and intestines.

From the variation in the amount of activity of a female serow aged 8 years old with its offspring, captured on 5 September 1984, a big drop in NMP and an increase in IMP in April was detected. It seems that this is due to the influence of fighting with her young to force it to separate prior to the birth of another kid from her. IMP also decreased sharply in January. From the variation in the amount of activity of a male serow age 4 years old from 23 September 1984 until December 1985, it was estimated that the young serow was always awake without indication of NMP patterns during the mating season from September to October. This was explained by the fact that the NMP pattern could not be determined because of damage to the aerial transmitter in August. The serow also displayed a decrease in NMP and an increase in IMP in May.

One male serow aged 6 years old captured on 29 July 1984 displayed an increase in IMP in the winter of 1985 (January). In one male serow aged 13 years old captured on 4 August 1985, a decrease in IMP and an increase in NMP were shown in winter of 1986 (February). Although, in general, activity patterns varied with each individual, the period from April to May was characterised by a marked drop in NMP and an increase in IMP.

ABOUT HOMING MOVEMENT

Japanese serows No. 1 and No. 6, captured in 1983, seemed to be unsettled after being released in remote places and attempted to return to their former habitats. Both serow died on their way back to these places. A male Japanese serow No. 15 (6 years old) captured in 1985 returned to his former habitat at a distance of 2.5 km in three days. These examples suggest how impractical it is to transfer serows from an overpopulated area to one with a sparse population in order to control their numbers.

10

A preliminary study on the ecology of Formosan serow *Capricornis crispus swinhoei*

Kuang Yang Lue

Department of Biology, National Taiwan Normal University, Taipei, Taiwan 117, Republic of China

INTRODUCTION

Due to the small surface area of Taiwan, there are only few species of big mammals on the island. Large herbivorous animals found in Taiwan include sambar deer (*Cervus unicolor*), sika deer (*Cervus taiouanus*), barking deer (*Muntiacus reevesii*) and Formosan serow (*Capricornis crispus swinhoei*). Formosan serow is a subspecies endemic to Taiwan. Morphologically there is little difference between Formosan serow and Japanese serow. In body size, the animal in Taiwan is smaller than the one in Japan. In addition, the Formosan serow is covered with shorter, dark brown hairs. The colour is light brown on the chin, throat and neck.

Since the end of the Second World War, untouched virgin forests in remote areas have been logged or planted with vegetables, and wildlife habitats, including that of the Formosan serow, have deteriorated seriously. Furthermore, illegal hunting by professional hunters and aborigines is common. The population size of the serow is shrinking. Only a few field studies of Formosan serow have been made since 1945, and data on the biology, behaviour, ecology and life history of the serow are needed.

METHODS

An attempt was made to cover as much of the island as possible during the survey. Droppings, footprints and browsing marks were examined and recorded. Food plants and bones were collected. Aborigines and professional hunters were interviewed. The survey started in 1982.

ZOOGEOGRAPHIC RELATIONSHIPS FOR FORMOSAN SEROW

Currently, there are only two species of serows recorded, the Japanese serow (*Capricornis crispus*), including the Formosan serow, and Sumataran serow (*Capricornis sumatraensis*). The Formosan serow and Japanese serow are found only in Taiwan and Japan, and the Sumatran serow lives only in India, south-east Asia and certain parts of Mainland China.

In geological history, Taiwan was connected to the Asian continent several times. The last glacial retreat occurred about 16000 years ago. From the evolutionary viewpoint, the time is too short for a big animal to evolve. Based on the most optimistic estimation, it could evolve only up to species level. Karyotype studies show that both Formosan and Japanese serows have 60 chromosomes (Soma 1981), and the difference between these two animals is very small. Lin and Lin (1983) and some Japanese researchers believed that the Formosan serow came from Japan and reached Taiwan by passing through the east part of Mainland China. After the last glaciation, these migrants were locked in this island and were isolated from their parental population in Japan. As a result of habitat differences, two subspecies of serows evolved.

DISTRIBUTION

From information obtained in the survey and interviews in the past three years, we discovered that the Formosan serow is still widely distributed all over the island, from the north to the south. (Figure 10.1). Animals were sighted or caught in 11 of Taiwan's 16 counties. These are Taipei, Ilan, Hualien, Hsinchu, Taichung, Nantou, Chia-I, Tainan, Kaohsiung, Pingtung and Tai-tung.

Records indicate that the altitudinal range of its distribution varies tremendously. (Figure 10.2) In Nan-jen Shan of Kenting National Park, near the southern tip of the island, the sighted location is only 200 m above sea level; whereas in the Yu-shan area it is about 3870 m above sea level. Figure 10.1 shows that droppings, browsing marks and footprints were found most frequently in forests at an elevation of 1000–3500 m. The range is wider than that found by Lin and Lin (1983). While Kano (1940) was doing his investigation before the war, he discovered that the altitudinal change for Formosan serow was about 3500 m. Our data showed that the range has not changed a lot since 1940. Dien (1964) conducted a parasitological study of wildlife in remote areas, and most of the serows he examined came from forests 1500 m above sea level. The results of a survey by McCullough in 1974 revealed that Formosan serow could be seen from the foot of a hill up to 3500 m, but were comparatively common between 1000 and 2000 m above sea level. Okada and Kakta (1970) studied Japanese serow and found the range for this animal to be 500–2500 m. It seems that the altitudinal distribution for Formosan serow is wider than that of Japanese serow. The difference is reasonable in view of latitudinal changes.

126

Figure 10.1: Distribution map of Formosan serow

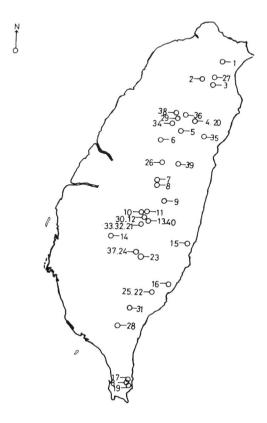

Based on footprints, droppings and browsing marks left by serows, the largest population sizes were found in remote areas, such as Nan-hoo-da-shan, the Watersheds of Tou-sai-shi, Old King-yang aboriginal village, the peak of Yu-shan, North-face Creek, Da-fen area, Pei-nan Mountain, Ba-lin, and Twin-ghost Lakes of Ping-tung (Figure 10.3). All areas are 2000 m above sea level. The vegetation includes coniferous and mixed deciduous forests.

HABITATS

McCullough (1974) stated that steep and rocky slopes are the favourite habitats of serow, and especially disturbed slopes. From our survey, we found droppings in alpine grassland consisting of *Yushania niitakaymensis*, on the forest floors of *Juniperus* sp. and on top of cliffs. This is a little

127

Figure 10.2: Altitudinal distribution of Formosan serow

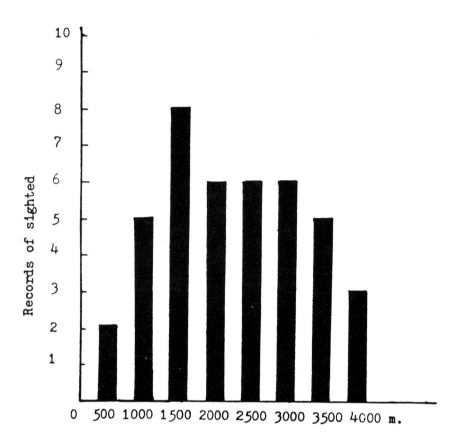

different from McCullough's findings. In Nan-hoo-da-shan, Yu-shan, Hsiuh-shan, Shou-ku-ran-shan and Pei-da-wu-shan areas, the vegetation where either droppings or footprints were found comprises *Tsuga chinensis, Abies kawakamii, Juniperus formosana* and *Juniperus squamata*. Among these, browsing marks were also found on coniferous plants. Apparently Formosan serows will eat the juvenile part of conifers, just as Japanese serow do (Okada and Kakta 1970). McCullough (1974) listed 17 species of feeding plants for Formosan serow, all of which were grasses and shrubs growing on disturbed slopes of early succession stages. *Miscanthus* sp. is the main dominant plant he mentioned. Information obtained by interviewing aborigines revealed that *Urtica fissa, Elatostema edule, Anisogonium esculentum, Begonia laciniata, Polygonum chinensis, Chamabainia cuspidata, Mussaenda parviflora, Perrottetia arisanensis* and *Pellionia arisanensis* are favourite plants for Formosan serow. It is worth mentioning that

128

Figure 10.3: Areas for larger populations of Formosan serow. 1,
Nan-hoo-da-shan. 2, Tou-sai-shi. 3, Old King-yang aboriginal village. 4,
Yu-shan. 5, North-face Creek. 6, Pei-nan Mountain. 7, Ba-lin. 8, Twin Ghost
Lakes. 9, Da-fen area. 10, Den-da area

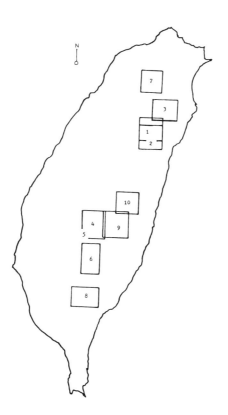

Urtica fissa is a poisonous plant, but it seems to cause no harm at all to
serows. In captivity, Formosan serows are fed with *Morus australis*,
Ipomoea batatas and *Ixeris chinensis*. The statements above show that
serow eats various types of plants. *Trimeresurus stejnegeri* is a poisonous
green snake with a very good protective coloration. Two cases of predation
by Formosan serow were sighted by local hunters.

In August 1981, one juvenile female serow was caught by the author
and colleagues near the edge of a rocky slope on Hon-da-la-shi about 1800
m above sea level. The slope was caused by a recent landslide. No vegetation was found. Above the slope Lauraceae and Fagaceae are the main
dominant plant families. Grasses, ferns and shrubs were found on the floor.
This is a typical mixed deciduous forest in Taiwan. In Nan-hoo-da-shan
and Yu-shan areas where serows live, there is little but mosses on the floor.

129

While we were doing our investigations, footprints and droppings were easily found near hiking trails in remote mountain ranges. At Shan-yang station of the Southern East–West cross highway, one serow came daily to eat corn that was provided for domestic chickens. We conclude that Formosan serow survives well with some human disturbance.

FAUNAL RELATIONSHIPS

In hills and mountain areas, Formosan macaque (*Macaca cyclopsis*), Formosan giant squirrel (*Petaurista grandis*), Formosan white-headed flying squirrel (*P. alborufus lena*), red-bellied tree squirrel (*Callosciuris erythraeus roberti*), Formosan black bear (*Selanarctos thibetanus formosanus*), wild boar (*Sus scrofa taivanus*), Formosan Reeves' muntjac (*Muntiacus reevesi micrurus*), Formosan sambar deer (*Cervus unicolor swinhoei*), Formosan weasel (*Mustela sibrica davidiana*), Formosan mountain rat (*Rattus culturatus*) and Formosan fieldmouse (*Apodemus semotus*) were found from various locations where the Formosan serow lives. Except weasel, all other animals cannot be predators of serows. Most of these animals are herbivorous. Among these, sambar deer and muntjac are potential competitors for Formosan serow. The relationships among these three big herbivorous animals in Taiwan are still unclear.

Weasel is a carnivorous animal, but it is much smaller than serow. There are only few chances for a successful predation by weasels. Several cases were reported by aborigines that weasels ate the trapped Formosan serows. From faeces we found that mountain rats and fieldmice are the main dominant food for weasels in Taiwan. Clouded leopard (*Neofelis nebulosa brachyurus*) might be the prime predator for serows in forests, but the author believes that this big cat has been exterminated in the wild.

Birds that can be found with Formosan serows are Mikado pheasant (*Syrmaticus mikado*), Swinhoe's blue pheasant (*Lophura swinhoei*) Formosan hill partridge (*Arborophila crudigularis*) and Formosan bamboo partridge (*Bambusicola thoracica sonorivox*). The competition between these birds and serows will not be as severe as with herbivorous mammals.

CURRENT STATUS OF FORMOSAN SEROW

Based on our observations, we found out that the population of Formosan serow is dwindling. Natural habitats for serows are deteriorating. Besides, illegal and excessive hunting occurs quite often in most areas. Both foot and neck snares are used most frequently. Traps and shooting are used occasionally. Most hunting is conducted in late autumn and throughout the winter, from November to March. Aborigines and professional hunters usually avoid the warmer seasons, because the trapped animals easily decompose.

Skeletons of lower jaws obtained from in the field were analysed (Table 10.1 and Figure 10.4). Two wildlife dealers interviewed said that they could get on average about 10 animals each month. A total of about 60 individuals are purchased every hunting season. This indicates that the hunting pressure in Formosan serow in certain areas is quite heavy. In remote areas, such as Dan-da, Jen-da and Da-fen of central mountain ranges, Nan-hoo-da-shan, and Old King-yang village, all above 2000 m with virgin vegetations, serows held their population well there in the face of heavy hunting. There is no doubt that Formosan serow requires protection in most areas.

Table 10.1: Lower jaws of *Capricornis crispus swinhoei* collected in the field

Location	Date	Length (cm)	Width (cm)	Height (cm)
Kuang-kaou	June 1981	16.62	3.02	5.76
Yen-pin forest	November 1981	13.62	2.86	6.18
Show-ku-ran	April 1982	14.01	a	6.28
Pei-da-wu	September 1982	12.18	a	5.42
Pa-tong-kuang	August 1982	11.64	a	5.66
North-face Creek	August 1982	11.16	a	5.66
North-face Creek	August 1982	9.44	a	a
Little Ghost Lake	June 1983	11.52	3.38	5.58
Unknown		16.9	8.50	7.40
Pa-lin	February 1986	17.20	8.86	a
Unknown		16.82	8.24	5.54
Unknown		15.33	7.71	5.81
Unknown		15.16	5.56	6.24

a Unable to obtain the measurement

CONCLUSION

Currently the Formosan serow is distributed widely all over the island of Taiwan. The highest populations are found in alpine grasslands and virgin coniferous forests. The Formosan serow seems to adapt very well to disturbances, including ones from humans. It eats various types of plants. Although hunting is illegal, the pressure on serows is still very heavy in this island, and a comprehensive conservation programme for Formosan serow is required.

Figure 10.4: Lower jaws of various age groups collected in the field

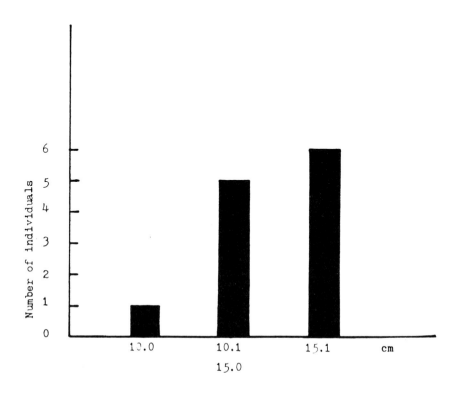

ACKNOWLEDGEMENTS

The authors wishes to thank the Council of Agriculture and Taipei Zoological Garden for financial support. I also thank students of the Biological Department and the Mountain Club of National Taiwan Normal University, who helped greatly with the fieldwork.

REFERENCES

Dien, Zen-Ming (1984) The Formosan serow. *Quat. J. Taiwan Mus. XVI* (1–2), 98–100
Kano, T. (1940) *Zoogeographical studies of the Tsugitaka mountains of Formosa.* 145 pp.
Lin, J.Y. and G.L. Lin. (1983) *Wildlife survey on Pa-tung-kuan area of Yu-shan National Park.* Tung-hai University, Taiwan 150 pp.

McCullough, D.R. (1974) *Status of larger mammals in Taiwan.* Tourism Bureau, Taiwan, 36 pp.

Okada, I. and T. Kakta (1970) Studies on the Japanese serow, *Capricornis crispus* (Temminck). An investigation report of Japanese serow. 10th Annual publication of preservation of the serow at Mt Suzuka, pp. 1-15

Soma, H., H. Kada, K. Matayoshi, M.T. Ysai, T. Kiyokawa, T. Ito, K.P. Wang, B.P. Chen and S.C. Chen. (1981) Cytogenetic similarities between the Formosan serow (*Capricornis swinhoei*) and the Japanese serow (*Capricornis crispus*). *Proc. Japan. Acad., Sci. 57*, 254-9

11

Social behaviour of Japanese serow, *Capricornis crispus crispus*

Mitsuko Masui

Ueno Zoological Gardens, 9-83 Ueno Park, Taito-ku, Tokyo 110, Japan

INTRODUCTION

In the Japanese serow, some form a family consisting of a female and her young (or including a male) in a certain area of territory, and others live alone. Animals observed in the field stand still in most cases, which gives an inactive impression. This chapter reports the findings of a study of the Japanese serow in which the behaviour observed in the field and in captivity were compared with those of other ungulates kept in Tama Zoological Park.

STUDY AREA AND METHODS

The area chosen for the field observations is Nibetsu Japanese Cedar Plantation, which is located at the foot of Mt. Taihei in the ENE part of Akita Prefecture (39° 40′ N, 140° 30′ E). It covers a space of 600 ha and is 300 m in altitude. Several streams run through the slopes, and afforested and deafforested lands form a mosaic. Observations with the naked eye or with binoculars (7×50) were made from 12 May 1975 to 12 November 1979 in the field, and amounted to 380 h 38 min. Observations from 8 March 1975 to 26 February 1976 were made in Tama Zoological Park, and totalled 140 h 31 min.

SOCIAL BEHAVIOURS

A total of 93 encounters were recorded over the period of 380 h 38 min of field observation. Figure 11.1 shows the behaviours during the corresponding time. The behaviour observed most frequently was staring fixedly. Animals were mostly on high stumps, so it seemed that they could easily see others from a distance (Figure 11.2). In some cases two animals

134

Figure 11.1: Behaviours at the time of encounters in Japanese serow

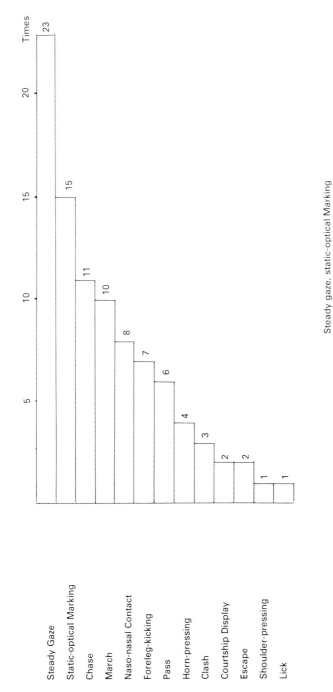

Figure 11.2: Boundary encounter between family consisting of a pair and young (left) and an adjoining solitary one (standing on the stump)

mutually recognised and came nearer to each other. The one was observed with its neck stretching low, and the other showed a slightly erect posture. Such an approach led to naso-nasal contact which caused courtship-like display, or optical marking, or agonistic behaviours such as clashing and horn-pressing (Figure 11.3).

THREAT AND DOMINANCE DISPLAY

Courtship-like display was observed from a male to a female, but examples from a male to a male seem to be seen. Agonistic behaviours are assumed between the same sex, but instances from a female to a male could be also observed. Violent chases were observed 11 times, and were presumed to be aimed at driving out wandering animals. There is similarity between court-ship display and dominance display. The mouflons show behaviours such as foreleg-kicking, twisting and neck low-stretching, followed by butting between males (Geist 1971, Masui and Nagai 1979).

Those behaviours are also observed among Japanese serow, not only during the oestrous period but also when they encounter one another. Whereas a dominant male shows active behaviours, a female or a submissive male stays still with its haunch lowered. This kind of display is also made

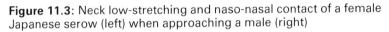

Figure 11.3: Neck low-stretching and naso-nasal contact of a female Japanese serow (left) when approaching a male (right)

towards an animal with which the same home range is shared, or towards one in the adjoining home range which the displaying animal knows well. It is also considered as a reconfirmation of dominance.

COURTSHIP DISPLAY

Japanese serow come into heat in autumn, and two examples of courtship display have been observed (Masui 1978). Table 11.1 shows a comparison between these two examples and other ungulates. Behaviours such as sniffing rear, flehmen, foreleg-kicking and anterior chest-pressing were frequently observed. The adoption of an erect posture was not so obvious, and the behaviour of foreleg-kicking looked like scratching in comparison with that of the mouflons or the oryxes (Figure 11.4). Behaviours involving the rubbing of a secretion from the preorbital glands against the horns and ears of a female were also frequently observed, and are similar to such behaviours among gazelles (Leuthold 1977, Walther *et al.* 1983, Walther 1984).

Table 11.1: Distribution of male courtship display in several ungulates

	Japanese serow *Capricornis c. crispus*	Moulfon *Ovis orientalis*	Scimitar oryx *Oryx dammah*	Nilgai *Boselaphus tragocamelus*	Reticulated giraffe *Giraffa c. reticulata*	Sika deer *Cervus n. yakushimae*
Erect posture	+	+	++	+	+++	+
Foreleg-kicking	++	+	++	−	−	−
Neck low-stretching	±	+++	−	+	−	+
Cajoling a doe by rubbing his chin on her rump	±	±	−	+	+	+
Anterior chest-pressing	+	+	−	+	++	+
Lick	+	+	±	++	+	+
Horn-rubbing	+	−	−	−	+	−
Horning	++	−	−	−	−	+
Marking a doe with his preorbital glands	++	−	−	−	−	−

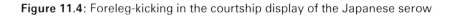
Figure 11.4: Foreleg-kicking in the courtship display of the Japanese serow

AGONISTIC BEHAVIOURS

Sociable ungulates frequently show the behaviour of sparring, but quite a few examples of sparring are observed among the unsociable Japanese serow. If many young are kept together, they play forehead-pressing or horn-pressing. On the other hand, in the field, young about the same age seldom meet, and the behaviour with grown-up animals is rarely observed. The young bucks play horning against stumps and other material.

Many animals that seem to have been injured by horning are found among weak and rescued Japanese serow. Table 11.2 shows the relation between frequency in dominance display and lethality in a range of animals. Those with big horns, or with striking parts in regions such as shoulder or rump, get off with a slight injury in many cases by making a threat display which shows these striking parts. As regards the Japanese serow, they have short and unimposing horns and have no striking parts. There is no sexual dimorphism between both sexes as far as the observations in this report are concerned, and the horns are also developed as weapons to the same degree in both sexes.

Japanese serow often get stabbed in the hypochondrium. There is evidence that the skin of regions that easily get stabbed has become thick (Geist 1971, Masui 1976). Figure 11.5 shows the thickness of the skin of a few species of animals. Japanese serows have a thin skin all over the body

139

Table 11.2: Relations between threat and dominance display and lethality in several ungulates

	Sociability	Threat and dominance display	Frequency of sparring	Main injured region	Lethality
Japanese serow *Capricornis c. crispus*	±	±	±	Hypochondrium	+++
Mouflon *Ovis orientalis*	++	++	+++	Face	
Scimitar oryx *Oryx dammah*	++	++	+++	Dorsum	
Nilgai *Boselaphus tragocamelus*	+	+	+	Hypochondrium	+++
Reticulated giraffe *Giraffa c. reticulata*	+	++	+++	Rump	
Chinese muntjac *Muntiacus reevesi*	±	±	±	Hypochondrium	+++
Japanese wild pig *Sus leucomystax*	++	++	+++	Shoulder	

Figure 11.5: Dermal thickness in six species of ungulates. Dotted line denotes the region of thickening. (From Masai 1976)

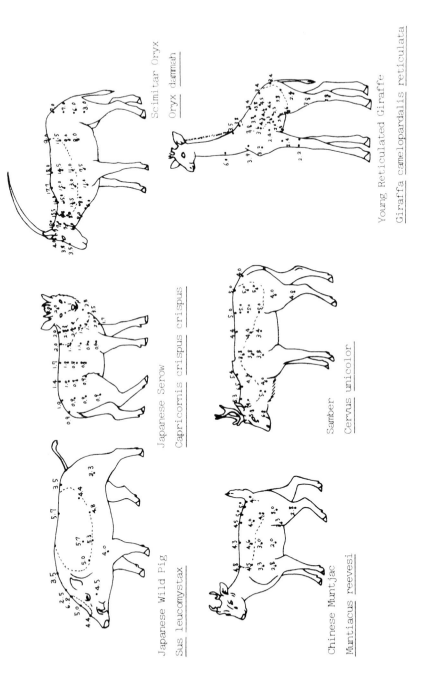

Japanese Wild Pig
Sus leucomystax

Japanese Serow
Capricornis crispus crispus

Scimitar Oryx
Oryx dammah

Chinese Muntjac
Muntiacus reevesi

Samber
Cervus unicolor

Young Reticulated Giraffe
Giraffa camelopardalis reticulata

(except the regions around the neck and the head, which are slightly thicker than other parts). They will therefore easily suffer a mortal wound if stabbed with sharp horns. They do not exhibit the behaviour of sparring, and their way of defending is quite primitive, which seems to cause their serious injuries.

MARKING BEHAVIOURS

Other noticeable behaviours of the Japanese serow include scent marking and object aggression (Figure 11.6). Both males and females exhibit behaviours such as scent marking, object aggression and optical marking, but a difference is seen between both sexes in the frequency when they live in family groups in certain areas. Observations of markings in a family group of a pair and young over the period of 140 h 31 min and the frequency of scent marking are shown in Table 11.3. A female also exhibits such behaviours frequently when it is not with a male. When a female shares the same home range with a male, the animal tends to leave the responsibility for marking with the male.

Among the behaviours related to territoriality that are exhibited by Japanese serows are scent marking by means of the preorbital glands; object aggression; statue-optical marking; and the depositing of dung piles. In optical marking, Japanese serows sometimes stand still on small protuber-

Figure 11.6: Preorbital-gland marking in the Japanese serow

ances such as stumps and rocks. The posture is useful for mutual recognition because it can be seen even from a distance. In some cases two animals stand on the same place in parallel or in a reverse position, which may correspond to the broadside display of other ungulates (Figure 11.7)

Table 11.3: Frequency of markings

	Object aggression	Secretion marks
Buck	134	685
Young buck	9	1
Doe	16	44

Figure 11.7: Two animals stand on a stump in a reverse position, which may correspond to broadside display by other ungulates

CONCLUSION

Japanese serows do not form large herds. Usually they live alone, in pairs or as a family. The ceremonial trend of agonistic behaviour is not obvious, which can easily lead to a serious fight, but dangerous encounters are effectively avoided by mutual use of olfactory and visual signals, although they lack the conspicuous threat display seen among other sociable ungulates. They keep a safe distance from one another, and encounters seem to be relatively quiet even if animals happen to meet.

ACKNOWLEDGEMENTS

The author is most grateful to Dr H. Hiramatsu, Mr Y. Mezawa and other staff of Tama Zoological Park, Dr Y. Matsumoto and Dr T. Matsumoto, Mr Y. Mikami and eleven other students at the Nature Conservation Club of Nihon University, Mr A. Hirai, student of Azabu University, and Ms Y. Iwasawa, for their assistance in observation in the field and in the zoo. The author would also like to thank Mr K. Maita, Mr T. Akasaka and Mr N. Hama for providing information on field observations. Thanks are also due to Mr A. Takato who prepared the photographs, and to Ms H. Masui for her assistance in translating the manuscript into English.

REFERENCES

Geist, V. (1971) *Mountain sheep. A study in behavior and evolution.* University of Chicago Press, Chicago Ill.
Leuthhold, W. (1977) *African ungulates. Zoophysiology and ecology 8.* Springer, New York
Masui, M. (1976) Zoo lecture room, weapons and shields. *Animals and Zoos 25,* 338-40
Masui, M. (1978) *Some observations of courtship and mating behavior* in free-living Japanese serow, *Capricornis crispus crispus. J. Mamm. Soc. Japan. 7* (3), 155-7
Masui, M. and Nagai, S. (1979) Social behavior of the mouflon. Hierarchy and behavior patterns. *Animals and Zoos 28,* 6-10
Walther, F.R. (1984) *Communication and expression in hoofed mammals.* Indiana University Press, Ind.
Walther, F.R., Mungall, E.C. and Grau, G.A. (1983) *Gazelles and their relatives. A study in territorial behavior.* Noyes Publications, USA

Part Three:
Keeping and Breeding of *Capricornis*

12

Breeding of Sumatran serow at Dusit Zoo

Chira Meckvichai and Alongkorn Mahannop
Dusit Zoo, Bangkok, Thailand

INTRODUCTION

There are two subspecies of Sumatran serow in Thailand. *Capricornis sumatraensis sumatraensis*, with black legs, is found in the south, and *C. s. milneedwardsi*, with red legs, ranges from Tenasserim to the north and to the east.

Dusit Zoo used to keep a group of serows which proved to breed well but no records were kept. In 1974 an epidemic spread through the serow exhibit and killed all nine of them.

Our male came to Dusit Zoo from the wild in December 1975 when he was about a year old, and he was kept singly until a 5-year-old female came in March 1978. We immediately put them together in the same enclosure. Both of them are *C. s. milneedwardsi*.

CAGING

The enclosure is a wire cage 18×20 m in area. The wall is 2 m high except behind the artificial hill where it is 5 m in height. Inside there is an artificial cement hill of 15×17 m, the highest point of which is 5 m. This proved to be very suitable for climbing, and the animals stay up on the hill for several hours a day. The living quarters are 4×6 m with a concrete roof 2.5 m high. All four sides are open. A cement water burrow measuring 0.7×2 m is provided. The lawn is covered with carpet grass (*Axonopus compressus* Beauv.), which the serow sometimes nibble. There is also a raintree (*Samanes saman* (Jacq. Merr.) and a pine tree (*Casuarina aquisetifolia* Linn.)

FOOD AND WATER

In the wild, serow is more of a browser than a grazer. At Dusit zoo we usually give the serow five main kinds of natural food, sweet potato (*Ipomoea batatus* Lamk), banana (*Musa sapientum* Linn.), string bean (*Vigna sinensis* Savi), Para grass (*Brachira mutica* Stapf) and morning glory (*Ipomoea aquatic* Forsh.). The food is given in the living quarters twice a day at 0900 and 1300. Besides the food provided by the keeper the animals are also fed by visitors but the amount given in this way is relatively small. The animals also nibble raintree leaves that fall from the tree into the cage. The animals will not eat all the food at feeding time. Instead they take some and leave the living quarters for the hill to ruminate, and come back to eat several times a day including during the night. Kruahong (1983) reported that each serow in Dusit Zoo takes 7.92 kg of food in fresh weight or 1.47 kg in dry weight, which proved to be enough. Banana is the most favoured food and string bean the least. All foods are offered in natural form except sweet potatoes which are cut into small pieces. Mineral salt is provided for the animals to lick at any time they choose. No other supplement food is provided. Extra calcium is given to the female after she has delivered. Three grams of calcium is inserted into banana and given to the mother every day for two months. Water is given in a main burrow outside the living quarters as well as in a small burrow inside the living quarters. The animals take a very little water each day.

BREEDING

Since the pair were put together, the female was on heat several times but she did not invite the male. The male tried to copulate but the female refused even though they seemed to be compatible. Early in 1979 the female had shown interest in the male for the first time by staying close and licking his body, head and ear, and the male answered by the same gesture. The female showed some restlessness while the male stayed close to her; sometimes the male left the female to run around the enclosure rubbing its horn against roots or the tree, and also made a loud snort, raking his hoof with the ground. The first copulation was observed on 21 June 1979 but no pregnancy ensued. Copulation lasted only 5 seconds. Since the first occurrence no further copulation was observed until September 1980, after which the veterinarian found that the female was pregnant. The first male offspring was born on 19 January 1981.

Normally a pair live separately, except during the rutting period as stated above. After copulation the female pays no attention to the male, and lives separately again until the baby is born. Lekagul and McNeely (1977) said that the breeding season was in October and November and that the gestation period is 210–230 days, but our record showed that they breed all year round. From the fourth month of pregnancy, the female sleeps more and other acivities are performed noticeably much less than usual. The male is no longer allowed to come close even during feeding time, when they usually eat together.

BIRTH

Two weeks before birth the female became more restless. The teats became bigger and red, the flank reduced in size and the vagina became swollen. She started looking for a place in which to give birth. There is a small den at the foot of the hill which is also close to the living quarters, and this was chosen by the female to give birth to her first offspring and each of her subsequent five young. We provide straw or hay in the den. After the kid was born the mother ate all the placenta and licked the body clean until the hair was dry. The new born kid's weight ranged from 3.7 to 4.5 kg. The length is approximately 50 cm and the height is 45 cm. The kid started standing 45 minutes after birth while the mother used her nose to support it. Four hours after birth it could walk for a short distance alongside its mother. Climbing the hill was practised with close supervision from the mother when the kid was 1 day old. It could climb as high as 1 m at its first try, and within a few days it could climb alone without any support. During the first week the mother took her baby for a walk around the enclosure for a short period early in the morning when there were no crowds around. Most of the time they would stay in the den.

After birth, relations between the father and the kid were limited. The mother always chased the father away, keeping the kid with her all the time. The father also showed aggression to his young. There were times when the mother had to leave it alone when she was eating in the living quarters, and the father took the chance to sneak in and he gored the kid. No father-and-son behaviour was observed. The present practice is to move the father to another enclosure once the female has shown signs of giving birth. He will be back in the same enclosure when the baby is big enough and strong enough to escape whenever the father attacks.

The kid started taking milk as soon as it could stand and would feed three or four times in the daytime. At night it would feed four to six times. From 0100 to 0400 both mother and kid were sound asleep and no milk taking was observed. The first week the kid took only milk. It started nibbling morning glory leaves during its second week but still took milk most of the time. Other solid foods were taken after 1 month when single behaviour started to take place. The kid would stay alone and came to the mother only to take milk. The weaning age is about 5–6 months when the mother is pregnant again.

The mother will be on heat 3-4 weeks after giving birth. Our records show that the female can get pregnant again as early as 3 months after giving birth.

ABORTION AND COMPLICATIONS IN SEROW

The following are our records on a stillborn serow.

15 May 1981: Female showed sign of oestrus (vulval swelling, clear discharge from the vulval opening, no aggressiveness with the male), and the male tried to copulate with her (walking behind the female) all day. The male copulated with the female once in the evening.

18 June 1981: No oedematous swelling of vulva and the male was not interested in the female.

5 October 1981: Expanding of abdomen (larger than normal) and dropping so that the flank is conspicuous. Swelling of mammary gland so that all this part is larger in size; the teats begin to become erect and to swell.

20 October 1981: There was a fetal movement (wave) on the left flank. Teats and mammary gland were conspicuous. Body hair in the ventro-lateral area of both sides of the abdomen was soft and dense, and the animal was slow in walking.

21 October 1981: Hyperpnoea. Oedematous swelling and opening of vulva. Dilatation of nostrils. Conspicuous fetal movement. The tail was erected, and the animal walked slowly, feeding less than normal and licking the abdominal hair frequently.

22 October 1981: Normal respiration, fetal movement more frequent, more swelling of vulva, urination more frequent, sleeping on the hill all day except when feeding.

3 November 1981: Abdomen larger and dropping. Flank more conspicuous. Normal feeding. Mammary gland and teats more swollen. Fetal movement (wave) three or four times a minute. Respiration 30 times per minute. Walking around the enclosure, urination infrequent.

10 November 1981: Clear fluid discharge from the vulva, fetal movement normal.

11 Novembc 1981: 0430. Abortion (stillborn). Size 51 cm in length, shoulder height 40 cm, hip height 36 cm, ear length 8 cm, tail length 6 cm, weight 3.5 kg, amniotic fluid in both nostrils and anus, body soft.

Treatment given to the mother:

R   Penicillin–streptomycin 800000 units I/M

Oxytocin 20 units I/M

Nystatin–diiodo hydroxyquin–benzalkonium chloride (NDB) 100000 units, 1 tablet inserted in the uterus

The animal in good health, feeding normally; placenta dropped from the vagina 1000.

15 December 1981: White placental tissue (part) from the vulva. The animal was restrained on lateral recumbency and the remaining placenta and cotyledons (7) were pulled out.
Treatment given:
R   Oxytocin 20 units I/M
      Penicillin–streptomycin 800000 units I/M
      NDB 200000 units inserted into the uterus
Vulva showing sign of oedematous swelling; feeding normally.
16 December 1981: Vulval swelling but less than yesterday, diarrhoea (greenish-yellow stool). Anorexia.
R   Penicillin–streptomycin 800000 units I/M
      NDB 200000 units inserted into uterus
      Sulfamonomethoxine 1200 mg I/V
17 December 1981: Vaginal discharge but no fetid odour, vulva nearly normal size, normal feeding, normal rumination.
R   Chloramphenicol 1g I/M
      Sulfamonomethoxine 1200 mg I/V
      NDB 100000 units inserted into uterus
      Cortisone acetate 50 mg I/M
18 December 1981: repeat treatment.
R   Douching with Dettol solution
      NDB 200000 units inserted into uterus
      Vitamin K 40 units I/M
      Kanamycin 1 g I/M
      Iron dextran 200 mg I/M
      Kao–pectin–neomycin 40 c.c. P/O
19 December 1981: Uterine infection improved, only the intestinal infection treated:
R   Kanamycin 1 g I/M
      Vitamin K 40 units I/M
      Sulfamonomethoxine in drinking water
24 December 1981: Diarrhoea stopped
R   Gentamycin 160 mg I/M every day for 4 days
      Sulfamonomethoxine in drinking water
Appetite good.
28 December 1981: Faeces normal; feeding normal.

RECORD OF BIRTHS AND DEATHS

Our records on six young born to the same parents are as follows:

(1) Male born on 19 January 1981; weighed 4.00 kg. It was separated from the parents at 8 months old, and is now at Khao Khiew Open Zoo.
(2) Male stillborn 11 November 1981; weighed 3.5 kg.
(3) Male born 17 December 1982; weighed 4.2 kg. Died 21 March, 1983 of acute pneumonia.
(4) Male born 26 November 1983; weighed 4.2 kg. Gored by its father at the age of 3 months, which caused serious injury to the left hind leg. Never recovered, became paralysed and died 16 June 1984.
(5) Male born 5 October 1984; weighed 4.5 kg. Died 16 May 1985 of acute lung oedema.
(6) Female born 27 August 1984; weighed 4.2 kg. Died 11 January 1986 of anaemia and pneumonia.

DISCUSSION

Our records show that the serow in captivity can breed all year round. This differs from what occurs in the wild as reported by Lekagul and McNeely (1977) who state that the breeding season is October and November. This may be due to the fact that food in the wild becomes scarce in the dry season (December–April) and is abundant in the rainy reason (May–November). Serows will be healthy and ready to breed after the rainy season. Young are born at the beginning of the rainy season and start taking solid food in the middle of the rainy season when young leaves and buds are plentiful. By the time the mother is pregnant again, the baby will be strong enough to live separately. Temperature is believed to have no significance in this case since it varies very little each season. In captivity food is ample all the time, which makes the serow healthy all year. Conditions of food shortage never occur in the lives of the zoo animals.

There has been no report in regard to relations between young and parents in the wild. Our record is limited to only one family. It is still questionable, especially the father's habit of hitting its young. Whether this is an individual behaviour remains to be studied.

The mortality rate of baby serows is very high. Only one of all six survived. The causes of death vary, though some cases have features in common. Most animals died when they were 4–8 months old, and we suspect that this is perhaps when they are weaning and adapting to solid food. At this age resistance is low, and the animals easily become ill if they are stressed or if the whether and temperature change abruptly. Our records also show that we lost most of our wild-caught serows when they were about the same age as the zoo-born ones.

REFERENCES

Kruahong, C. (1983) Ecology and some behavioral characteristics of serow (*Capricornis sumatraensis*) in captivity. Unpublished Thesis, Bangkok University

Lekagul, B. and J.A. McNeely. (1977) *Mammals of Thailand.* Kurusapa Ladproa Press, Bangkok

13

Breeding and behaviour of Formosan serow at Taipei Zoo

Chen Pao-Chung

Taipei Zoological Garden, 66 Chung Shan N., N. Road, Section 3, Taipei, Taiwan

INTRODUCTION

The Formosan serow is a subspecies of *Capricornis* endemic to Taiwan. Its habitat is primitive coniferous forest and cliffs at elevations of 1000–3000. Populations of these animals in Taiwan have always, been small, and unfortunately their habitat has been diminishing gradually in recent years due to overexploitation of the forest, and because of the failure of the authorities to enforce the wildlife conservation law, resulting in unimpeded illegal hunting. The number of Formosan serow is therefore decreasing rapidly.

Taipei Zoo started its Formosan serow breeding programme in 1974. Because we fear that a demand for this species might cause large-scale hunting, thereby reducing the number of Formosan serows in the wild, we can only obtain the animals from local wildlife meat restaurants. Thus whenever we see a healthy one in a restaurant, we grab the chance and buy it. However, because these animals are collected only for eating purposes, the trapper seldom worries about catching them in such a way as to avoid damage to the animals. This has caused a very high mortality rate in keeping Formosan serow obtained in this way. Furthermore the natural habitat of this species is in the high mountain region, and so it seems to be very difficult for them to adapt to the warm and humid weather conditions that prevail on flat low-lying land. The task of breeding Formosan serow has shown very slow progress. There has been no record of successful breeding in Taiwan except the one from Taipei Zoo.

This chapter provides information regarding keeping, breeding and behaviour of the Formosan serow recorded in Taipei Zoo during the past eleven years. It is hoped that this record will serve as a reference for future breeding programmes and will help in educating the public and in conserving the animals.

HOUSING

At the very beginning we mixed the Formosan serow with Formosan sambar deer. The two species seemed to get along very well. The Formosan serow can jump without difficulty on to the fence, which is 2 m above the ground. It walks easily on top of the fence, which is as narrow as 3 cm, and if the animal is not being frightened, it always goes back to its enclosure. Three years later, we built an exhibit area specially for Formosan serow at the highest point of the zoo. The area is fan-shaped and about 350 m^2 (Figure 13.1). A house is located in the area. There are two isolated rooms inside the house. We also fence off a small outside area to give a more aggressive animal an opening separate from that used by the others.

Figure 13.1: The Formosan serow enclosure

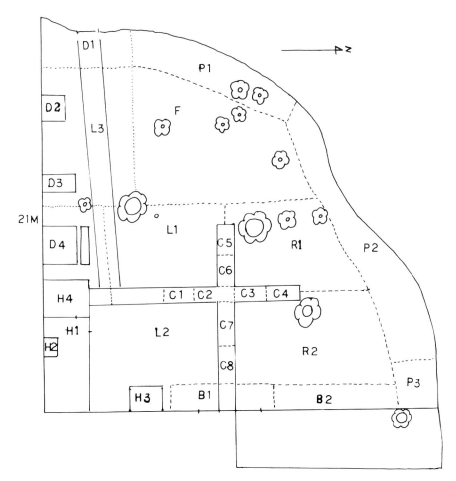

The highest part of the area is flat land which occupies half of the whole exhibit area. A cross-shaped concrete platform and a shelter are also provided there. The rest of the area has a 30° slope down to the ground. There are a few flamboyant trees with trunks 30 cm in diameter, and ten maple trees with trunk 7-8 cm in diameter occur randomly in the area. The total enclosure has a fence 160 cm high and 30 cm wide with an overhang inclined inwards at an angle of 45°.

It seems that the exhibit area is a little small for eight Formosan serows. A few aggressive ones always occupy the feeders, and the weaker ones are often cornered. Also, the area slopes only on one side which faces the cold winter wind. Shelters in the area are not strong enough to protect the animals from cold wind. Therefore, the health and skin condition of the animals usually become worse in the winter. The concrete platform in the area gives the animals a place in which to rest. Trees in the enclosure also provide the animals with shade in summer. One large tree inside especially helps the weaker animals to hide from being attacked by the aggressive ones. It is our observation that two weaker Formosan serows always jump up to a branch 3 m high and stay there for a long time.

The gate of the house is always opened, although the animals seldom go inside except when it rains or there is food. They spend most of their time (including night-time) outdoors. There was grass upon the ground when animals were first released into this section. However, there has been no more grass ever since then. We have tried to grow various leguminous plants there but without success.

FOOD

We gave each animal 0.4 kg of pellets (see Table 13.1), 0.5 kg of carrots, 0.8 kg of fresh napier grass and 0.6 kg of soybean curd pomace per day to start with. The nutritional condition did not look right. After careful observation and changes, we gradually came to use more natural food to feed them. At the present time, the menu for the eight animals per day is 1 kg of pellets, 4 kg of cabbage, 5 kg of fresh corn, 2 kg of bamboo shoots, 3 kg of carrots, 10 kg of mulberry branches and 4 kg of sweet potato leaves. The result seems pretty satisfactory.

The only matter that continued to concern us was that the animals suffered diarrhoea or soft stools from time to time. As a result of our investigations, it was found that those that suffered diarrhoea were those that ate more pellets. The diarrhoea situation was much improved after the quantity of pellets was cut down. Most of the Formosan serows have learned to beg food from visitors. They are fairly interested in biscuits, candies, various fruits, popcorn and bread which visitors offer to them.

Table 13.1: Ingredients of food pellets

Ingredients	% per kg
Corn, ground	47.4
Soybean meal	20
Sorghum	8
Wheat bran	10
Yeast	3
Napier grass, meal	6
Molasses, cane	3
Bone meal, steamed	2
Crude salt	0.6
Vitamin/trace-element mixture	0.06

BREEDING

One serow started mating at the age of 16 months. She gave birth to one lamb later but it died because its mother lacked colostrum. For the past eleven years, there are six records of mating. It happens from September to December, the time period when daylight is becoming shorter (Figure 13.2). When mating, the male will follow the female around and will touch her with his horns very gently, or he will use his front legs to touch the female's belly and hips intermittently. Then copulation starts.

Six births were recorded in total. The young were born during March to July. The gestation periods were recorded as 240, 211 and 210 days. We believe the latter two figures are more correct after comparing with Japanese serow. During the last phase of its pregnancy, the belly of the female shows a protrusion to one side. A few days before giving birth, it shows a little swelling of the teats from which white milk would come out if squeezed. The mother always gives birth at night. The weight of the newborn is approximately 1.3 kg. The lamb can stand up and walk on its first day of life, and starts consuming grass and solid food at the age of one month. The mother eats the placenta immediately it is dropped.

We are now successfully keeping three of the offspring. The other three died because of a lack of colostrum, from diarrhoea and from pneumonia. It is our belief that the death of the young has something to do with the mother's nutritional state. This is the problem we aim to conquer for future breeding, i.e. the nutritional condition of a pregnant female during the final stage of pregnancy must be closely monitored.

TERRITORIAL BEHAVIOUR

Among the eight Formosan serows (two female, six male), No. 10 is usually aggressive and has to be kept separately. The others get along well.

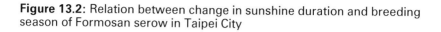

Figure 13.2: Relation between change in sunshine duration and breeding season of Formosan serow in Taipei City

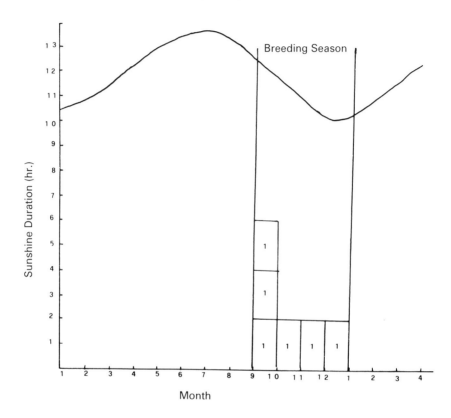

Fighting is rarely seen, but chasing and fighting for food happen quite frequently. Starting in October of 1985, one person was sent to the site to observe the animals' behaviour with both sampling of all occurrences of some behaviours and scan sampling (Altmann 1974) from 0800 to 1700 and at least three days a week. Altogether 249.5 h of observation until April of 1986 were recorded. From records of animals approaching and avoiding each other, an interaction table was made, and from this their order of social rank can be understood (Table 13.2).

We also observed that a few serows often use the preorbital gland and horns to rub the door frame, iron poles, the corners of the walls, or tree trunks. Table 13.3 and Figure 13.3 show the number of times that each animal rubs. On comparing this behaviour with the results shown in Table 13.2, we found the two are very closely related (with a coefficient of 0.873), i.e. the higher the animal's social status the more times it rubs with its preorbital gland and horns and vice versa. If we study the records

Table 13.2: The distribution of approach–retreat interactions among Formosan serow

Avoiding

Approaching

	1	8	2	6	7	5	9		
1	0	6	10	0	1	1	5		23
8	1	0	12	40	144	4	11		212
2	17	85	0	38	59	12	8		219
6	0	25	3	0	2	16	15		61
7	0	3	0	10	0	12	11		36
5	0	0	0	31	3	0	0		34
9	0	58	36	32	59	2	0		187
	18	177	61	151	268	47	50		772

regarding locations in which they rub their preorbital gland and horns (Figure 13.4), we can see that this rubbing behaviour happens mostly on the main path through which keepers or animals must pass every day, such as the exhibit entrance, the house entrance, the aisle to the isolation opening, and alongside the drinking pool. During the observation period, there was one change in the social ranking in the group. The number of times in which an animal that has just entered the higher social class rubs with its preorbital gland and horns increases greatly. We suggest therefore that the rubbing behaviour is used for the marking of territory.

OTHER BEHAVIOURS

Males tamed by keepers were found to show mating movements towards keepers, i.e. the animal will use its front legs to touch the keeper intermittently and will follow the keeper around. The movements of attack include

Table 13.3: Frequency, location and method of marking territories among Formosan serow

NO	Times	Location															Method		
		P1	P2	P3	D1	D2	D4	H1	H3	H4	B1	L1	F	R1	R2	Other	G*	H**	G***
2	719	167	22	18	61	23	40	95	27	10	125	30	24	29	14	34	624	61	34
9	400	87	18	17	24	6	36	50	18	11	47	9	12	22	2	41	373	25	2
8	52	7	2	2	0	0	2	12	5	0	2	3	5	5	2	5	41	9	2
1	45	15	3	2	8	4	1	3	0	0	0	0	2	0	0	7	26	14	5
5	16	5	1	0	0	0	0	5	0	0	0	0	1	1	0	3	6	10	0
6	32	28	1	0	0	0	1	0	1	0	0	0	0	0	0	1	28	3	1
7	2	0	0	0	0	0	0	0	0	0	0	0	1	0	0	1	0	2	0
Total	1266	309	47	39	93	33	80	165	51	21	175	42	45	57	18	92	1098	124	44

Figure 13.3: Frequency of marking territories among Formosan serow

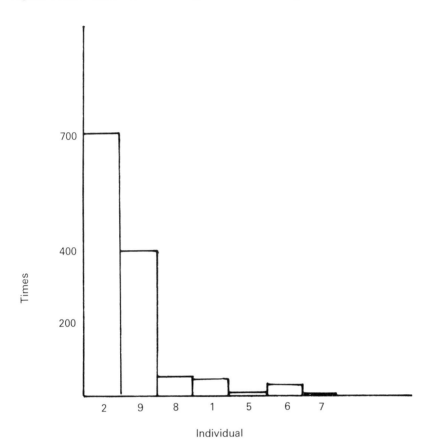

facing each other with horns lowered or tapping the floor heavily, whistling very sharply. All Formosan serows go to a specified place for defaecation. For a while this was a high but flat area (L1), but later all of them moved down to lower places (P2 and P3) for the same purpose. A small amount of urine is frequently seen to be excreted right after chasing one another. Some will urinate when they are marking the territory.

We divided their daily activities into three categories, feeding, walking and resting. From focal-animal sampling records, we can tell that they spend a lot of time walking (an average of 32.83 per cent over a total observation time of 122 h). Figure 13.5 shows the percentage of time they spent at each daily activity. Food was provided between 0830 to 0900 each day. Two observations were made during night-time, and feeding and walking were seen from time to time.

161

Figure 13.4: Frequency of marking territories in various locations

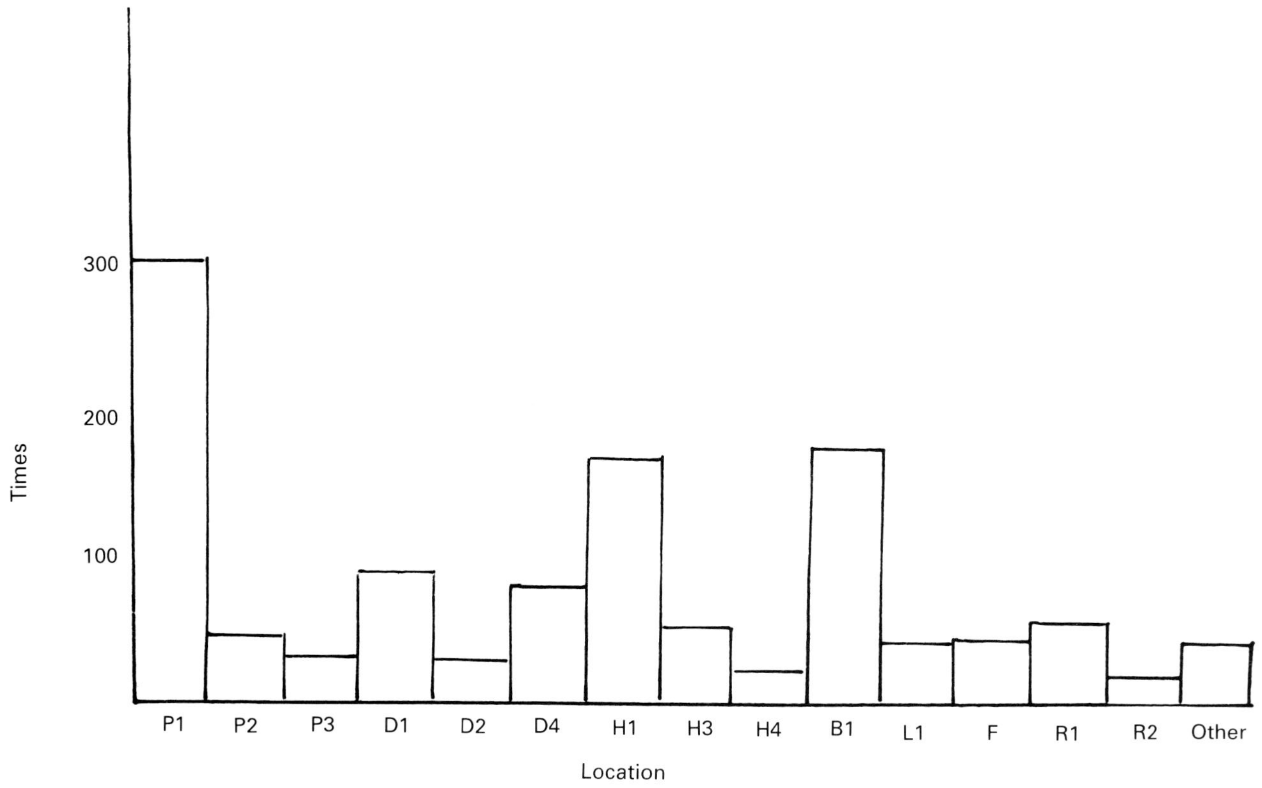

Figure 13.5: The pattern of daily activities of Formosan serow

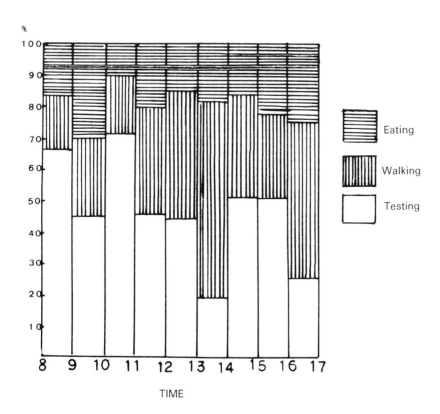

They are often seen gathering in groups of two to three animals up on the concrete platform to rest. On hot days, they will find a shady area and do the same. A couple of low-ranking serows are usually chased up to a branch 3 m high and are forced to stay there for a few hours.

DISCUSSION AND CONCLUSION

An area of 350 m² is rather crowded for keeping eight Formosan serows. Scenes of animals chasing each other are therefore frequent. It is very important to provide the animals not only with shady areas but also with shelter from the winter cold. A high platform for resting is also a must in the enclosure. A fence 160 cm high with a 30 cm overhang is safe enough to keep Formosan serows in. We suggest feeding them with more natural food, e.g. sweet potato leaves, mulberry branches, bamboo shoots cabbages, etc. Not only do the animals like it, but it also has a good nutri-

163

tional effect. The distribution of the food must also be very careful to avoid low-ranking animals going hungry.

The general behaviour of the Formosan serow is about the same as that of the Sumatran serow (West 1979) and the Japanese serow (Yamamoto 1967). The habit of using the preorbital glands and horns to mark its territory is very close to that of muntjacs, as described by Barrette (1977)

Warm and humid weather conditions may have a negative effect on breeding of Formosan serows, judging by the absence of successful breeding records on level land in Taiwan and the rare breeding records from Taipei Zoo (temperature and humidity of Taipei City are shown in Table 13.4). It may therefore be worth trying to breed Formosan serows in a higher region where the temperature is cooler, or in a temperature-controlled environment.

Table 13.4: The mean temperature and relative humidity of Taipei City

M	1	2	3	4	5	6	7	8	9	10	11	12
T.°	14.8	15.4	16.7	20.4	24.8	27.5	29.1	28.6	27.7	24.6	20.9	16.7
R.H. %	84	87	87	80	82	82	77	80	80	79	79	80
T / R.H.	17.6	17.7	19.1	25.5	30.2	33.5	37.7	35.8	34.6	31.2	26.4	20.9

REFERENCES

Altmann, J. (1974) Observational study of behavior: sampling methods. *Behaviour* 49, 227–67

Barrette, C. (1977) Scent-marking in captive muntjacs, *Muntiacus reevesi. Anim. Behav. 25*, 536-41

West, J. (1979) Notes on the Sumatran serow *Capricornis sumatraensis* at Jakarta Zoo, *Int. Zoo Yrbk 19*, 252

Yamamoto, S. (1967) Breeding Japanese serow *Capricornis crispus* at Kobe Zoo. *Int. Zoo Yrbk. 7*, 174

14

Behaviour of the Japanese serow (*Capricornis crispus*) at the San Diego Wild Animal Park

Judith K. Berg

Zoological Society of San Diego, San Diego Wild Animal Park, Route 1, Box 725E, Escondido, California 92025, USA

INTRODUCTION

The designation of the Japanese serow as a 'natural monument' in 1934 and as a 'special natural monument' in 1955 gave this fascinating species of the subfamily Caprinae complete protection from legal hunting and capture (Akasaka 1974). Because of this protective status, only three facilities outside Japan are fortunate to house selected individuals of this species: the Peking Zoo in China, and the San Diego Wild Animal Park and the Los Angeles Zoo in the United States. The Wild Animal Park received a male and a female serow from the Japanese Serow Center, Gozaisho Mountain, Mie Prefecture, Japan. This pair was housed in an open-air mixed-species exhibit. However, due to the larger size of the enclosure and the elusive nature of the animals, very little was known about their behaviour, which created a concern to Park Management for their welfare. Therefore, in 1983, I undertook a project whose purpose was to determine the serows' individual, interspecific, and intraspecific behaviour and their use of space at the Park.

MATERIALS AND METHODS

Setting

Two Japanese serows were observed from January 1983 until April 1985 in a 14 ha open-air mixed-species exhibit at the San Diego Wild Animal Park in California. The serows shared this enclosure with ten other Asian species (approximately 100 animals). All species had continual access to all areas of the exhibit. The landscape of this hilly, semi-rocky area was vegetated with indigenous plants plus introduced trees and grasses. A man-made stream flowed through the north-west half of the exhibit, emptying into a 0.5-ha pond. As a result of interspecific conflicts and of there being no successful births, a new enclosure was constructed for the serows. While awaiting the completion of their new area, in April 1985 they were moved into 0.5-ha fenced section of the large enclosure. This section was a portion of the exhibit that the serows had primarily utilised. They remained in this area until September 1985 at which time they were relocated to their new 1-ha fenced exhibit. The terrain of this new area was semi-rocky with sloped banks which led down to a stream bed in the centre of the area. The landscape was densely vegetated with indigenous species of low shrubs, herbaceous plants, grasses and trees.

The average monthly temperature during the daytime hours of the study ranged from a minimum of 5° C (January) to a maximum of 36° C (August). The extreme temperatures at the Park ranged from a minimum of 0° C (December) to a maximum of 40° C (July).

The serows were fed a special diet twice a day at 0730 and 1330. Their provided food consisted of small high fibre herbivore pellets (crude protein 14 per cent, crude fat 3 per cent, crude fibre 21 per cent, ash 10 per cent, added minerals 2 per cent); small low-fibre herbivore pellets (crude protein 17 per cent, crude fat 3 per cent, crude fibre 13 per cent, ash 8.5 per cent, added minerals 2 per cent); amalene (compressed alfalfa pellets, rolled oats and corn, molasses); yams; carrots; apples; a *Camellia japonica* bush; and a flake of alfalfa. They had unrestricted access to their own food, in addition to the feeders for the other animals within the large enclosure. Their primary human contact was with their keepers, who, after feeding them, remained with them while they ate. There was no public access to the animals except by an electric monorail which travelled around the outer perimeter of the exhibit every 15 to 30 minutes during the open hours of the Park. The same monorail affected their new enclosure; however, since this area was chosen to fulfil the serows' requirements, they could, and usually did, remain out of view from the public.

Subjects

The subjects consisted of a male and a female Japanese serow which came from the Japanese Serow Center, Gozaisho Mountain, Japan. The male, Ken, was born on 3 May, 1976 and arrived at the Park on 18 March 1980. The female, Mariko, was born on 19 July 1980 and arrived at the Park on 15 January 1982.

Procedure

Observations were recorded one day each week during a 3-year period which began in January 1983 and extended until December 1985. The

166

serows' behaviour was documented throughout a 7-hour period during each day of observation beginning at 0700 and extending until 1400. In addition, eleven periods of night-time observations were made during full moon beginning at 1430 and extending until 0630. The method employed was systematic recording, in which written descriptions documented consecutive 5-minute intervals of the animals' individual behaviours, social interactions, movements within the enclosure, and the presence of any other species in their area. Observations were made from the periphery of the enclosure; no attempt was made to conceal the observer since the animals were already habituated to field keepers and became habituated to the observer.

The behavioural categories examined in this study were lying down (resting and sleeping), feeding (eating provided food and foraging on naturally occurring vegetation), locomoting (walking, trotting and running), standing, scent-marking and their other behaviours. An additional category, 'other', included eliminating (urinating and defaecating), drinking, horn rubbing, vocalisations, sexual behaviour (naso-genital testing, flehmen, courtship and mating), naso-nasal, butting, mutual butting, intraspecific rubbing and interspecific interactions. All of the above behaviours, except for the last two, are described by Kishimoto (1981).

Data analysis

The percentage of total occurrences for each category of behaviour was used for the data analysis. To determine if there were inter-individual differences between the behaviour of the male and that of the female, the data were analysed using the t-test. To determine if there were intra-individual differences within each category of behaviour among the three years, the data were analysed using the Kruskal-Wallis one-way analysis of variance. The level of significance for all tests was set at 0.05.

RESULTS

There were 140 days of observation throughout the three-year period comprising 10 686 five-minute recording intervals. During these five-minute periods of observation the male engaged in a total of 14 060 behaviours and the female in a total of 12 909 behaviours. The eleven periods of night-time observations were analysed separately and included 1342 five-minute periods for the male and 1153 five-minute periods for the female. During these five-minute periods of observation the male engaged in a total of 1567 behaviours and the female in a total of 1328 behaviours.

General behaviour patterns

The primary behaviours in which the serows engaged from 0700 to 1400 during the three-year period were lying down (M 46 per cent; F 49 per cent), locomoting (M 16 per cent; F 17 per cent), feeding (M 13 per cent; F 18 per cent), and scent marking (M 11 per cent; F 2 per cent). Standing accounted for the least number of total behaviours (M 6 per cent; F 4 per cent), except for the female's scent marking (2 per cent). The remaining activities, which included 14 behaviours, were lumped into the category of 'other', and accounted for 9 per cent of the male's total behaviours and 10 per cent of the female's (Figures 14.1 and 14.2). During the eleven periods of night-time observations from 1430 to 0630, these same behaviours accounted for the major proportion of each animal's total activities (Figure 14.2). Standing was not observed during the night-time hours.

t-tests were performed to compare the data for the male with those for the female for each category of behaviour during each year of observation. The results are shown in Table 14.1. The only statistically significant differences occurred in the behaviours of feeding and scent marking. The female performed significantly more feeding behaviour than the male during 1983 and 1984. The male consistently performed more scent-marking behaviour than the female during all three years. The remainder of their activities showed similar trends during each year (Figures 14.3–14.5).

To compare the behaviour of each animal within each category among the three years, the Kruskal-Wallis one-way analysis of variance was performed. The results are shown in Table 14.2. The only significant differences occurred in the 'other' behavioural category for both the male and the female. This is understandable since this category contains the accumulation of 14 different behaviours, some of which are directly affected by external variables (e.g. the introduction of other species into the exhibit).

Hourly patterns

Figure 14.6 shows the hourly patterns of behaviour that occurred during the combined eleven periods of night-time observations. The male's behaviour was chosen to illustrate these patterns because he was on view during more hours than the female; however, when both animals were seen, their behaviours were similar. Lying down accounted for 70 to 100 per cent of each hour's behaviour from 1900 until 0600. The serows were most active from 1600 until 1900 (feeding, locomoting, scent marking and performing other behaviours) and then again at 0600. Some activities occurred throughout the other hours, except at 2100 and 0400 when lying down accounted for 100 per cent of the total behaviours.

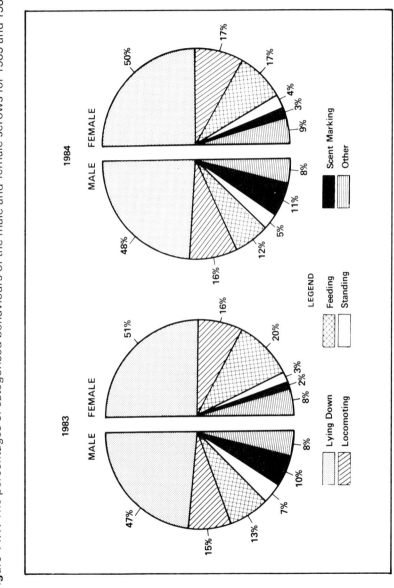

Figure 14.1: The percentages of categorised behaviours of the male and female serows for 1983 and 1984

Figure 14.2: The percentages of categorised behaviours of the male and female serows for 1985 and night-time

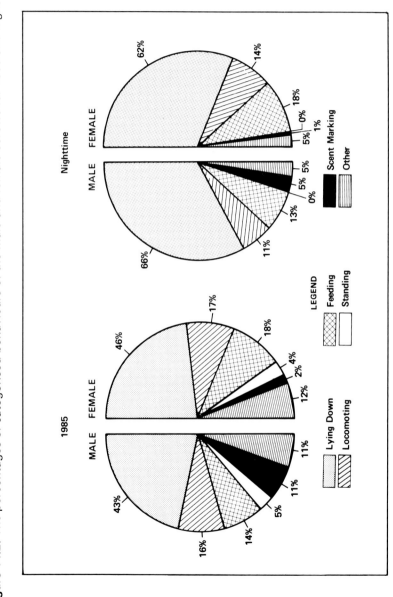

Table 14.1: *t*-tests used to compare the percentage of each behaviour of the male serow with that of the female for each of the three years

| | Male | | Female | | |
	Mean	SD	Mean	SD	*t*
Lying down					
1983	48.50	9.48	51.33	4.35	0.94
1984	49.50	9.46	50.00	6.78	0.14
1985	43.80	7.53	47.27	7.72	1.06
Locomoting					
1983	14.58	2.71	15.66	2.42	1.03
1984	15.58	3.11	17.50	4.42	1.22
1985	15.90	1.92	18.36	3.61	1.98
Feeding					
1983	13.08	2.15	20.16	3.12	6.46*
1984	11.75	2.73	17.75	4.07	4.23*
1985	13.81	3.60	17.18	3.81	2.12
Standing					
1983	6.16	3.56	3.08	1.72	2.69
1984	4.91	4.60	3.08	3.98	1.04
1985	5.09	2.80	3.45	2.65	1.40
Marking					
1983	9.83	3.35	1.58	1.08	8.11*
1984	9.83	3.71	2.50	1.97	6.03*
1985	10.27	2.00	1.81	1.16	12.08*
Other					
1983	7.83	2.94	8.16	2.48	0.29
1984	8.41	2.64	8.91	2.99	0.43
1985	11.09	1.81	11.90	3.04	0.76

$*P > 0.05$

Table 14.2: Kruskal-Wallis *H* values for male and female serows comparing three years of each category of behaviour

	Lying down	Locomoting	Feeding	Standing	Marking	Other
Male	2.53	1.99	2.50	1.13	0.88	9.82*
Female	1.51	3.17	5.19	1.44	1.50	8.28*

*Value required for significance at the 0.05 level, df = 2, is 5.99

Information from a six-month analysis of the daytime periods (0700 to 1400) during months of wide temperature variations at the Park showed that the serows engaged in a varying number of activities throughout all hours (Berg 1984). Lying down accounted for 50 to 75 per cent of each hour's total behaviours between 0900 and 1300 during the warmer months and between 1000 and 1400 during the cooler months. Feeding and loco-moting peaked between 0700 and 0900 during the warmer months and between 0800 and 1000 during the cooler months. Another peak in their

Figure 14.3: The percentage of total behaviours for each category of behaviour for the male and female serows during each month of 1983

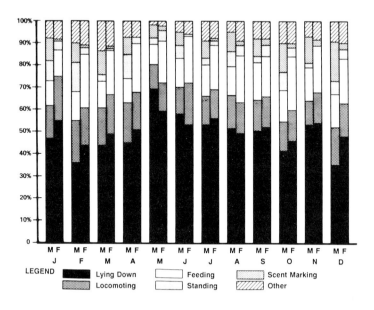

Figure 14.4: The percentage of total behaviours for each category of behaviour for the male and female serows during each month of 1984

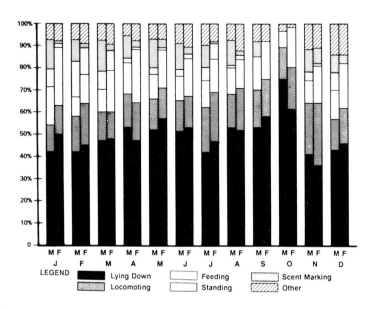

Figure 14.5: The percentage of total behaviours for each category of behaviour for the male and female serows during each month of 1985

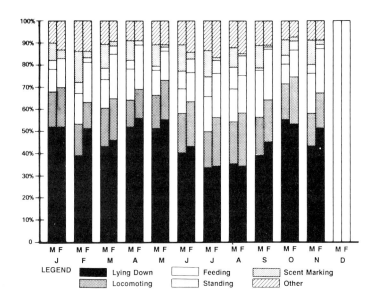

Figure 14.6: The hourly behavioural patterns of the male serow for the combined eleven periods of night-time observations

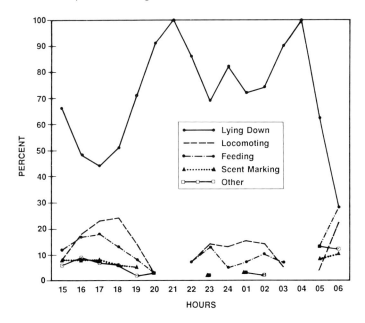

173

feeding behaviour occurred between 1300 and 1400. The hours spent eating corresponded to the times they were fed by their keepers. A further breakdown of their feeding time budget showed that the serows spent 70 per cent eating provided food and 30 per cent foraging on natural vegetation. This contrasts with the night-time periods when they spent 44 per cent eating provided food and 56 per cent foraging. (The time spent foraging in their new enclosure is above 30 per cent during the daytime.)

Space usuage

The serows spent 91 per cent of the total daytime periods of observation within a well-defined 1-ha area at the north-west end of the large exhibit. In this area they lay down in ten rest spots, defaecated in three dung piles, and scent-marked in twelve locations (Figure 14.7).

When the serows were found lying down at the beginning of the observation period, 55 per cent of the time they were found in either rest spot (RS) 1 or 2. These two rest spots were the highest locations in their area and those from which they could see most of the enclosure. When resting during the day, the serows moved in relation to shade. RS 1, RS 3 and RS

Figure 14.7: Space utilisation by the male and female serows in their 1-ha area over the 3-year period. The counts are based on the total number of 5-minute periods for both animals

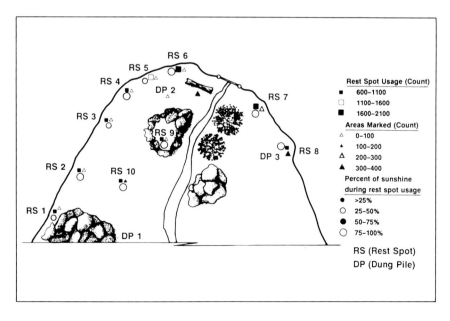

5 were not shaded. The sun shone only 33 per cent of the time when the serows used these three rest spots (Figure 14.7). RS 6 and RS 7 were shaded by a cement tram bridge during the morning and mid-day, whereas RS 2, RS 4, RS 8 and RS 9 were shaded by overhanging bushes or trees throughout most hours of the day. RS 10 was a shelter. In addition, RS 6, RS 7 and RS 9 were most affected by winds which blew through the stream bed during mid-day, cooling these areas 6° C lower than other locations. The serows' behaviour of choosing shaded locations during the day at which to rest has persisted in their new enclosure. The exception to this occurred during the cooler morning hours and at night when they were observed lying on top of a large flat boulder or in other open high-ground locations.

Three areas were used by the serows for elimination (defaecating and urinating). Dung pile (DP) 2 was the area most frequented. The serows usually eliminated twice a day, primarily between 0700–0800 and 1200–1400.

Marking with the preorbital gland was directed towards specific inanimate objects (fence posts, a gate lock, shelter), stems of bushes, edges of rocks and branches of a fallen tree, primarily on the outer perimeter of the serows' area. The locations most often marked were those closest to their most frequented rest spots (RS 6 and RS 7) and the area where they were most often fed (RS 8) (Figure 14.7). Although both the male and the female scent-marked, most marking events were performed by the male (Figures 14.1 and 14.2) while locomoting around their area (Table 14.3). The male also marked the female, primarily on her ears and less often on her ano-genital region. Pawing the ground with a forefoot, thus marking with the interdigital gland, occurred infrequently.

Table 14.3: Scent marking of the male serow. Each value is the percentage of the 1486 total observed marking events

	Marking objects	Marking female
When other species not in the area, while:		
locomoting	36	
standing	8	7
approaching female	3	11
prior to lying down	2	
When other species were in the area, while:		
locomoting	14	
standing	4	3
approaching female	1	8
prior to lying down	3	

Sexual interactions

The behaviour of the male scenting and/or licking the female's anogenital region was observed during each month of the year. Flehmen was observed only 25 per cent of the time following this behaviour. However, there may have been a subtle form of flehmen which wasn't always seen. During the observed flehmen behaviours, 45 per cent occurred following the naso-genital testing and 55 per cent occurred testing the female's urine flow or from fresh urine on the ground. This behaviour was also observed during each month of the year.

Mating behaviour was observed from September to March; 75 per cent of these behaviours occurred during September to November with a peak in October (Figure 14.8). The sequences of mating behaviour (courtship and mounting) lasted 5 to 35 minutes. More than one sequence often occurred during the same day. The components and postures of the serows' courtship and mounting behaviours were described by Kishimotor (1981). In addition to the described components which included naso-genital testing, flehmen, foreleg kicks and butting, the male marked the female's ears and/ or anogenital region during 50 per cent of these sequences. There follows a description of a mating sequence. The male approached the female from the rear and performed naso-genital testing followed by flehmen. The male then directed foreleg kicks between the female's rear legs followed by mounting. The male marked the female's anogenital region with his

Figure 14.8: Sexual behaviour of the male and female serows for each month of the combined 3-year period. The percentages are based on the total number of all behaviours

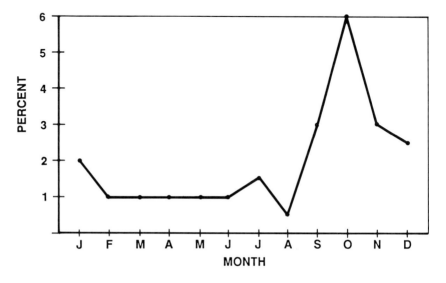

preorbital gland followed by more foreleg kicks. The female turned and stood in a reverse parallel position. The male moved to the female's rear, mounted, then performed more foreleg kicks. The female turned to face him and again moved into a reverse parallel position. The male moved perpendicular to the female, then directed his foreleg kicks to her flank. He moved to her rear and performed two consecutive mounts. The female walked from the male, who then followed her. The two animals ate. The entire sequence lasted five minutes. During some of the mating sequences, low-intensity vocalisations were heard. The durations of the timed mounts were 2–3s.

The female was first observed soliciting from the male in November 1984 at the age of 4 years. This solicitation included the female standing in front of the male, or walking in front of him as they locomoted with her tail raised, shaking from side to side. Also during two sequences of this same breeding season, the female butted, marked, then mounted the male. The male responded to her behaviours with courtship and mounting. When the male performed courtship behaviours and the female walked or ran from him (i.e. would not stand for him), the serows' snort sound was vocalised.

The result of the serows' mating behaviour has been two stillbirths, in May 1983 and July 1984. Each infant was a female. The necropsy of the first infant showed possible asphyxiation from mucus, although the actual cause is undetermined (Nielson and Benirschke 1983). The necropsy of the second infant found no abnormalities to indicate the cause of the stillbirth (Anderson 1984).

Non-sexual interactions

There were a total of 300 social interactions which did not appear to be related to the serows' sexual behaviours. These included naso-nasal, intraspecific rubbing, butting, and mutual butting. Except for mutual butting, they occurred during each month of the year. Mutual butting was not observed during the same months as mating. Naso-nasal occurred during 58 per cent of the total non-sexual interactions followed by intraspecific rubbing (20 per cent), butting (12 per cent), and mutual butting (10 per cent). Naso-nasal occurred when one serow approached the other or while they stood together, especially during feeding. Sometimes the initiator also licked the nose and mouth area of the other serow. Intraspecific rubbing occurred when one animal brushed against the other, usually with the head or neck. Although butting occurred during mating behaviour, it also occurred in other contexts. Butting was observed when both serows were performing horn rubbing at the same bush (Kishimoto 1981), following horn rubbing, prior to mutual butting and in unassociated contexts. Butts were usually directed towards the rear or flank of the other animal. During mutual butting, the initiator lowered its head towards the other serow who responded with the same behaviour. The two animals then knocked their horns. This behaviour lasted only a few seconds during each encounter. Following butting and mutual butting, running or chasing would sometimes occur. During chasing, snort sounds were vocalised.

Interspecific interactions

The number of species and individual animals who shared the large exhibit with the serows varied throughout the years as introductions and removals occurred. The serows appeared to be intimidated by most of the other species, especially the larger deer (*Cervus elaphus, Dama dama,* and *Cervus nippon*). Their responses to approaches of the larger deer species were marking, withdrawal, flight and what Kishimoto (1981) described as an 'alarm posture'. Most of the confrontations with the deer occurred at the serows' food tubs. Although a keeper remained with the serows during their first feeding of the day, when the keeper left, and throughout the other hours of the day (and night), other animals would often move into the serows' area to eat their special diet. The serows would not defend their food against these other animals. They also would not defend their resting places. If they were approached by a deer while lying down, they would stand, then move away. The serow would, however, tolerate muntjac (*Muntiacus reevesi*) at both their food and rest spots.

Usually only an approach and/or visual display was directed towards the serow by the larger deer species. However, on a few occasions a deer was seen chasing a serow who responded by running into a rocky area and emitting snort sounds. Although physical encounters were rarely observed, during one of these encounters in June 1984 a female deer stood on her hind feet and came down on top of the female serow while she was urinating. The female deer had hidden her fawn in the serows' area; although the serows were curious about the fawn, they were not aggressive towards it.

On occasion a serow would stomp its feet against the ground and lower its head towards an approaching animal but if the other animal responded with an aggressive display, the serow would withdraw. The only time a serow was seen chasing another animal occurred when two Central Chinese goral (*Nemorhaedus goral*) were introduced into the large exhibit. The gorals' initial encounter with the serows resulted in the serows' first scenting then chasing the gorals. Some chasing of the gorals by the serows continued throughout the next two days. The gorals sought refuge by crouching down among rocks or inside the serows' shelter. This behaviour decreased in frequency during the succeeding days, even though the gorals spent most of their time in the serows' area. Although the four animals adapted to each other's presence, the serows remained dominant to the gorals. The gorals displayed submissive postures and/or withdrawal in response to an approach by a serow.

DISCUSSION

The general behaviours of free-living Japanese serow — lying down, loco-moting, feeding, scent marking, standing and 12 other individual and social behaviors (Kishimoto 1981) — were similarly observed in the present study. In addition to those behaviours, intraspecific rubbing and inter-specific interactions were also documented. The male and female serows engaged in most of their behaviours during the same time period, either while together or in close proximity to each other; thus, their monthly behavioural patterns showed similar trends. The exception to this was the male's performance of significantly more scent marking and less eating than the female. The serows were most active, especially feeding and loco-moting, at 0600–0900 and at 1600–1900. They spent 50–75 per cent of each of the other daytime hours and 70–100 per cent of each of the night-time hours lying down. Their pattern of feeding during the morning, late afternoon and evening hours, and primarily resting during the other hours of the day, is similar to that of the free-living genera of their tribe: *Capricornis*, *Nemorhaedus*, *Oreamnos*, and *Rupicapra* (Allen 1940, Akasaka 1974, Nowak and Parodiso 1983, Pachlatko and Nievergelt 1985).

In their natural state, the serow residing in a defined range remain within that area throughout the year (Kishimoto 1981). The Parks' serows, as those in the wild, also performed most of their activities within a well defined area. This 1-ha area seemed to fulfil all their terrain and behav-ioural requirements. It was semi-rocky and sloped for them to climb and take refuge; the rest spots at which they were most often found in the morning were in the highest spots in their area, and the ones from which they could see all possible approaches of other species; there were shaded areas to which they moved during the day; and naturally growing grasses, shrubs and trees fulfilled their foraging requirements. Of the total time they were observed in the large enclosure, only 9 per cent was spent outside of this area. This appears to show evidence of the required components of a captive serow environment.

The serows' behaviour of scent-marking specific objects within their defined range was similar to that of their free-living relatives. As in the wild, the serows at the Park performed this behaviour during each month of the year (Akasaka 1974, Kishimoto 1981). In their natural state serows rest while lying down on flat surfaces near trees, on flat stumps, on terraces of steep slopes (Kishimoto 1981), or in caves (Akasaka 1974). The serows at the Park chose flat surfaces of earth or boulders near overhanging bushes or trees, or under a cement cover. In their new enclosure they also retreated into a thicket of trees and bushes to rest during the day, as their free-living relatives retreat into forested areas (Allen 1940, Akasaka 1974). Having shaded areas plus areas in which to retreat should be a consideration for a captive serow environment.

The serows supplemented their provided food with naturally occurring species of grasses, low growing plants, twigs, and, in their new enclosure, the leaves of trees and shrubs. Although the species of vegetation are different, this is the type of forage they eat in the wild (Okada and Kakuta 1970, Akasaka 1974, Kishimoto 1981, Nowak and Paradiso 1983). Because foraging on natural vegetation accounted for 30 per cent of their daytime and 56 per cent of their night-time feeding behaviour, this is an obviously necessary component of a captive environment.

Even though the serows were forced together into a captive environment, by living in a large enclosure they could have separated at any time, but they instead remained together in the same area. Kishimoto (1981) found that in free-living serows the adult male's range overlapped with that of an adult female in the majority of his resident animals. Thus, a pair could live within the same area during all months of the year.

The serows' peak mating season, September to November, corresponded with that of their free-living relatives (Kishimoto 1981). The continuation of mating into other months may have been due to their captive state (Sugimura *et al.* 1983). The serows' courtship rituals and postures during these behaviours were the same as those described by Kishimoto (1981). The exact reason(s) why there were two stillbirths, with undetermined physiological explanations as to the cause, remains a mystery. However, from the observations during this 3-year study, it was concluded that the stresses placed on the serows by the interspecific competition in the exhibit may have been a major contributor to this problem. These shy, elusive animals would not normally encounter such an array of other animals as they did at the Park. Thus, they now reside in a 1-ha enclosure which is as close to their natural environment as possible for southern California. The only other animals who shared this area with the serows are two muntjac, a species the serows tolerated in the large enclosure. This should be a consideration for mixed-species exhibits that include serows.

CONCLUSION

The unique nature of the San Diego Wild Animal Park has enabled the conduct of behavioural research on a pair of Japanese serows who lived in a simulated environment which most closely approaches that of their free-living brethren. The data gathered from this research furnished Park management with information that was used to provide the serows with the best possible living circumstance for their species, and, hopefully, for their successful perpetuation. Because their behaviour, while living in this environment, closely resembled that of their free-living relatives, the use of this information similarly for the provision of other captive serow environments as to be encouraged. Only by learning as much as possible about the behaviour of this species in both its free-living and captive states can we hope to preserve them.

ACKNOWLEDGEMENTS

I am grateful to Dr John Phillips and Dr James Dolan for their review of the original manuscript. I extend my gratitude to Dr James Dolan and Larry Killmar for encouraging me to conduct this research at the Park, and to Rich Massena, Randy Rieches, and the Field Keepers for their assistance and cooperation during this project. I thank Dennis Gjertson for the graphics and I extend a very special thank you to David Berg for his editing, helpful suggestions and encouragement during this project.

REFERENCES

Akasaka, T. (1974) Japanese serow in the wild. *Wildlife 16*, 452–58

Allen, G. (1940) The even toed ungulates: serows. In *The mammals of China and Mongolia, Part II*, American Museum of Natural History, New York, pp. 1227–38

Anderson, M. (1984) Necropsy report of *Capricornis crispus*. Zoological Society of San Diego

Berg, J. (1984) The behavior of the Japanese serow (*Capricornis crispus*) in captivity. Paper presented at the American Society of Mammalogy Conference, Arcata, CA

Kishimoto, R. (1981) Behaviour and spatial organization of the Japanese serow (*Capricornis crispus*). MS thesis, Osaka City University

Nielsen, N. and K. Benirschke. (1983) Necropsy report of *Capricornis crispus*. Zoological Society of San Diego

Nowak, R. and J. Paradiso. (1983) *Walker's Mammals of the World*, 4th edn. Johns Hopkins University Press, Baltimore, MD

Okada, Y. and T. Kakuta. (1970) Studies on the Japanese serow, *Capricornis crispus*. In *Soc. Preserv. Japanese Serow, Suzuka Mountain (Tenth Anniversary Volume)*, pp. 1–15

Pachlatko, T. and B. Nievergelt. (1985) Time budgeting, range use pattern and relationships within groups of individually marked chamois. In S. Lovari (ed.) *The Biology and management of mountain ungulates.* Croom Helm, London pp. 93–101

Sugimura, M., Y. Suzuki, I. Kita, Y. Ide, S. Kodera and M. Yoshizawa. (1983) Prenatal development of Japanese serows, *Capricornis crispus*, and reproduction in females. *J. Mammal. 16*, 302–4

Part Four:
Ecology and Breeding
of the Rupicaprini

15

Keeping and breeding of chamois (*Rupicapra rupicapra rupicapra* Linné 1758) at the Alpine Zoo, Innsbruck/Tirol

Helmut Pechlaner

Alpine Zoo, Austria

INTRODUCTION

The alpine chamois is the only animal of the *Capricornis* group living in the European Alps. Generally dwelling in timberline regions, it moves to higher zones in summer.

The alpine chamois shares its habitat with the alpine ibex (*Capra ibex ibex* Linné 1758), which stays at higher altitudes even in winter. The natural enemies of the chamois are of hardly any consequence even though the bearded vulture (*Gypaetus barbatus aureus*), which mostly feeds on bones, is reported to attack weak and sick chamois with its strong wings, pushing them from crags. The golden eagle (*Aquila chrysaetos*) goes after newly born chamois, but a strong rock-doe will always be able to ward it off successfully. Only the deserted kid of a chamois is easy prey for the golden eagle. Because of their behaviour, ibex were much easier to hunt for than chamois; and the mighty antlers of the males topped a successful kill with a proud trophy.

Chamois are much shyer. Their behaviour and build are well adjusted to the conditions of life in Alpine regions. For millennia the naked rock in summer and the deep snow in winter made for a severe selection. Chamois are long-legged animals, and an average body length of 120 cm and a body height of 1 m gives them an almost square frame. The average weight is about 25 to 30 kg, and the bucks are heavier by approximately 20 per cent.

In summer the coat shows a reddish-brown to grey–yellow colour; it is dark brown to black in winter. The black stripe on the back, also referred to as the list, as well as the short tail and the extremities covered with black hair, produce an attractive contrast in summer. The face and the longish ears in particular are characterised by clearly bound stripes and markings; the pupils remain horizontal, no matter in which way the head is positioned. The black stripe at the side of the face, the so-called mastax, is very clearly defined during the first half of life, but it loses its sharp outline with age.

The growth of the horny substance commences during the first weeks of life. Towards the end of the first year of life the uppermost layer is rubbed off. The curvature of the horn tip also develops in the course of the first year of life. The structure of the antlers is the same as that of other horned animals, and the respective curvatures are created by differing intensities of growth at the sides.

The male always carries much stronger antlers than the female. The tip of the horns of bucks is bent so strongly that it points down to the base of the ear. With females the tip points backwards, parallel to the back. When rivals are engaged in a fight, they use their antlers as weapons. Ritualised marking may already be observed in young animals.

The biology of the chamois is dealt with in excellent fundamental and sophisticated publications such as Knaus and Schröder (1983) and Meile and Bubenik (1979).

BREEDING OF CHAMOIS

Approximately 130 years ago man first succeeded in breeding chamois. Notwithstanding that considerable period of time, zoological gardens still regard the chamois as a very delicate fosterling and so it is a rarely seen animal. In recent decades, however, there has been an increase in the keeping of chamois.

The Alpine Zoo in Innsbruck, the capital of the Tirol, is perched on the hillside near Weiherburg, and is a late Gothic castle. Located in Alpine country, this highest zoo of Europe has a favourable climate and so is well suited to the keeping and breeding of Alpine species.

Its field is limited to animals still indigenous to the Alps (such as the marmot (*Marmota marmota*)), or that have been native to the Alps at some time during the past few centuries (such as the bald ibis (*Geronticus eremita*)).

Starting with the foundation of our zoo in 1962, our staff was concerned with the keeping and breeding of the chamois. During the early years it was quite impossible to get *Chamois offspring* from other zoos, and the mere thought of catching grown wild animals and acclimatising them had to be ruled out for various reasons.

Once in a while hunters brought us young animals in the first winter of their life, which they had found exhausted in the deep snow. Their mothers had been killed by either avalanches or erroneous hunting. These animals were usually in a state of health that would have made them a victim of natural selection, but intensive treatment against parasites and bacterial infection primarily of the lung saved a small number of them.

Occsionally the Alpine Zoo received young chamois which had been found as kids by farmers or hunters and were then bottled-fed at home.

186

These animals generally proved to be in a good state of health, but the male specimens were disturbed, i.e. imprinted on human beings. Having adopted false patterns of behaviour these bucks regarded every human being as a rival and opposed him with incredible aggression. Since this would endanger both keepers and visitors to the zoo on the one hand, and would entail self-mutilation at the grid barrier on the other hand, such animals cannot be kept. In some enclosures the horn tips are so to speak disarmed with rubber hoses or plastic balls. It may become necessary to use such bucks for covering the does for one or two years. Then, however, they are replaced by their sons who have been raised under natural conditions by their mothers.

DIET

It took the Alpine Zoo a few years to find out, try and enforce the right feeding method. Based on a misinterpreted love of animals and out of laziness, keepers tend to feed much too generous amounts of whole meal, oat flakes and grain mixtures. In general, ruminants will be able to digest such highly concentrated food only if they are supplied with large amounts of stringy food at the same time. Otherwise, acidosis, indigestion with diarrhoea, and a grave disturbance of the nutrient balance will ensue. If these animals do not die of false feeding, they will certainly lose weight and get an inflated belly. Exaggerated growth of the hooves may be another consequence. Today we know that a maximum of 0.25 kg of mixed grain feed may be fed per grown chamois and enclosure.

As the principal food, good hay must be available at all times, accompanied by an all-year supply of branches of non-toxic deciduous and coniferous trees. The Alpine Zoo mainly provides spruce (*Picea abies*), pine (*Pinus silvestris* and *Pinus nigra*), hazel (*Corylus avellana*), willow (*Salix* sp.), ash (*Fraxinus excelsior*), sycamore (*Acer pseudoplatanus*), birch (*Betula* sp.), and alder (*Alnus viridis*). Even young animals love needles, leaves and buds, but they just as greedily feed on bark. Large branches are also used for marking and coat care. It goes without saying that fresh water must also be supplied during the cold season, but the animals will still like to lick up fresh snow. Also, they readily accept carrots and fodder beet.

In 1970 the Alpine Zoo celebrated the first birth of a chamois in the zoo; from 1972 on, chamois were bred as a matter of routine.

Housing

Feeding, however, is not the decisive factor: rather it is the structure of the enclosures. At the Alpine Zoo the chamois enclosure has a total size of approximately 500 m³, and one-quarter of it can be partitioned off with a sliding door to provide an alternative enclosure. From the beginning of April till about the end of July the male is kept in the smaller enclosure.

On principle, a chamois breeding group has only one sexually mature male. Otherwise rival fights would be unavoidable and they might easily entail the death of an animal. In the wild, too, females and males go their different ways prior to the time of birth, so there is even more reason to give them their own territories within an enclosure.

The whole chamois enclosure is divided and subdivided again vertically and horizontally so as to provide the animals with places to which they can retreat during their resting phases without seeing or being seen by the other animals. This structure becomes particularly important during the mating season.

In the case of rock-goats, the highest-ranking male virtually prances on the spot when courting the doe and the animals copulate straight afterwards. The chamois, just as the roe deer, however, require a long period of driving before mating. The buck grunts and bristles its long hair on the back in the region of the sacrum. For several days he chases the doe over several hours. In between, the rock-doe tries to hide, which is another reason for providing a large enough enclosure. When, on the day of the climax of the rutting time, both animals are hot from the chase, the doe invites the buck to start copulation.

In our enclosure practically the whole surface is lined with natural stones. This makes for optimum natural wear of the hooves, which is to say that the hooves of these animals require no additional care. Because of the use of this type of flooring, the droppings of the animals cannot cause any reinfection with parasitic or bacterial diseases: after every rain the water runs off immediately, and the excrement dries up. Twice a week the enclosure is first 'dry-cleaned' with a broom, then sluiced with running water. This is a further reason for the fact that routine veterinary-medical care in the Alpine Zoo remains restricted to the supply of vitamins and minerals, and to anaesthetising the animals prior to catching them.

The chamois enclosure also features a natural cave which is always dry and has a sand-and-earth-filled floor. This popular hiding place is used by the various specimens in accordance with heirarchy and social structure. In spring the rock-does retreat there to give birth to their young.

In the Alpine Zoo chamois kids are usually born during the first half of May, which is about two weeks earlier than in their natural habitat in our high mountain regions.

The enclosure of the chamois is located approximately 700 m above sea level whereas the regular habitat of the chamois in the Alps is at an elevation of 1500 m above sea level. There have never been twins at the Alpine Zoo, and we do not know of any births of twins in other enclosures. Even though rock-does with two kids have been seen in the wild, and twin births occasionally do happen, they are quite exceptional. In the wilderness six-month-old kids who had lost their mother were observed attaching themselves to other rock-does.

SUCCESSFUL FOSTERING OF AN ORPHAN CHAMOIS

At this point I should like to report on a successful experiment the Alpine Zoo was forced to try in 1985. Somewhere above the timberline hikers had found a rock-doe killed by falling rock. A kid of only a few days was with the dead animal. The hikers took the kid home and bottle-fed it for five days, using an artificial preparation for the first days of the life of human babies. On the fifth day they brought the kid to the Alpine Zoo.

It so happened that one of our rock-does had given birth to a kid the night before. So, hoping the mother would accept the young animal, we simply put it into the enclosure. The kid, however, ran to the next best doe and that female hurled it away with her antlers — among other reasons because it did not have a kid of its own.

Human beings entering the enclosure were immediately accepted by the kid as substitute parents since it acted on its experience of the previous five days.

We reacted by removing the rock-doe and its newly born kid into the small enclosure and added the newcomer. The doe quickly walked to the kids and started licking her own. Our new kid exploited the situation to drink from the tightly filled teats of the wet nurse.

The acclimatisation proceeded very well in this way and the two kids grew up as pseudo-twins. The successful experiment not only supplied interesting observations but had an additional advantage: the newcomer was a female but was still able to become a full integrated member of the group. Unfortunately, the size of the chamois enclosure at the Alpine Zoo does not permit us to incorporate a grown female into a working breeding group. Some years ago, even a six-month-old female kid was rejected by the group. So far, we have been forced to introduce new blood to the genetic potential by exchanging the respective breeding buck.

CONCLUDING REMARKS

Today the keeping and breeding of the alpine chamois do not anymore present a problem. Healthy young animals are available from other zoos, and with suitable enclosures and optimum feeding as well as eventual veterinary-medical prophylaxis the chamois develop beautifully. Ethological work designed to provide the basis for comparative outdoor work can be carried out without difficulty.

REFERENCES

Knaus, W. and W. Schröder. (1983) *Das Gamswild.* P. Parey, Hamburg
Meile, P. and A. Bubenik. (1979) Zur Bedeutung sozialer Auslöser für das Sozialverhalten der Gemse, *Rupicapra rupicapra* (Linné, 1758) *Säugetierkundl. Mitt.*
27, 1–42

16

Experiences of keeping and breeding saiga antelope at Tierpark, Berlin

Claus Pohle

Tierpark Berlin, Am Tierpark 125, DDR-1136 Berlin, Democratic Republic of
Germany

INTRODUCTION

Certainly one of the most important conservation successes of the present century was the saving of the saiga antelope (Figure 16.1) from extinction. As a result of a complete hunting ban in the Soviet Union, the saiga was able to revitalise itself from a population of only approximately 1000 animals at the beginning of the twentieth century. Within a period of 40 years the animal had recolonised areas where it had ceased to exist, and today, with a population of about two million animals, saigas are the most common hooved animals in the USSR.

Variations in the population occur through great losses when there are dry periods in the summer or deep snow and ice in the winter. These losses, however, are quickly overcome due to the high reproductive potential of these animals. Since the beginning of the 1950s (1951 on the west bank of the Volga, and 1954 in Kazakhstan), saigas have been hunted under licence for economic use. The number of animals that can be taken yearly depends upon the size of the population and varies between 20 000 and 500 000 saigas. This produces large amounts of meat and valuable leather, as well as the horns which are exported to eastern Asia where they are used as folk medicine. The end of the 1950s saw the high point in the distribution and population size of the saiga. Increasingly since the 1960s, negative human influences have been registered relative to the population of the saiga antelope. Irreversible changes have resulted in the loss of habitat for the saiga through agricultural use of the land and the construction of villages. Of particular importance with reference to the loss of saigas has been the construction of fences, canals, heavily travelled highways, and also diseases transmitted by domestic animals, such as foot and mouth disease. The present goal in the USSR is the stabilising of a basic population of from 1 to 1½ million saigas (250 000 to 350 000 on the west bank of the Volga, and 700 000 to 850 000 in Kazakhstan). These numbers must maintain a high reproductive capacity in order to overcome losses in bad years and the exploitation of the population for economic reasons (Shirnov 1982).

Figure 16.1: A male saiga 3 years old. (Photo: Klaus Rudloff)

One should also attempt to build stable populations of saigas in zoo-logical gardens as has already been done with a number of hooved animals. We are, however, a long way from achieving this goal with the saiga. After the first saiga was exhibited in the London Zoo over 120 years ago, these animals were seldom seen in the following decades, and for the most part did not live very long in captivity (Mohr 1943). It was only after the animals began to be used for economic reasons in the Soviet Union that it was possible to obtain saigas in large numbers for zoos abroad. These exported animals were captured as newborn lambs and were bottle-fed. All of these saigas originated from the population living on the west bank of the Volga where the Astrakan Forestry Department busied itself with the artificial rearing of saigas (Bannikow 1963).

The first saigas arrived in Tierpark Berlin in 1958 (Petzold 1963). As a transit point and quarantine station in the East/West animal trade, Tier-park Berlin has kept 332 Saigas exported from the Soviet Union. These animals, which arrived in the months from August to June, were almost all young animals of the breeding season (3 to 14 months old) (Figure 16.2). The majority arrived as 6-month-old lambs in the months of October/November. I have previously described our first experiences in keeping and breeding these difficult animals (Pohle 1974).

FEEDING

The feeding of saigas presents no difficulties. The animals are easily raised on high-energy mixes which are normally used for antelope and deer as well as crispbread, rolled oats, and vegetables. Hard foods such as carrots or potatoes should be cut into very small pieces. We lost two animals as a result of suffocation after they had eaten large pieces of potato. Hay and green food are freely available to the animals in the summer. Drinking water and salt blocks are always available to the animals.

HOUSING

Of great importance in the successful keeping of saigas is an enclosure of sufficient size. The artificially raised imported animals as well as those already born in the enclosure can suddenly break into a panic and dash themselves against the fences. Deaths due to trauma are therefore not a rarity. Saigas which are inhabitants of open steppes are apparently not able to avoid fences when panicked. This can lead to injuries due to the restric-tive conditions in zoos. Our enclosure for the saigas at present consists of three units which are in a row parallel to one another and measure 1050 m², 1800 m² and 800 m². The smallest enclosure is for the male group. The two larger enclosures are for the breeding groups. There are two 30-m-long

Figure 16.2: Saiga herd exported from the USSR at about 7 months old. (Photo: Klaus Rudloff)

alleys between the enclosures (2.7 m and 2.2 m wide respectively) which are closed at the ends and are connected to the adjoining enclosures by means of doors. When necessary, an animal can be isolated in these alleys. They are also ideal for trapping an animal that has to be caught by hand. Closed stalls are not provided. A roofed area is provided to protect the food from rain. In both of the large enclosures, the feeding areas are fenced and are connected to the enclosure by means of two doors. In this way during the rut an aggressive male can be enclosed, which then allows one to work within the other enclosure.

BREEDING

The males are only placed with the females from November to the beginning of April. The first saiga birth in Tierpark Berlin occurred on 19 May 1969 when two females each produced a single lamb. At this time the male was kept within the herd. At the onset the lambs were of no interest to the buck, but four hours later he suddenly attacked the young animals so aggressively that he had to be separated from the group. In order to avoid any risks we have always separated the breeding males from the group before the lambing season. In the wild, a separation occurs between the sexes in the spring although some males (2–4 per cent of the entire population) do remain on the lambing grounds among the pregnant females which congregate in large numbers (Shirnov 1982).

When the adult breeding males are returned to the group during the autumn, the approximately 6-month-old males must be taken out of the herd. Young females are left in the breeding group as they can be successfully impregnated at the age of 7 months.

It is absolutely impossible to keep more than one adult male (older than a year and a half) together with a group of females. In order to avoid injury to the female, the horns of the males are covered with rubber tubing (Figure 16.3). To prevent damage to the horn, the tubing is removed after the bucks have been separated from the females. Within the male groups, even outside the rutting period the males can often interact aggressively. Injuries can be avoided by cutting off the tips of the horns, but it is better to isolate the individual bucks.

The rut begins during the second half of December and continues until January. Cold, dry weather results in the height of the rut being during the last ten days of December. According to Shirnov (1982) the rutting period in nature is approximately 40 days, although the greatest proportion of females are bred within the first five to ten days after the rut. If impregnation does not occur during the first cycle, a second cycle follows for the females after 16 to 19 days. According to this, the males must have a renewed period of fertility in May and June as females have been taken

Figure 16.3: A younger male saiga (1.9 years old) with winter fur and with tubes put on horns. (Photo: Klaus Rudloff)

with well developed embryos in October and November. Since the males are removed from the group in Tierpark Berlin during May and June, we have not observed that phenomenon here. The latest successful breeding among our saigas occurred in the middle of February, based on a gestation period of 4.5 months. A female which was placed with our male on 19 January produced a lamb on 28 June.

During the rut the males pursue the females at a trot. The head is held forward and the nose is inflated, and loud calls, grunting, and nasal rumbling can be heard. Our saiga enclosures are outside the area for visitors. With the appearance of humans, the buck begins to drive his harem together and to force them away from observers.

As already recorded, the first saiga birth occurred in Tierpark Berlin in 1969, and 116 lambs (63 males, 53 females) have been born at the time of writing. In the year 1984 alone, 34 saigas were born in Tierpark Berlin. Of the 90 births, there were 26 pairs of twins (6 × 2♀♀, 8 × 2♂♂, 12 ×♂♀).

In the Kalmuck steppes whence the zoo saigas originated, the birth period begins earlier than in Kazakhstan, around 15 April. Most births occur between 3/4 and 8/9 May. Although the birth period extends for over a month, the greater majority of females (not less than 80 per cent) produce their lambs within a short period of from three to five days (Shirnov 1982).

Saigas have never been born in April in Tierpark Berlin. Our 90 births occurred in a period between 3 May and 28 June, of which 66 took place in the period up to 20 May and only 13 in June (Figure 16.4). As already mentioned, the birth on 28 June was an exception. The latest birth, by a female that was consistently with a buck from the beginning of the rut, was on 15 June. In the various years, most of the young were produced within a period of eight to ten days (Figure 16.5).

The birth weights of 77 live-born lambs varied between 1960 g and 4400 g. The lightest animal was weak and an attempt was made to rear it artificially, but it died when it was two days old. A mother successfully raised a lamb that was one of a twin pair which weighed only 2340 g. The median weight for 77 live-born saigas is 3250 g. There is not sufficient material to make any accurate statements regarding weight differences between single births and twins as well as males and females. The variations within a single group are large, and the lightest as well as the heaviest lambs were single females. In single births the males on average are heavier than the females, but in the case of our twin births the situation was just the reverse:

Single ♂♂ (n=15): 2800–4000 g (median 3480 g)
Single ♀♀ (n=20): 1960–4400 g (median 3310 g)
Twin ♂♂ (n=22): 2200–4200 g (median 3130 g)
Twin ♀♀ (n=20): 2340–3800 g (median 3160 g)

Figure 16.4: Temporary distribution of saiga bred in a single year

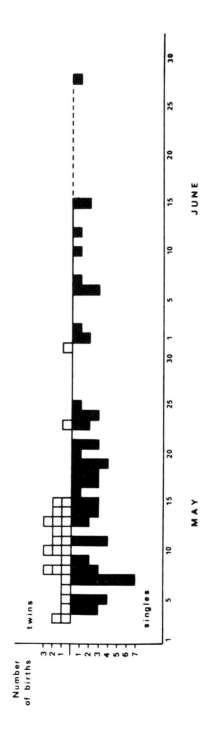

Figure 16.5: Temporary distribution of 90 saiga bred at Tierpark Berlin

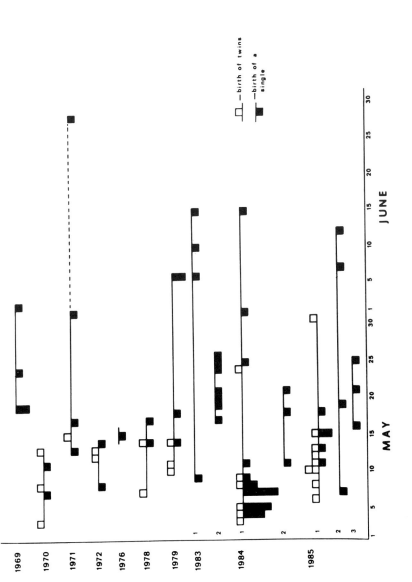

During its first few hours, a newborn lamb remains in close contact with its mother, who immediately follows it if the lamb leaves her side. The important olfactory recognition of the young animal is formed during this phase. If there are complications during the birth of twins, and the firstborn separates itself from the mother who is not in a position to follow it, the contact between mother and lamb can be destroyed. In one of these cases, a newborn lamb was adopted by a female who had produced a stillborn lamb. The lamb was successfully raised by its adoptive mother while the second-born youngster was raised by its natural mother. The adoption of a newborn animal by another who had lost her lamb was also reported by Zimmermann (1980).

Later the young animals distance themselves from the mother. The lamb will even go out through the fence and rest outside the enclosure in locations without visual contact with the herd. This behaviour is also known in deer and other antelope (Dathe 1966). Initially, twins rest in different locations. As the lambs become stronger in their ability to run, they tend to lie closer together and, in fact, build regular kindergartens.

The choice resting places for young animals are those areas where their backs are protected, for example along fences and under the roof of the feeder as well as in open areas where the adult animals have dug holes in the earth (Figure 16.6). This particular preference for such depressions resulted in a day-old lamb entering the burrow of a wild rabbit (*Oryctolagus cuniculus*). In attempting to free itself it had only pushed itself deeper into the burrow, and it was only by accident that we discovered it and were able to dig it out. However, it was so exhausted that a few hours after being freed it died.

The mothers recognise their lambs olfactorily. When a lamb bleats a number of mother animals will go to it and only walk away after they have checked it olfactorily. If a lamb that is only a few days old attempts to nurse from a female which is not his mother, she feigns biting or bucking. When the lambs are over three weeks old, they are so persistent that the females are not able to repel them, and three lambs have been observed nursing from the same female (Figure 16.7). Since females who have lost their young are prepared to adopt lambs, and it is also possible for older lambs to nurse from females that are not their mothers, this would seem to indicate that in nature many lambs are saved whose mothers have died. Because there is less space available in zoological gardens, this practice, which may enable lambs to survive in the wild, can have fatal results in captivity. If lambs are born weeks after the majority are born, the chances of survival are reduced because the older lambs will nurse from their mothers. In such cases, one should attempt to separate the newborn lambs and their mothers from the herd.

The artificial rearing of lambs in Tierpark Berlin has been successful on a number of occasions. Cows' milk is an excellent substitute, as has been

Figure 16.6: Saiga lamb one day old close to the fence. (Photo: Klaus Rudloff)

Figure 16.7: Three saiga lambs 3–4 weeks old drinking together from a mother saiga. (Photo: Klaus Rudloff)

substantiated by Bannikow (1963) and Orbell and Orbell (1976). The fat content of 100 g of full milk (2.2 per cent) is increased by the addition of 50 g of heavy cream (10 per cent fat). To this mix is added 80 g of linseed slime and 20 g vital serum. The amount fed four times a day increases up to the third day from 50 g to approximately 150 g.

A few years ago we began to mark newborn saigas with ear notches but since the ears are so heavily covered in hair it is not possible from a distance to read the notches. However, upon the death of an animal,or when it is caught for veterinary reasons or to be transported, one can then identify the animal.

Of the 116 newborn saigas, 12 were stillborn. Of these, nine were single lambs of mothers who were primiparae. In the case of twin births to older females there is only one case of a single stillborn young, and in another instance, full-term stillborn twins were delivered by Caesarian section.

LONGEVITY

Of the saigas born in Tierpark Berlin, 13 were shipped to other zoos and 20 are still in the collection. Of the remaining 71 births, 63 did not reach their second year. By the time they were 6 months of age, 41 had already died. Only six (two males, four females) reached a longevity of over 3 years. The oldest female died at 7 years 9 months. A year before it had stopped producing lambs. Presently, we have eleven females that were born in 1982 and imported in October of the same year. Of the animals imported earlier, six (two males, four females) lived to be over 4 years of age. The oldest, a female, was 6 years old and had lived for 5 years 8 months in Tierpark Berlin. Dolan (1977) reported that an imported female lived for 7 years 10 months in the San Francisco Zoo and was possibly over 9 years old at the time of death. Jones (1982) gives the greatest longevity in captivity as 8 years. For saigas living on the Kalmuck steppes, a longevity of from 9 to 10 years has been given, though the majority of the animals live only 4 to 5 years (Shirnov 1977). Females older than 4 years produce only single lambs. Shirnov (1982) characterises the peculiarities of the saiga antelope as follows: 'In the ecological system of the hooved animals one can characterise the saiga as the ephemeron of its clan characterised by early maturity and a high reproductive potential compared with the short longevity of the individual generations.'

REFERENCES

Bannikow, A.G. (1963) *Die Saiga-Antilope.* Neue Brehm-Bücherei Vol. 320. A. Ziemsen Verlag, Wittenberg Lutherstadt.

Dathe, H. (1966) Zum Mutter-Kind-Verhältnis bei Cerviden. *Beitrag zur Jagd und Wildforsch. V*, 83–8

Dolan, J.M. (1977) The saiga, *Saiga tatarica*: a review as a model for the management of endangered species. *Int. Zoo Yrbk 17*, 25–30

Jones, M.L. (1982) Longevity of captive mammals. *D. Zool. Garten (NF) 52*, 113–28

Mohr, E. (1943)Einiges über die Saiga, *Saiga tatarica* L. *D. Zool. Garten (NF) 15*, 175–85

Orbell, E. and J. Orbell. (1976) Hand-rearing a saiga antelope, *Saiga tatarica*, at the Highland Wildlife Park. *Int. Zoo Yrbk 16*, 208–9

Petzold, H. -G. (1963) Im Tierpark Berlin 1958 erstmalig gehaltene Tierformen. *Milu, Berlin 1*, 177–202

Pohle, C. (1974) Haltung und Zucht der Saiga-Antilope (*Saiga tatarica*) im Tierpark Berlin. *D. Zool. Garten (NF) 44*, 387–409

Shirnov, L.V. (1977) Sajgak In: *Kopytnye zveri*, Lesnaya promyšlennost, Moscow, pp. 79–118

Shirnov, L.V. (1982) *Vozvraščennye k žizni*. Lesnaja promyšlennost, Moscow

Zimmermann, W. (1980) Zur Haltung und Zucht von Saiga-Antilopen (*Saiga tatarica tatarica*) im Kölner Zoo. *Z. Kölner Zoo 23*, 120–7

17

Breeding of Mongolian gazelle at Osaka Zoo

Minoru Miyashita

Osaka Municipal Tennoji Zoo, 6-74 Chausuyama-cho, Tennoji-ku. Osaka 543, Japan

INTRODUCTION

On 6 September 1974, as part of an animal-friendship exchange between Japan and the People's Republic of China, the Osaka Municipal Tennoji Zoo received a pair of Mongolian gazelle from Beijing Zoo. Although the Mongolian gazelle (Figure 17.1) is classified in the genus *Procapra* together with the Tibetan gazelle *P. picticaudata* and the Przewalski's gazelle *P. przewalskii*, there was no detailed report about its ecology and breeding, so we started to keep them without any prior knowledge.

As far as we know, these gazelles are the only representatives of this species outside China, and the birth of a calf in 1977 (Figure 17.2) was the first birth in captivity (Miyashita and Nagase 1981). Following this birth, we succeeded in breeding in 1979 and 1982, and we were able to record such information as breeding behaviour in captivity.

HOUSING AND FEEDING

At our zoo there is an enclosure for antelopes measuring about 3700 m², where five species of antelopes and gazelles have been kept mostly in small groups or pairs. They are groups of Beisa oryx, eland, springbok, Thomson's gazelle and blackbuck. As we were concerned that the valuable Mongolian gazelles might be injured if housed with the other species, we partitioned off a part of this enclosure. This new enclosure for the Mongolian gazelles has an area of about 120 m² and is surrounded by a wire fence 2 m high. It also contains two individual stalls, each measuring 3 × 4 × 2 m high, and individual access into the enclosure.

At first the pair co-existed peacefully, but as the male grew he frequently attacked the female and in August 1974 we decided to separate them. They were allowed access to the outdoor enclosure on alternate days, until the following May when it was fenced down the middle to give them individual outdoor as well as indoor enclosures.

Figure 17.1: Male Mongolian gazelle

The gazelles are unaffected by cold but they appear to be sensitive to the heat and dampness of Japan, so in summer we set up a sun- and rain-shelter roofed with reeds in the enclosure.

We feed them on a pelleted herbivore diet, chopped carrots, sliced potatoes, cabbages, bread, hay cubes, fresh hay, wheat bran, salt and bone meal, as shown in Table 17.1.

Faecal samples are examined regularly and soon after their arrival we detected the eggs of some parasites, including *Strongyloides trichuris*. Since treatment with Parbendazole for intestinal worms, there has been no recurrence of parasitism by helminths.

OESTRUS AND MATING

Our female first came into oestrus on 17 December 1975 and again on 15 January and 13 February 1976. Thus we found that the oestrus cycle appeared to be a regular 29 days, but the actual oestrus is very short and only lasts between a half to one day. We recognised the female's swollen

Figure 17.2: Female with her 3-month-old calf born at Osaka Zoo in 1977

Table 17.1: Diet of the Mongolian gazelle

Foodstuff	Quantity per head per day(g)
Pellets	500
Carrot	500
Potato	300
Cabbage	300
Bread	200
Hay cubes	200
Fresh hay	500
Wheat bran	100
Bone meal	30
Salt	30

and pinkish genitalia and frequent tail wagging as signs of oestrus. On 13 February we put the pair together, but although the male showed interest, the female appeared nervous and ran from him. As he continued to chase and butt her, without attempting to mate, we quickly separated them again.

From the third oestrus onwards, we perceived no further evidence of oestrus until the following winter when the female appeared receptive on 1 and 30 December. Assuming a 29-day cycle, we planned to put the pair together on 28 January and we had already covered the male's horns with rubber tubing to give the female some protection. As expected, on the morning of the 28th the female showed signs of oestrus by frequently wagging her tail. They were put together and when the male approached and smelled her, she remained still, standing close to him with tail raised. He attempted to mount three times within a few minutes, and at the third attempt intromission was achieved. Immediately the mating was completed the female ran off, and as the male became aggressive again, they had to be separated. The female showed no further signs of oestrus that season and we assumed she was pregnant.

GESTATION AND PARTURITION

Based on comparisons with gazelles of similar size, we assumed a gestation period of 170–180 days. By mid-July the female's udders were noticeably swollen and they continued to increase in size until a week before delivery. At the end of July, pregnancy was confirmed by fetal movements. On 2 August 1977, after a gestation of 186 days, a female calf was born.

DEVELOPMENT OF THE YOUNG

The calf was able to stand after 20 minutes and first suckled two hours after birth. It measured about 53 cm in body length and was 42 cm high. As it struggled violently when handled, we did not attempt to weigh it. At first the calf, which spent most of the time lying down, was shut into the stall with its mother, but after a week the dam and the calf were released into the outside enclosure. At 35 days old we observed the calf picking up hay in its mouth and at 40 days it was eating solid food. By 3 months of age its body length was about 85 cm and it was 52 cm high. By 8 months it was as tall as the adult.

SECOND AND THIRD BIRTHS

The male was kept apart from the female, which remained with her young until June 1978. From June 1978 onwards, the adult pair were often put together again. However, they had to be separated when the male showed signs of aggression.

From 11 to 13 December the female showed some signs of oestrus, so we put the pair together in the daytime over three days. We did not however, observe the mating. Although she did not appear to have further oestrus, we were not confident of pregnancy until the June of the following year, when swelling of the udder and movements of the fetus were seen. She was then confined to her stall, while her two-year-old offspring remained in the paddock. On 18 June 1979, 187–189 days after the only observed oestrus, a second female was born.

On 24 February 1982, we noticed the female's oestrus and we put the pair together. We observed the mating. On 30 July, 187 days after the mating, the third birth occurred. The calf was the first male to be born.

DEATH OF THREE MONGOLIAN GAZELLES

We had five Mongolian gazelles at one time, but we then lost three. The adult female died on 12 September, 44 days after her third delivery. During the third pregnancy, her left mandibula had been swollen since the beginning of July 1982, and continuous slavering and masticating were recognised. We therefore suspected dental inflammation or lumpy jaw, but we did not plan to give any antibiotic therapy until weaning because we feared it would affect the pregnancy. Fractures of the left mandibula and sequesturm were found in autopsy findings. In addition, dentes decidui was absent and the subpostmolar dentes was fractured. Our final diagnosis was that, probably, periodontitis led to bronchopneumonia and oedema, and to general septicaemia.

The third calf had been kept with its two elder sisters since his mother died. As the young male began to butt his sisters a little before his first birthday, we isolated him in a new enclosure from the females in June 1983. On 16 September 1983, the young male seemed to be surprised at something and smashed into the fence. His cervical vertebra was fractured and he died instantly.

The old male had been kept apart from his mate and offspring since his mate was pregnant with the third calf. The sire was left with the young females several times in their oestrus, but no mating was observed. About July 1984, we noticed a tumour in his lower gingiva. It swelled gradually, and finally we observed bleeding from the tumour. Although we administered antibiotics, haemostatics and vitamins to the old male, he died on 26 October. He was found by the autopsy findings to have osteosarcoma.

CHROMOSOME STUDIES

Blood and biopsy samples were obtained from the sire and first-born female and investigated by the Department of Obstetrics and Gynaecology at Tokyo Medical College. Chromosome preparations stimulated with phytohaemagglutinin (PHA) were obtained from lymphocyte cultures incubated for four days and from skin cultures after 21 days. Following the cultures, the air-dried preparations were stained by Giemsa in the conventional method (Figure 17.3).

The chromosome numbers and karyotypes found in this species were reported and described in detail by Soma *et al.* (1979, 1980). The Mongolian gazelles were found to have a diploid number of chromosomes of $2n=60$, consisting of uniform acrocentrics.

DISCUSSION

When the pair of Mongolian gazelles from Beijing arrived at Osaka after one month's quarantine, we had estimated that they were about 2 years old on the basis of size, colour and length of the male's horns, but we

Figure 17.3: Chromosomes of male Mongolian gazelle (bottom) and karyotype (top)

did not really know until the male calf became about 1 year old. When the young male reached his first birthday, the length of his horns was the same as that of his father's horns on arrival. We therefore corrected our estimate, and we believe they were about 1 year old on arrival. Therefore, I believe the Monogolian gazelle reaches sexual maturity at 3 $\frac{1}{2}$ years old. The first sign of oestrus was found at 2 $\frac{1}{2}$ years old in the paired female, and at 4$\frac{1}{2}$ years old in our zoo-born females.

According to Nowak and Paradiso (1983), the mating season is at the end of autumn and birth occurs in June. Twins are commonly produced, but the gestation period is not known exactly. We found that the mating season was from November to the following February. Our female has always had only one calf per birth on each of the three occasions, and her gestation period is from 186 to 189 days.

We noticed the dam's oestrus nine times and observed her oestrus cycle five times from 1975 to 1982. Her oestrus cycle appeared to be a regular 29 days except in one case of 26 days. We oberved the first-born female's oestrus cycle only once in 1984. It was a 29-day cycle the same as her mother's. The dam's oestrus was easy to recognise, but with the two young females it was not so clear because of timidity and nervousness. (See Figure 17.4.)

The sire could not be kept with the females continually. One reason is that the enclosure is too small for them. In the wild, the males and the females form separate herds in the summer, and they remain apart until the

Figure 17.4: Mongolian gazelles' oestrus

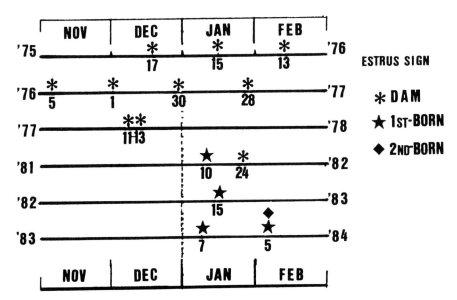

211

mating season late in the autumn. In the winter, they form mixed herds and migrate over great distances to better pastures and to drop the calf (Walther 1972). It may be that there is no necessity to keep the male and the female together continually in captivity.

At present, only two females are on display, so we have asked about acquiring a male from some zoological gardens in China for the past two years. We hope to do so in the near future, and to breed the Mongolian gazelle again.

ACKNOWLEDGEMENTS

The author wishes to express his deepest appreciation to Professor H. Soma, of the Department of Obstetrics and Gynaecology, Tokyo Medical College, for his kindness in providing me with the opportunity to present these findings, and for his comments on the manuscript. Thanks are also due to the staff of Osaka Municipal Tennoji Zoo for their advice and cooperation.

REFERENCES

Miyashita, M. and K. Nagase. (1981) Breeding the Mongolian gazelle at Osaka Zoo. *Int. Zoo Yrbk 21*, 158–62

Nowak, R.M. and J.L. Paradiso. (1983) *Walker's mammals of the world.* Johns Hopkins University Press, pp. 1286–8, Baltimore

Soma, H., T. Kiyokawa, K. Matayoshi, I. Tarumoto, M. Miyashita and K. Nagase. (1979) The chromosomes of the Mongolian gazelle, a rare species of antelope. *Proc. Japan Acad. 55 (B)*, 6–9

Soma, H., H. Kada, K. Matayoshi, T. Kiyokawa, T. Ito, M. Miyashita and K. Nagase. (1980) Some chromosomal aspects of *Naemorhaedus goral* (goral) and *Procapra gutturosa* (Mongolian gazelle). *Proc. Japan Acad. 56 (B)*, 273-7

Walther, F. (1972) The gazelles and their relatives. In B. Grzimek (ed.) *Grzimek's animal life encyclopedia 13*, Van Nostrand Reinhold, New York, pp. 438–42

18

Nemorhaedus cranbrooki Hayman

Cizu Zhang

Shanghai Zoological Garden, 2381 Hong Quai Road, Shanghai, Kiangsu, People's
Republic of China

INTRODUCTION

Red goral (*Nemorhaedus cranbrooki* Hayman) is a Class 1 protected
animal in the People's Republic of China. In 1979 and 1982, the author
looked for red gorals in various parts of Tibet (Mainling, Nyingchi, Bomi,
etc.), observed their ecology, caught them, and bred them. Red gorals in
Gyaca, Nangxian, Mêdog and Zayii Prefectures were also investigated, and
ten skull specimens and pelts were obtained. There are currently nine red
gorals in Shanghai Zoo.

GEOGRAPHICAL DISTRIBUTION

Nemorhaedus cranbrooki Hayman is found in remote mountainous wood-
lands of south-east Tibet and north-west Yunnan, People's Republic of
China; Assam, north-east India; and northern Burma. In south-east Tibet,
it is distributed in southern Gyaca Prefecture, all over Nangxian Prefecture,
south to Yarlung Zanbo Jiang in Mainling Prefecture, all over Mêdog
Prefecture, drainage of Dongju River in Nyingchi Prefecture, drainage of
Pêlung River, Bomi River in Bomi Prefecture, drainage of Yigong River,
lower Zayii in Zayii Prefecture, and extending south-east up to Nujiang
River in Yunnan Province. The principal habitats are in four prefectures,
namely Bomi, Nyingchi, Mainling and Mêdog. This is the largest primitive
coniferous woodland in Tibet, located on the eastern boundary of Mount
Himalaya and the southern slopes of Kaqin Glacier, Nyaingentanglha
Shan, which is very heavily forested and where the geomorphology is
extremely complex. Drops from high peaks to deep valleys may reach
several thousand metres. Rapid streams whirl among steep cliffs, most of
them going north–southward, and converge into the big turning segment of
Yarlung Zanbo Jiang. Here, affected by a warm wet air current from the
Indian Ocean, the climate is warm and there is abundant rainfall: *c.*
2000 mm annually. The rainy season is from May to August.

ECOLOGY AND BEHAVIOUR

Nemorhaedus cranbrooki Hayman lives in mountainous forests at 2000 to 4500 m altitude, and is a typical woodland-dwelling mammal. The animals migrate perpendicularly according to the seasons. In summer, they stay around the upper margins of the forest, and roam about in meadows and shrubby lands above the woods. In winter, with the lowering of the snow line, they come to the lower part of the forest, grazing in coniferous and deciduous mixed forests as well as in meadows and bushes at lower altitude.

Generally, they drink from the streams in the morning and then look for food. During the day, they graze on the sunny grassy slopes, and stay overnight on rocky cliffs. By observation, it was found that they feed mainly on *Usnea* and other lichens, grasses and tender bushy stems and leaves.

Red gorals are excellent springers. They always jump up to the cliffs in the event of danger. Their principal predators are leopards and jackals. From their faeces, either from the wild or after taming, the presence of ascarids, nematodes and tapeworms is always detected. Furthermore, almost every red goral is severely attacked by *Ixodes* sp. Most of them attach themselves to soft tissues (in the ears, eyelids, the vagina or the anus) and suck blood.

Except during the mating period, the males generally live alone and the females live with their young. Occasionally, there are families of over three members. These are either a mother with her child, temporarily staying with a male; or a female with two lambs, born in the current and the previous year. For example, at 0900 on 1 September 1979, the author observed with a telescope at the summit of Eastern Peak (4500 m) near Laiguo Bridge, Mainling Prefecture, and saw three groups of red gorals climbing and grazing: 200 m beneath, a female with two lambs, born in that year and in 1978 (Figure 18.1); 500 m eastward, a mother with one lamb; and one red goral again 800 m further eastward. The group of three beneath were not afraid of me. They did not run away even when I got within about 40 m.

The copulation period of *Nemorhaedus cranbrooki* Hayman is around December, and the animals copulate frequently during oestrus. The males utter a 'zer ... zer' call to ask for coupling. The lambs are born around the following June, and there is usually only one. However, the author has caught a young lamb in April 1982, born in early spring of that year (Figure 18.2).

The principal wildlife in the same region with red goral consists of forest musk deer, takin, Fea's muntjac, mainland serow, black bear, leopard, jackal, Asiatic golden cat, lesser panda, leopard cat, weasel, Rhesus macaque, and even tiger.

214

Figure 18.1: Female with lambs, on Eastern Peak (4300 m) near Laiguo Bridge, Mainling Prefecture

Figure 18.2: Early spring lamb (Bayu, Yigong drainage)

MORPHOLOGICAL CHARACTERISTICS

Table 18.1 summarises the morphological features of nine red gorals. The size of red gorals is similar to that of common gorals, but there is no mane, more curvature on two antlers, and a much shorter tail. The females are a little larger than the males. Both sexes have two antlers, reddish-brown hair (darker in the dorsal region and lighter at the throat, thorax and abdominal regions). A line of black hair in the middle of the thorax, and a stripe of black hair (width *c.* 2 cm) extends along the back from the post-occiput to the tip of the tail. In a young animal, an obvious white spot can be seen in the middle of the forehead that fades as it matures. However, there are mature animals with a trace of white hairs, and lambs with an indistinct white spot. Furthermore, individuals without antlers are also discovered. (See Table 18.2.)

CATCHING AND TAMING

A relatively successful method for catching a red goral is to drive it with experienced hounds to the edges of a cliff, then lasso it around its head (together with a foreleg) with a rope. It is important to remember to control the hounds in time to prevent them from biting the animal after it has been lassoed.

On arriving at the breeding base, all the *Ixodes* sp. attached on its soft tissues should first be removed. Its wounds should be examined and, first aid performed as necessary. Often, as a result of the pursuit by hounds, red gorals become frightened and overexcited, and the heart and lungs may be injured through overstraining. Thus, a three-day treatment with strepto-mycin and penicillin is necessary for therapeutic and prophylactic purposes.

Table 18.1: Body weight (kg) and dimensions (mm)

	Specimen number								
	1	2	3	4	5	6	7	8	9
Sex	Adult	Adult	Adult	Adult	Adult	Adult	Adult	½ year	½ year
	♂	♂	♂	♀	♀	♀	♀	♂	♂
Body weight	22.6	25.0	28.6	30.6	28.5	29.5	27.5	15.8	16.8
Body length	930	970	960	980	1030	990	950	810	840
Shoulder height	590	570	580	605	600	610	590	460	530
Hind foot	270	260	275	280	260	280	260	245	250
Antler length	150	160	125	140	Broken	118	150	48	35
Ear length	109	115	117	106	105	103	102	95	95
Tail length	103	105	118	104	102	112	120	75	100
Chest measurement	700	685	780	810	810	790	800	580	630
Abdomen measurement	710	770	815	835	830	860	890	660	680

Table 18.2: Skull, teeth and antler dimensions (mm)

Specimen	Sex	Skull length	Base length	Nose-bone length	Zygomatic breadth	Upper teeth	Lower teeth	Antler-bone length	Antler base diameter	Date and place of collection
82109	Adult ♂	197	175	59	87	68	67	—	—	March 1981. Mêdog, Xinge Jiang
82111	Adult ♂	189	173	66	90	64	69	76	19	April 1982. Nyingchi, Dongju, Pêlung
82125	Adult ♂	206	184	61.5	92	57	60	91	22.5	December 1982. Bomi, Yigong
83170	Adult ♂	210	—	75.5	95	54	56	79	23	February 1983. Bomi, Yigong
83171	Adult ♂	208	184	63	92.5	60.5	62	95	23	February 1983. Bomi, Yigong
81026	Adult ♀	205	181	60	95	59	63	—	16	September 1979. Mainling, Laiguo Bridge
82100	Adult ♀	195	171.5	57	86	66	69	55	15	February 1982. Bomi, Yigong
82110	Young ♀	163	139	43	72	45	51	—	—	April 1981. Mêdog, Xinge Jiang
83172	Young ♀	157.5	134	34	71	44.5	53	—	—	February 1983. Bomi, Yigong
83173	Young ♀	169.5	148.5	49	76	55	56	30	12.5	February 1983. Bomi, Yigong

Newly captured red gorals should be kept in a dark wooden cage with enough space to turn the body freely and to raise the head. A little light should come in at the front only for feeding and drinking. The animal must be kept as calm as possible for one week. Then, after the goral has recovered its normal daily rhythm of feeding and drinking, it should be transferred into a dimly lit room for two weeks. Outdoor activities for the goral may then be allowed in open yards. As red gorals are excellent springers, care must be taken to prevent them from escaping. The mesh of the wire net used to surround the yard should be small enough to prevent the animals from hurting their feet.

FARM BREEDING

Red gorals are relatively easily tamed herbivores, and soon become accustomed to frequent contact with human beings. In Shanghai Zoo, seven wild animals are breeding in two groups: two males with one female in Group A, and one male with three females in Group B. During the mating and oestrous periods, because of mutual fighting, we have to separate the two males in Group A. Females at the lambing stage should be kept alone.

Female No. 4 in Group A was mated in December 1983 to January 1984, and a lamb was born on 19 July 1984. The animal was again mated in December 1984 to January 1985, and a lamb was born on 13 July 1985. Both lambs died as a result of inadequate maternal care.

Female No. 5 in Group B was mated in December 1984, and a lamb was born on 14 June 1985. It survived.

Female No. 6 in Group B was mated in December 1984, and a lamb was born on 18 June 1985 which survived.

The two surviving lambs were being weaned $3^1/_2$ months later, feeding in another cage. The fodder consists mainly of grasses and leaves. Fresh grasses and fresh leaves of *Ulmus pumila*, *Poulownia fortunei* and *Sophora japonica* are fed from May till December. Dry grasses and dry mulberry leaves are fed from November until the following April, together with wheat seedlings, carrots, and privet leaves. In addition, 300 g of mixed cereal powdered fodder are given per head daily all the year round.

CONSERVATION OF *NEMORHAEDUS CRANBROOKI* HAYMAN

There are still a number of red gorals in the above-mentioned regions, but hunting by the natives is not insignificant. Furthermore, the rapid development of lumbering industries has aggravated the situation. Thus, the author suggests that a conservation are of 200 km diameter should be established with its centre at the big turning point of Yarlung Zangbo Jiang, which includes part of four Prefectures: Mainling, Nyingchi, Bomi and Mêdog. The altitude in this region ranges from South Jiabawa Peak ($>$ 7700 m) down to south Yarlung Zangbo Jiang ($<$ 1000 m). The whole region is affected by warm wet air currents from the Indian Ocean, and characterised by tropical rainforest, sub-tropical, temperate zone, and highland cold-zone climates. Among all the mountainous perpendicular natural habitats in the People's Republic of China, it is the most undamaged and most complete one, with very abundant and complex vegetation types as well as flora and fauna components, together with various species of rare and precious animals. The establishment of such a natural conservation area on the Tibet Plateau would be of particular significance.

SUMMARY

(1) The main distribution areas of red goral (*Nemorhaedus cranbrooki* Hayman) in the People's Republic of China are in four Prefectures of Tibet, i.e. Mainling, Nyingchi, Bomi and Mêdog. The northernmost boundary is the northern part of Yigong river, Bomi Prefecture, i.e. woodland regions on the southern slopes of Kaqin Glacier, Nyainqen-tanglha Shan.

(2) *Nemorhaedus cranbrooki* Hayman is a forest-dwelling animal, mainly living in mountainous regions at 2000–4500 m altitude in evergreen coniferous forests. It lives alone or in small family groups. The mating period is around December, and usually one lamb is born the following June.

(3) The morphological characteristics are reddish-brown hair, with a longitudinal black dorsal stripe. In young animals, there is an obvious white spot on the middle of the forehead.

(4) A successful catching method is to drive with hounds and lasso with ropes. First-aid wound treatment must be given after capture and medical care may be needed to prevent heart and lung failure due to over-straining. Once captured, the goral should be kept in the dark for one week, in a dimly lit place for two weeks, after which the amount of light can be increased gradually.

(5) Herd-breeding in farms is possible, and animals should be fed with grasses and leaves. Lambing females should be fed individually. In the mating period males should be fed separately.

(6) The principal habitats are located in the most untouched and typical regions of Chinese mountainous areas, with perpendicularly distributed natural habitats, abundant flora and fauna components, and many rare and precious animals. The establishment of a natural conservation area in such locations would be particularly beneficial.

ACKNOWLEDGEMENTS

The author wishes to thank Associate Professor Sheng Helin, who approved this chapter, and Associate Professor Wang Song for his help in its preparation. Zhou Jianhua, Yu Minhua and Zhou Huilin have also taken part in this investigation.

19

Breeding of goral, Formosan serow and chamois

Takeyoshi Ito
Japan Serow Center, Gozaisho-dake, Komono-cho, Mie-gun, Mie Prefecture
510-12, Japan

SEROWS

In April 1959 the 2200 m-long Gozaisho Ropeway was opened at Mt Gozaisho-Dake, standing at 1210 m above sea level, and in July 1960 the first Scientific Research Meeting for the preservation of Japanese serows in the Suzuka mountain range was held. Since then, the Japan Serow Center has established the breeding of Japanese serows at the top of Gozaisho-Dake. A pair of Japanese serows were captured in April 1972 from the wild and given protection there, and in August 1965 we had the first successful birth of Japanese serow (a female) in captivity. They have bred every year since then. As a result we have been able to determine that their gestation period is about 213 days.

December 1973 saw the establishment of the Japan Serow Center, and in November 1980 Gozaisho Natural Museum, affiliated to the Gozaisho Alpine Zoo, was opened. Rocky Mountain goats were transported from the USA in an exchange for Japanese serows, and in this way the Serow Center was responsible for the founding of the Gozaisho Alpine Zoo.

GORALS

Gorals were sent from Peking Zoo in December 1978 as an animal friend-ship-exchange. The goral has a number of unique characteristics as follows: the hair on their backs is long, they have no suborbital glands, and their tails are long like those of horses. They have a white spot on their throat. Defaecation is localised, and the animals form small groups. Moulting takes place between April and May.

They consume a large amount of leaves all through the year and on Mt. Gozaisho like to eat *Sasa albo-marginata*. At the Center the main food for one animal consists of carrots, potatoes, bran (900 g), corn (150 g), hay and grass (3 kg), and soft leaves. They are in good health and have never suffered from diarrhoea.

In winter, when there is a snow cover up to 1 m deep, their movement is active, especially in the morning and in the evening. In the daytime they sit and sleep on a raised area. Gorals can spring to a height of 2 m, and climb large trees with an inclination of 30°.

The breeding season is from November to December. During the rutting period the male becomes aggressive. He is also seen to rub himself in the female's urine. The male frequently kicks the female with his forelegs, and copulation lasts for about 20 seconds. The gestation period is about 180 days, and a single lamb is born. The male usually pays no attention to it. As the lamb hides itself in the daytime, it is often impossible for us to find it.

Since the first lamb was born in June 1980, there have been nine further births. Three gorals have been transported to San Diego Zoo as an animal friendship-exchange, and in April 1986 a female goral was sent from Peking Zoo in order to avoid inbreeding.

FORMOSAN SEROWS

In November 1983 three Formosan serows (one male and two females) were transported from Taipei Zoo to the Japan Serow Center. They are smaller than the Japanese serow to which they are closely related, the kids being 85–90 cm long and about 60 cm high. The pelt is short and not curly, and they have a dark yellow spot on the throat; their ears are big. In appearance they resemble gorals quite closely, but they have suborbital glands.

In their first winter at the Center they were put in the pen at night and went outside only during the daytime. Later, even in winter, they were more active, but slept on a bedding of straw at night.

The main food for one animal consists of carrots, potatoes, bran (800 g), pellets (150 g), hay and grass (3 kg) and leaves. Like the gorals their health is good and diarrhoea has never been a problem. They have localised defaecation, and they like to rest in small groups consisting of three serows and to sit in high places or in areas where they have walls at their backs.

The breeding season is from October to November. Two females were separated from the others, but a male was exchanged for one of them after a short period. For one week before giving birth the female was kept in isolation, and in June 1985 the first male kid was born, the first birth in captivity outside Taiwan.

CHAMOIS

A female chamois (born in May 1980) and a male (born a month later), were sent in December 1980 from San Diego Zoo via Helsinki as an animal friendship-exchange. These animals are the most adept of all the serows at rock climbing. The body of the chamois is about 80 cm long and 70 cm high. They weigh about 20 kg and are dark yellow with a white spot on the face.

Two chamois with which we were familiar were active in deep snow, the male performing the courtship display every month. In November 1981 the male chased the female repeatedly and they copulated for 15-20 seconds. In May 1982 the first birth (a female, born to a 2-year-old mother) occurred, after a gestation period of 168 days. The male attacked the kid after its birth, injuring it. The female and her kid were therefore separated from the male. The young chamois recovered, and at 5 months its horns measured 2 cm.

In May 1983 a male kid was born, after a gestation period of 178 days. Because of the father's aggressiveness, we were obliged to keep him separately except during the breeding season. A further male kid was born in May 1984 after a gestation period of 180 days, and since then there have been five more births.

Chamois defaecate in a defined area. They do not have suborbital glands, and moulting occurs between April and June. As they are weak in summer they have to be carefully watched.

CONCLUSION

The Japan Serow Center now has 20 serows comprising eight subspecies. In the past, they have been seriously threatened with extinction, but they have survived. We therefore wish to do our best to ensure that they continue to breed.

Part Five:
Anatomy of *Capricornis*

20

Morphological characteristics of Japanese serow, with special reference to the interdigital glands

Makoto Sugimura, Yoshitaka Suzuki, Yasuro Atoji, Toshiko Hanawa and Koji Hanai

Department of Veterinary Anatomy, Faculty of Agriculture, Gifu University, Gifu 501-11, Japan

INTRODUCTION

The Japanese serow, a special natural monument in Japan, has been preserved since 1955. Present investigators had an opportunity to survey anatomically more than a thousand Japanese serows which were harvested in Gifu Prefecture from 1979 to 1985, with the permission of the Agency of Cultural Affairs. In this report, the anatomical features of the animals are described in comparison with those of domestic bovine ruminants, paying special attention to the morphology and functional meaning of the interdigital gland, which has been well known since the early 1900s as one of the scent glands of ruminants (Pocock 1910).

ANATOMICAL SURVEYS OF JAPANESE SEROWS

Skeletal and muscular systems

Adult Japanese serows are 30 to 45 kg in weight, and the horns are 12 to 16 cm in length. No sex dimorphism has been detected in the size of the body and the skeleton (Matsuo *et al.* 1983, Matsuo and Morishita 1985). In detailed morphometrical studies of the skeleton, however, the width between sacral wings, the transverse diameters and several other measurements of the os coxae are larger in the female than in the male (Sugano *et al.* 1982, Matsuo *et al.* 1984, Morishita *et al.* 1984). In the skull, the basifacial axis, the height of the occipital region and the length of the horncore are significantly larger in the male, and the lateral length of the premaxilla and the length of the upper and lower diastemas are larger in the female (Tsuchimoto *et al.* 1982). The muscular system appears to be similar to that of the small domestic ruminant (Figure 20.1). Further detailed examinations are needed to clarify the differences between the serow and

Figure 20.1: Superficial muscles of Japanese serow. The *m. cutaneus trunci* and *m. cutaneus omobrachialis* are removed

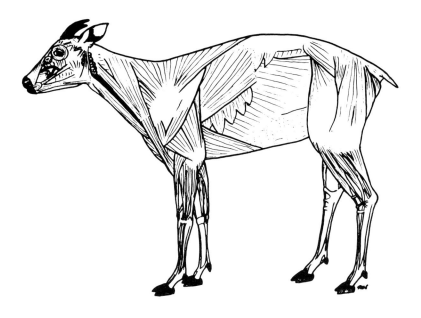

other bovine species. As Atoji *et al.* (1984) have pointed out, the *musculus scalenus dorsalis* tended to be more rudimentary than in the domestic ruminants.

Digestive and respiratory systems

Surveys of digestive organs have revealed basically the same results as those of small domestic ruminants. The morphology of the tongue was described as being similar to that of the goat on scanning electron microscopy of lingual papillae, but to have more undeveloped conical papillae (Funato *et al.* 1985). The stomach has four compartments, and the rumen is relatively small, 4–6 litres in capacity. The small intestine is 11–17 m, and the large intestine including the discoid–spiral colon is 5–7 m in length. The liver, 500–800 g in weight, shows indistinct lobation and often has a prominent papillary process. The pancreas has only a pancreatic duct confluent with a bile duct, and no accessory pancreatic duct like the goat. The presence of the well developed zygomatic salivary gland was noted in the animal (Figure 20.2), because the gland is not detected in domestic ruminants (Tsuchimoto *et al.* 1984). The bronchial ramification and lobular divisions of the lung are the same as those of the goat (Nakakuki 1986).

Figure 20.2: Deep section of the head. The left mandible is removed. A, Infraorbital gland; B, zygomatic gland; C, mandibular gland; D and E, mono- and polystomatic sublingual glands; F, parotid gland (almost removed); G, parathyroid gland; 1, major sublingual duct; 2, mandibular duct, 3, zygomatic duct (Tsuchimoto *et al.* 1982)

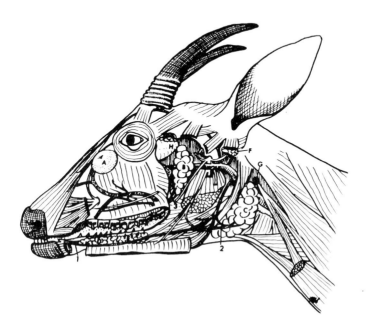

Urogenital organs

The kidneys are smooth externally, and have a common papilla and a renal pelvis (see Figure 20.8). The weight of the kidney is approximately 70 g in adult animals. The reproductive organs are similar to those of the goat, though there are slight morphological differences: for example, there is no sigmoid flexure of the penis in the serow (Figure 20.3). The female reproductive organs, ovaries, oviducts, bicornuate uterus and vagina, are basically the same as those of domestic ruminants (Figure 20.4). However, the absence of the vestibular glands and the presence of the preputial glands of the clitoris consisting of large hepatoid sebaceous glands are considered characteristic of the serow (Figure 20.5). The preputial gland seems to act as an odour gland under the control of sex hormones in females (Uno *et al.* 1984). The placenta is a cotyledonary semi-placenta with about 100 cotyledons opposite the uterine caruncles. Anatomical and histological observations of the testis and the ovary have provided abundant information on reproduction in females (Sugimura *et al.* 1981, 1983b, 1984, Ide

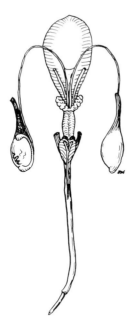

Figure 20.3: Male reproductive organs

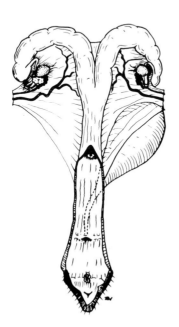

Figure 20.4: Female reproductive organs

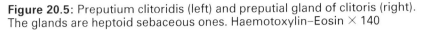

Figure 20.5: Preputium clitoridis (left) and preputial gland of clitoris (right). The glands are heptoid sebaceous ones. Haemotoxylin–Eosin × 140

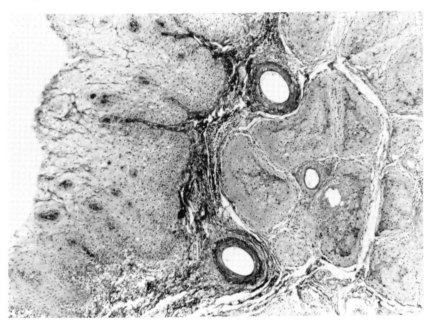

et al. 1982, Kita *et al.* 1983a, Ito *et al.* 1984, Yasuki *et al.* 1984), and in males (Tiba *et al.* 1981), as reported in separate papers.

Circulatory system and blood-forming organs

The shape of the heart and the distribution of the arteries are basically similar to those of small domestic ruminants. In the serow, however, there is a facial artery as found in the cow unlike the goat (Figure 20.6). The weight of the heart is about 300 g, and the spleen is 170 g on average. It has been noted that the relative splenic weight is about twice as great as in small domestic ruminants. The serow has also haemal nodes in addition to the usual lymph nodes. The morphology of the thymus has been reported to be similar to that of the goat (Figure 20.7; Sugimura *et al.* 1983a).

Endocrine organs

The thyroid gland has right and left lobes without a distinct isthmus, and its total weight is 2–4 g. The outer parathyroid glands are found at the dorsal

231

Figure 20.6: Superficial portion of the head. A, Infraorbital gland; B, parotid gland; C, buccal gland, 1, A. and V. faciales, 2, parotid duct

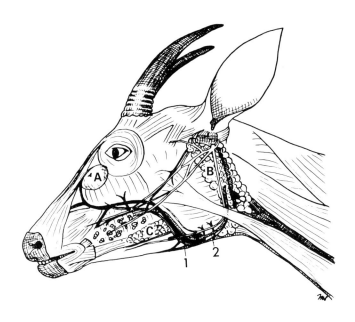

Figure 20.7: Organs of the neck and the thoracic cavity. A, thymus; B, left lung; C, thyroid gland (Sugimura *et al*. 1983a)

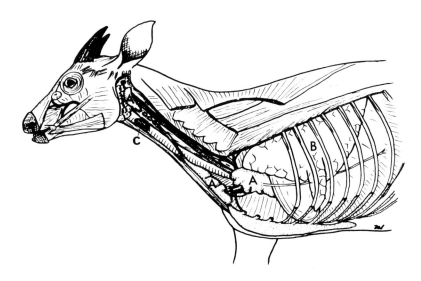

end of mandibular salivary glands, or at the dorsocranial end of the thymic cervical lobes if the animal is young, but the inner ones cannot be ascertained at present (Figure 20.2). The adrenal gland is approximately 2–5 g on either side, and is heavier in the female than in the male; they average 2.4 g and 3.3 g each in weight. The location of the glands is shown in Figure 20.8. Accessory adrenal cortical nodules are prominent in aged serows. The hypophysis and the pineal glands have not been examined.

Nervous system and other systems

The distribution of peripheral nerves seems to be basically similar to that seen in small domestic ruminants. The brachial and lumbosacral plexuses are reported in separate papers (Atoji *et al.* 1987a,b). The brain is 130 to 150 g in weight. The animal has four mammae as in the cow.

Figure 20.8: Kidneys (A) and adrenal glands (B) in ventral view

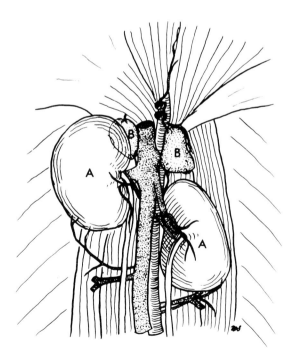

STRUCTURE AND FUNCTION OF THE INTERDIGITAL GLAND

The Japanese serow has three types of special skin glands, infraorbital and interdigital glands (Pocock 1910), in addition to the preputial glands described earlier (Uno *et al.* 1984). The infraorbital gland is considered to play an important role in scent marking in bovids (Gosling 1985). In the gland of the serow, sex dimorphism is clearly detected in the sebaceous gland, which consists of large lobules of hepatoid glands in the female unlike the usual ones in the male (Kodera *et al.* 1982). The development and lipid analysis of the gland have also been reported (Yokohata *et al.* 1985, Yokoyama *et al.* 1985). The infraorbital gland of the animal will be dealt with in Chapter 21.

It is generally accepted that the interdigital gland releases a substance on to the ground that has a characteristic scent which an isolated animal can follow back to the herd in the gregarious species of bovids (Walker 1975). Since the Japanese serow, which is a solitary species, has well developed glands on all four feet (Pocock 1910, Schaffer 1940), the structure of the gland is interesting from the viewpoint of comparative anatomy and function. In this section, the pre- and postnatal histological structures of the gland are described, with a consideration of its function.

Development of the interdigital gland

Serow fetuses, 2–34 cm in crown–rump length (CRL), were used as the material. The fetuses used are approximately 30 to 170 days of gestation (Sugimura *et al.* 1983b). The anlage of the organ was histologically found to be a bud of epidermal cells at the anterior level of the pastern joint (Figure 20.9) in an early embryo of CRL 2.5 cm; it suggests that the formation of the organ may also be earlier in the evolutionary history of the animal. The cellular bud rapidly extends between the two digits, and becomes stomach-shaped in a CRL 18 cm fetus. The cavity is formed by desquamation of superficial keratinised cells of the wall in a 33 cm fetus. The anlages of hair, sudoriferous and sebaceous glands appear in CRL 12, 16 and 18 cm fetuses, respectively. The anlages of both glands and hairs do not increase in more than CRL 19 cm fetuses, and consequently the anlages of both glands are about one-fifth the size of those of digital skin. Lamellar sensory corpuscles, which are distributed abundantly around the wall of the organ in adult serows, originate in nervous fibres of a CRL 10 cm fetus.

Figure 20.9: Anlage of interdigital gland in CRL 5 cm embryo, H–E, × 150

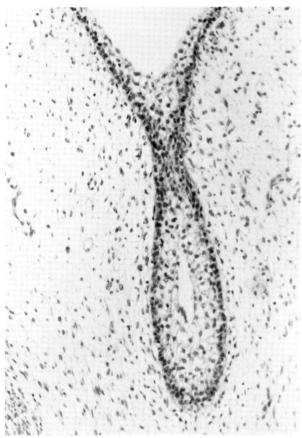

Postnatal histology of interdigital glands

Interdigital glands of 72 serows of both sexes were observed. The gland is a stomach-shaped blind sac, 2.5 to 4 cm in its longest dimension, with a thin wall of connective tissue. The sac opens through a short duct at the anterior side of the pastern joint (Figure 20.10). The duct has no sphincter muscles. No secretory substance is found in the sac, but a small amount of brown or black cerumen-like substance often adheres to the inner wall. The wall of the sac is covered with stratified squamous epithelium. Both sudoriferous and sebaceous glands of the organ are smaller than the digital skin, as detected in the prenatal study (Figures 20.11 and 20.12). The sudoriferous glands are but a half of the size of the digital skin ones, and the sebaceous

Figure 20.10: Lateral view of interdigital gland (I) in left pelvic limb. The fourth pedis is removed

Figure 20.11: Digital skin. Numerous hair follicles with sebaceous and sudoriferous gland are seen. H–E, × 20

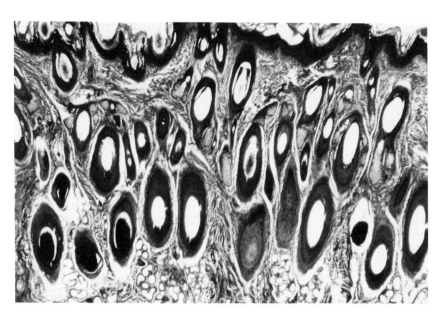

Figure 20.12: Interdigital gland. Note less developed hairs and glands. H–E, × 20

glands are less than one-tenth in comparison with the glands of the skin, as shown in Figures 20.13 and 20.14. In the Japanese serow, therefore, the organ is better called the 'interdigital pouch or sac' rather than the 'gland', based on its histological structure. This may imply that the organ of the animal has no active role as a scent gland. The sudoriferous glands may, however, release a small amount of specific secretory substance different from that of usual skin, because they are branched or have a complex tubular form unlike the simple tubular form of the digital skin.

In addition, it was noted that lamellar sensory corpuscles are widely distributed near the sac wall of the organ in the present study. The corpuscles are similar to the Vater-Pacinian corpuscles in general structure, but are long and club-like in shape, 500 μm in length and 230 μm in diameter (Figure 20.15 and 20.16). The sensory corpuscles are densely distributed near the lateral and bottom sides of the organ (Figure 20.17).

The organ of bovine gregarious species generally has well developed sebaceous and/or sudoriferous glands, such as in chamois, sheep and red duiker (Schaffer 1940, Mainoya 1978). Accordingly, it is generally accepted that the interdigital gland releases some odoriferous substances for social communication (Walker 1975), although the gland may be not employed for any specific marking activity, as stated by Gosling (1985). In the Japanese serow, however, the findings obtained suggest that the interdigital glands act to ensure surefooted movement of the hooves rather than as scent glands.

237

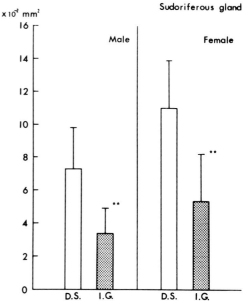

Figure 20.13: Comparison of sudoriferous glands of interdigital gland and digital skin

Area of glands per 1 mm of skin surface-line on section.

D.S.=digital skin, I.G.=interdigital gland. Bar represents SD.   ** : P < 0.01

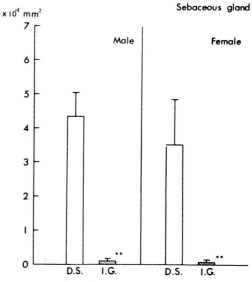

Figure 20.14: Comparison of sebaceous glands of interdigital gland and digital skin

Area of glands per 1 mm of skin surface-line on section.

D.S.=digital skin, I.G.=interdigital gland. Bar represents SD.   ** : P < 0.01

Figure 20.15: Lamellar sensory corpuscles (arrows) in subcutaneous tissue of interdigital gland. H–E × 60

Figure 20.16: Two lamellar sensory corpuscles. H–E × 160

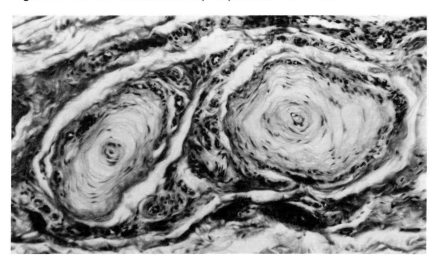

Figure 20.17: Distributions of lamellar sensory corpuscles. Small points show corpuscles

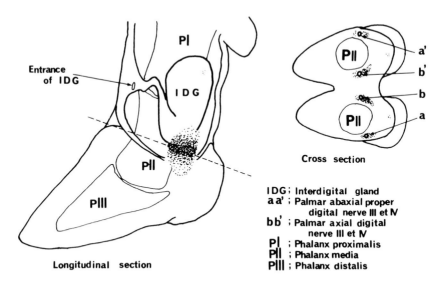

REFERENCES

Atoji, Y., Y. Suzuki, and M. Sugimura. (1984) Musculus scalenus dorsalis of the Japanese serow, *Capricornis crispus. Res. Bull. Fac. Agr. Gifu Univ. 49*, 197–200

Atoji. Y., Y. Suzuki, and M. Sugimura. (1987a) The brachial plexus of the Japanese serow (*Capricornis crispus*). *Anat. Anz. 163*, 25–32

Atoji, Y.,Y. Suzuki, and M. Sugimura. (1987b) The lumbosacral plexus of the Japanese serow, *Capricornis crispus. Anat. Anz.* (in press)

Funato, H., Y. Atoji, Y. Suzuki and M. Sugimura. (1985) Morphological studies on the tongue of wild Japanese serows, *Capricornis crispus. Res. Bull. Fac. Agr. Gifu Univ. 50*, 205–19

Gosling, L.M. (1985) The even-toed ungulates: order Artiodactyla. In R.E. Brown and D.W. Macdonald (eds) *Social odours in mammals*, 2, Clarendon Press, Oxford, pp. 550–618

Ide, Y., M. Sugimura, Y. Suzuki, and I. Kita. (1982) The morphology of female reproductive organs and their changes caused by parturition in Japanese serows. *Res. Bull. Fac. Agr. Gifu Univ. 46*, 193–203

Ito, Y., I. Kita, T. Tiba, and M. Sugimura. (1984) Histological observations on the regression of corpus luteum graviditatis in the Japanese serow (*Capricornis crispus*). *Res. Bull. Fac. Agr. Gifu Univ. 49*, 273–82

Kita, I., M. Sugimura, Y. Suzuki and T. Tiba. (1983a) Reproduction of wild Japanese serows based on the morphology of ovaries and fetuses. *Proc. Vth World Conf. Anim. Product (Tokyo)*, 2, 243–4

Kita, I., M. Sugimura, Y. Suzuki and T. Tiba. (1983b) Reproduction of female Japanese serows, *Capricornis crispus*, based on pregnancy and macroscopical ovarian findings. *Res. Bull. Fac. Agr. Gifu Univ. 48*, 113–19

Kodera, S., Y. Suzuki and M. Sugimura. (1982) Postnatal development and histology of the infraorbital glands in the Japanese serow, *Capricornis crispus. Japan. J. Vet. Sci. 44*, 839–43

Mainoya, J.R. (1978) Histological aspects of the preorbital and interdigital glands of the red duiker (*Cephalophus natolensis*). *East Afr. Wildl. J. 16*, 265–72

Matsuo, S., Y, Morishita and K. Ohshima. (1983) Studies on the skeleton of Japanese serows (*Capricornis crispus*) I. Bone of the thoracic limb. *J. Fac. Agr. Shinshu Univ. 20*, 173–92

Matsuo, S., Y. Morishita and K. Ohshima. (1984) Studies on the skeleton of Japanese serows (*Capricornis crispus*) II. Bones of pelvic limb. *J. Fac. Agr. Shinshu Univ. 21*, 59–90

Matsuo, S. and Y. Morishita. (1985) Studies on the skeleton of Japanese serows (*Capricornis crispus*) IV. Bones of the head (cranial and facial bones). *J. Fac. Agr. Shinshu Univ. 22*, 99–138

Morishita, Y., S. Matsuo, and K. Ohshima. (1984) Studies on the skeleton of Japanese serows (*Capricornis crispus*) III. Bone of the trunk (vertebral column, ribs and sternum). *J. Fac. Agr. Shinshu Univ. 21*, 120–48

Nakakuki, S. (1986) The bronchial tree and blood vessels of the Japanese serow lung. *Anat. Anz. 161*, 61–8

Pocock, R.I. (1910) On the specialized cutaneous glands of ruminants. *Proc. Zool. Soc., London* 840–986

Schaffer, J. (1940) *Die Hautdrüsenorgane der Saügetiere*, Urban & Schwarzenberg, Berlin and Vienna, pp. 248–370

Sugano, M., N. Tsuchimoto, M. Sugimura and Y. Suzuki. (1982) Morphometrical study on the skeleton of Japanese serows. I. Vertebral column and appendicular skeleton. *Res. Bull. Fac. Agr. Gifu Univ. 46*, 205–14

Sugimura, M., Y. Suzuki, S. Kamiya and T. Fujita. (1981) Reproduction and prenatal growth in the wild Japanese serow, *Capricornis crispus. Japan. J. Vet. Sci. 43*, 553–5

Sugimura, M., Y. Suzuki, Y. Atoji, M. Sugano and N. Tsuchimoto. (1983a) Morphological studies on the thymus of Japanese serows, *Capricornis crispus. Res. Bull. Fac. Agr. Gifu Univ. 48*, 113–19

Sugimura, M., Y. Suzuki, I. Kita, Y. Ide, S. Kodera and M. Yoshizawa. (1983b) Postnatal development of Japanese serows, *Capricornis crispus*, and reproduction in females. *J. Mammal. 64*, 302–4

Sugimura, M., I. Kita, Y. Suzuki, Y. Atoji and T. Tiba. (1984) Histological studies on two types of retrograde corpora lutea in the ovary of Japanese serows, *Capricornis crispus. Zool. Anz. Jena 213*, 1–11

Tiba, T., M. Sugimura and Y. Suzuki. (1981) Kinetik der Spermatogenese bei der Wollhaargemse (*Capricornis crispus*) I. Geschlechtsreife und jahreszeitliche Schwankung. II. Samenepithelzyklus und Samenepithelwelle. *Zool. Anz. Jena 207*, 16–24 and 25–34

Tsuchimoto, N., M. Sugano, M. Sugimura and Y. Suzuki. (1982) Morphometrical study on the skeleton of Japanese serows. I. Skull. *Res. Bull. Fac. Agr. Gifu Univ. 46*, 215–21

Tsuchimoto, N., M. Sugano, Y. Atoji, Y. Suzuki and M. Sugimura. (1984) Zygomatic salivary glands in Japanese serows, *Capricornis crispus. Japan. J. Vet. Sci. 46*, 593–6

Uno, K., M. Sugimura, Y. Suzuki and Y. Atoji. (1984) Morphological study on vagina, vestibule and external genitalia of Japanese serows. *Res. Bull. Fac. Agr. Gifu Univ. 49*, 183–95

241

Walker, E.P. (1975) *Mammals of the world,* vol. 2, 3rd edn, Johns Hopkins University Press, Baltimore, pp. 647–1500

Yasuki, D., I. Kita, T. Tiba and M. Sugimura. (1984) Atresia of ovarian follicles in Japanese serows (*Capricornis crispus*), especially in final structural changes of vesicular follicles. *Res. Bull. Fac. Agr. Gifu Univ. 49,* 283–90

Yokohata, Y., M. Sugimura, Y. Suzuki, T. Nakamura and Y. Atoji. (1985) Histology and lipid analysis of sebaceous portion of infraorbital gland in female Japanese serows. *Res. Bull. Fac. Agr. Gifu Univ. 50,* 185–91

Yokoyama, H., Y. Atoji, Y. Suzuki and M. Sugimura. (1985) Prenatal development of infraorbital glands in Japanese serows, *Capricornis crispus. Res. Bull. Fac. Agr. Gifu Univ. 50,* 193–203

21

Histology and lipid analysis of the infraorbital gland of Japanese serow, and functional considerations

Yasushi Yokohata*, Shuhei Kodera, Harumi Yokoyama, Makoto Sugimura, Yoshitaka Suzuki, Takao Nakamura and Yasuro Atoji
Department of Veterinary Anatomy, Faculty of Agriculture, Gifu University, Gifu 501-11, Japan

INTRODUCTION

The infraorbital gland, *Gl. sinus infraorbitalis*, of the Japanese serow is so developed and characteristic that it is thought to have an important function, such as territory marking (Kiuchi *et al.* 1979). The gross morphology of the gland was briefly reviewed by Pocock (1910), but little information is available on the histology of the gland. A preliminary analysis of the secretion of the gland has been reported (Asahi *et al.* 1979) and a histological study and lipid analysis have been performed to elucidate the function of the gland (Kodera *et al.* 1982, Yokohata *et al.* 1985, Yokoyama *et al.* 1985) and are reviewed herein.

MATERIALS AND METHODS

For histological study, infraorbital glands were obtained from 222 males and 218 females, including 15 male and 12 female fetuses smaller than 34.0 cm in crown–rump length (CRL), captured in Gifu Prefecture during December through March in 1979 to 1983. The organs were weighed and then fixed in 10 per cent formalin or Bouin solution. Tissue blocks from the middle of the organs were embedded in paraffin, sectioned at about 8 μm and stained with Haematoxylin–Eosin and periodic acid–Schiff (PAS) reaction. On the fetuses, the blocks were sectioned serially at 5 μm. For histoplanimetry, the areas of each histological structure in the sections were measured with an image analyser (Kontoron MOP-10), and the weight of each structure was estimated.

---

*Present address: Hokkaido University (Department of Parasitology, Veterinary Faculty), Sapporo 060, Japan

Materials for lipid analysis were obtained from 12 males and 53 females captured in Gifu Prefecture during December to March 1982–83. The organs were separated into the outer sudoriferous gland portion containing only the sudoriferous glands, and the inner sebaceous gland portion composed of the sebaceous glands, the residue of the sudoriferous glands and the secretion in the internal sinus. Total lipid extraction, performed in glass homogenisers at −60°C and at room temperature, was repeated three or four times with ethyl ether. These total lipids were separated into six fractions by chromatography on silicic acid columns using the solvents linked in Table 21.1. Mallincrodt's silicic acid, reagent grade, was used.

Table 21.1: Solvents used in separation of total lipids

Fraction	Solvent
Hydrocarbons	Hexane
Sterol esters	Hexane:benzene (85:15, v/v)
Triglycerides	Hexane:ethyl ether (95:5, v/v)
Sterols	Hexane:ethyl ether (80:20, v/v)
Mono- and diglycerides	Ethyl ether
Phospholipids	Methanol:chloroform: water (75:25:5, v/v)

These fractions were weighed after solvent removal under a stream of nitrogen. For comparison, the muscles of some individuals were analysed in the same way.

Gas chromatography (GC) of each fraction esterified with diazomethane (Thiessen and Regnier 1974) was used for analysis under the following conditions:

Carrier gas: helium, 50 ml/min
Column: glass column (2 m × 2 mmϕ), 5% SP-2300, ChW (AW, DMCS)
Temperature: Injector 210°C
            Detector 250°C
            Column 40–240°C (5°C/min)
Detector: Flame ionisation detector

The age of each individual was assessed by teeth eruption and wear of incisors. They can be classified into 0, 1 and over 2 years of age. Those over 2 years of age can again be classified into five grades, I, II, III, IV and V, as evidenced by the wear of incisors (Sugimura *et al.* 1981, Sugano *et al.* 1982, Miura and Yasui 1985). The size of the largest ovarian follicles, the presence or not of the corpus luteum, and the CRL of fetuses of the pregnant individuals among the 77 females were recorded.

RESULTS AND DISCUSSION

Histological structure

The infraorbital gland formed a swelling in the inferio-cranial portion of each eye (Figure 21.1). The gland weighed 4 to 24 g in adult serows. Each gland had a deep, hairy sinus in which secretion accumulated (Figure 21.2). The parenchyma of the organ was divided into two zones, the narrow whitish inner one consisting of sebaceous glands, and the wide brownish outer one consisting of the sudoriferous glands. The sudoriferous secretory cells were cuboidal to columnar in shape, depending on their activity. A layer of myoepithelial cells was located between the secretory cells and the basement membrane. Cytoplasmic blebs often protruded from the apical surfaces of secretory cells into the lumen of the tubule, and the secretion was apocrine in type. Several lipofuscin-like granules were usually observed under the nuclei. The granules in old serows were larger than in young serows. The highly branched acini of sebaceous glands were superimposed on the zone of the sudoriferous glands. Two types of sebaceous gland were noted; type I was small and the mode of secretion was holocrine (Figure 21.3); type II was hepatoid-like and stained more intensely with Eosin (Figure 21.4), similar to the so-called 'hepatoid Drüsen' (Schaffer 1940).

Figure 21.1: Infraorbital gland of an old serow (facial skin removed)

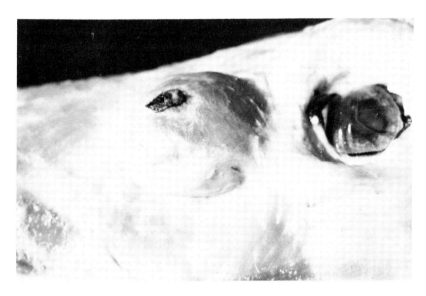

Source: Kodera *et al.* 1982

245

Figure 21.2: Tranverse sections of infraorbital glands. Samples are in order of age

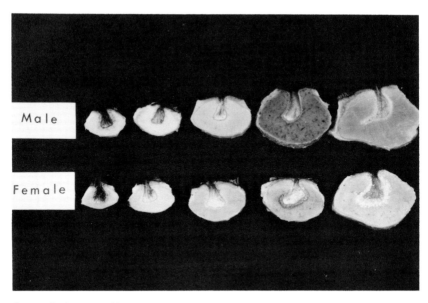

Source: Kodera *et al.* 1982

Pre- and postnatal development

The anlage of the infraorbital gland has been identified as a subcutaneous cellular projection of the integument. All anlages of hair follicles, sudoriferous and sebaceous glands appeared later (10.4–18.3 cm in CRL) in the gland than in the outer integument. The sudoriferous and sebaceous glands in the gland continued to develop faster than those of the integument during 17 to 20 cm in CRL, and the glands in female fetuses showed a tendency to develop better than in male ones. In pregnant serow, dramatic regression of the corpus luteum was observed histologically just as the CRL reached 17 cm (Ito *et al.* 1984), suggesting that the development of the gland is dependent on a sex hormone such as progesterone. Complete development of sebaceous gland type II seems to be in the later period of pregnancy or early after birth because it was not detected in a CRL 34 cm fetus.

The body weight of serows continues to increase until they are about 2 years old when they reach sexual maturity (Ito 1971, Komori 1975, Sugimura *et al.* 1981, 1983, Kita *et al.* 1983a,b), and becomes constant in serows thereafter, whereas the weight of the gland rises continuously after 2 years of age (Figure 21.5). The estimated weight of the sudoriferous

246

Figure 21.3: Sebaceous gland type I in an adult male. Sudoriferous glands are also seen on the lower right-hand side of the figure H–E stain × 100

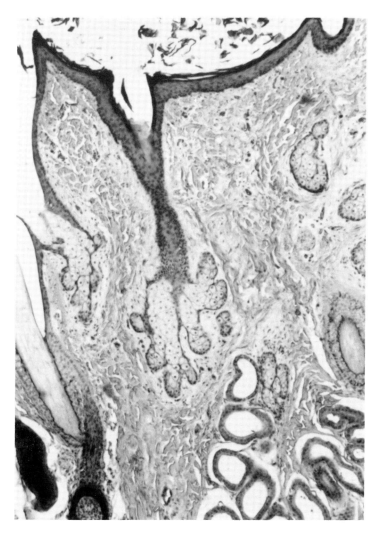

Source: Kodera *et al.* 1982

portion also increases with age, and the weight of the organ is dependent upon the weight of the sudoriferous glands. Mykytowycz (1965) showed that in Australian rabbits the chin glands were larger in dominant old males than in subordinate ones. In the serow, there also may be a correlation between the weight of the infraorbital glands and social status.

Figure 21.4: Sebaceous glands type I (arrows) and type II with one cyst in an adult female. H–E stain × 100

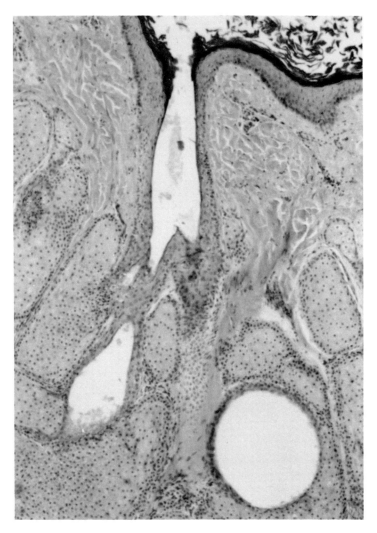

Source: Yokohata *et al.* 1985

Sexual dimorphism and cyst formation of the sebaceous gland

The infraorbital gland of adult males had a small amount of sebaceous glands type I, while the gland of males less than a year old and all females had both sebaceous glands type I and II. Especially in the female, the organ contained a large amount of sebaceous glands type II. The estimated

248

Figure 21.5: Body weight and weight of infraorbital glands in serows

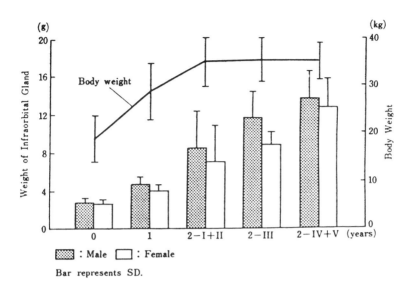

Bar represents SD.

Source: Kodera *et al.* 1982

weight of the sebaceous glands in males was constant at any age, whereas in females it increased until age class 2-III (Figure 21.6).

A lot of cysts were observed in the acini of the sebaceous glands type II (Figures 21.4 and 21.7). The histological patterns of the cysts varied, but the presence of the cysts, which were specific in sebaceous glands type II, suggested that their role was one of storage and the acceleration of secretion. Cyst formation was more remarkable in adult females with large ovarian follicles and/or corpus, as well as in females less than a year old, than in males less than a year old, 1-year-old females and adult females with only small follicles without corpus luteum (Figure 21.8). Acceleration of sebaceous-glands secretion by female sex hormone has been reported in many animals (Niimura *et al.* 1979, Vandenbergh 1983, Hayashi 1984). In Japanese serow, the preputial gland of the clitoris which is also a hepatoid-like sebaceous gland was suggested to be stimulated by oestrogen (Uno *et al.* 1984). Thus, a similar acceleration of secretion may be possible in the sebaceous glands type II. In addition, a high concentration of oestradiol was detected in the blood of a female less than 1 year old (T. Nakamura, unpublished), and concurred very well with the observed results.

Figure 21.6: Weight of sebaceous gland portion of infraorbital gland

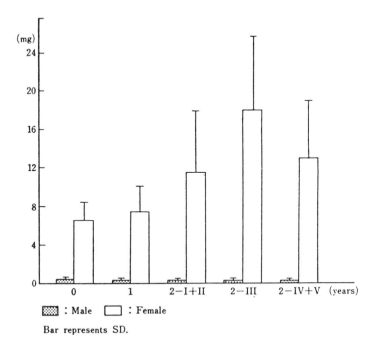

☒ : Male   ☐ : Female

Bar represents SD.

Source: Kodera *et al.* 1982

Lipid analysis

Analysis of relative lipid distribution (Figure 21.9) showed a high concentration of hydrocarbons and sterol ester fractions in the sebaceous gland portion as compared with those in the sudoriferous gland portion and the muscle. The sterol esters are related to the complex of steroids that contain known sex attractants in other animals (Kloek 1961; Sato *et al.* 1981). A specific complex of odoriferous substances has been detected in the hydrocarbon fractions of the anal gland extracts of female rabbits and chin gland extracts of male rabbits, respectively (Goodlich and Mykytowycz 1972).

A characteristic complex of peaks was detected by GC (Figure 21.10) from fractions of the extract of the infraorbital gland of serow at 20–30 min retention time. In a non-pregnant adult female with a large follicle and without corpus luteum, the complex was most remarkable in the sterol ester fraction of the sebaceous gland portion, and it was also much more obvious in all the six fractions of the sebaceous gland portion as compared with the fractions from the sudoriferous gland portion or from the muscle.

250

Figure 21.7: Two acini of the sebaceous gland (type I upwards). Acini of type II contain large cysts. Adult female H–E stain × 200

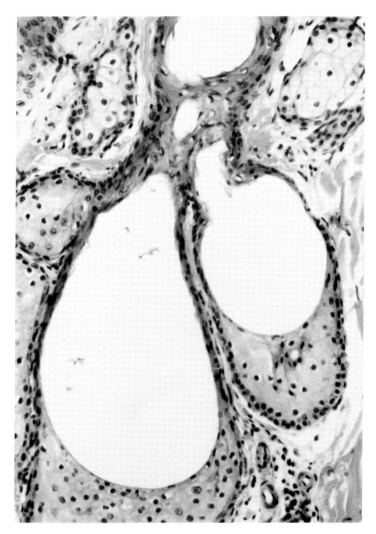

Source: Yokohata *et al.* 1985

Since the odoriferous substances of many mammals are composed of a number of compounds (Müller-Schwarze 1969, Brownlee and Silverstein 1969, Goodlich and Mykytowycz 1972, Stoddart 1973, Müller-Schwarze *et al.* 1974, Thiessen and Regnier 1974, Sasada *et al.* 1982), the complex isolated from the infraorbital gland of the serow may be no exception. The serows are solitary, and their mating season is limited to 2–3 months (Ito

Figure 21.8: Area of cysts per 1 mm^2 of area of acini of sebaceous gland type ll in infraorbital gland according to sex, age, size of follicle, presence of corpus luteum and size of embryos. 0, none of age group; 1, one of age group; S.F., adult group having small follicles (less than 4.8 mm in diameter) without corpus luteum; L.F., adult group having large follicles (more than 4.8 mm in diameter) without corpus luteum; C.L., adult group with corpus luteum without embryo; S.E., adult group with small embryo (less than 17 cm in CRL); L.E., adult group with large embryo (more than 17 cm in CRL). Bar represents SD

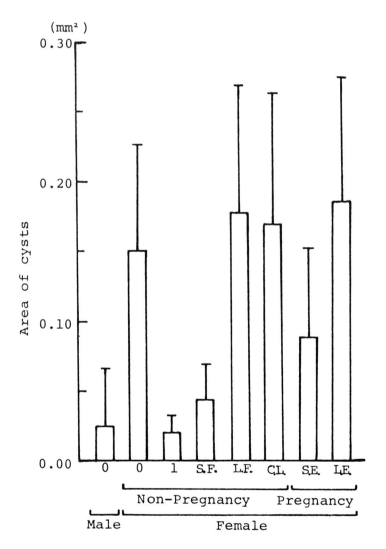

Source: Yokohata *et al.* 1985

252

Figure 21.9: Relative lipid distribution in infraorbital gland and muscle

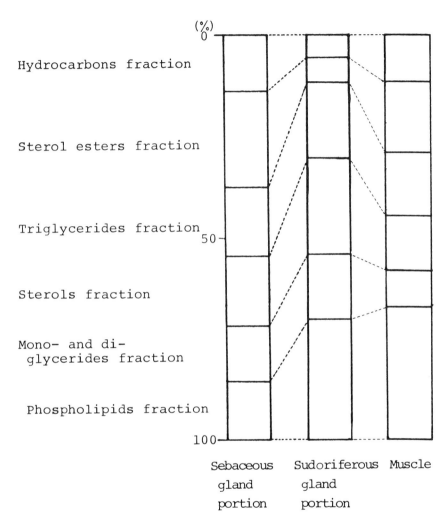

Source: Yokohata *et al.* 1985 and unpublished data

1971, Komoro 1975, Sugimura *et al.* 1981, 1983, Kita *et al.* 1983a, b). In captivity, a female in oestrus was observed to increase its frequency of scent marking with the infraorbital gland (Kanomata and Izawa 1982). From these features, it may be concluded that the secretion of sebaceous glands type II may include a kind of sex attractant that can inform the male of the female's own sexual condition. However, the complex becomes more

Figure 21.10: Gas chromatograms of lipid fractions of infraorbital gland. (a) Sebaceous gland portion (obtained from an adult, non-pregnant female with large follicle and without corpus luteum). (b) Sudoriferous gland portion (obtained from an adult, non-pregnant female with large follicle and without corpus luteum). (c) Sebaceous gland portion (obtained from 1-year-old female). (d) Sebaceous gland portion obtained from pregnant female with large embryo)

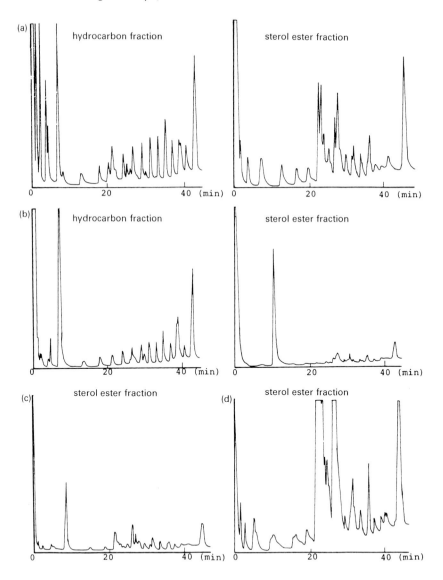

Source: Yokohata *et al.* 1985

254

remarkable in pregnant serows. Presumably, pregnant individuals send different information with other patterns of the complex compounds, depending on sex hormones such as progesterone, though we have no data at present to support this hypothesis.

As such, our results suggest that the infraorbital gland may have multiple functions which include not only the so-called territory marking but also sex attraction and other social behaviours.

REFERENCES

Asahi, M., K. Fujitani, M. Kozuka and T. Inoue. (1979) An attempt on the survey of bioactive principal of infraorbital gland of Japanese serow. *Rep. Natl. Conserv. Soc. Japan 56*, 117–26 (in Japanese)

Brownlee, R.G. and R.M. Silverstein. (1969) Isolation, identification and function of the chief component of the male tarsal scent in black-tailed deer. *Nature (London) 221*, 284–5

Goodlich, B.S. and R. Mykytowycz. (1972) Individual and sexes differences in the chemical composition of the rabbit, *Oryctolagus cuniculus. J. Mammal. 53 (3)*, 540–8

Hayashi, S. (1984) Sex attractants of mammals. *Taisha, 21*, 145–50 (in Japanese)

Ito, T. (1971) On the oestrous cycle and gestation period of the Japanese serow, *Capricornis crispus. J. Mammal. Soc. Japan, 5*, 104–8 (in Japanese)

Ito, Y., I. Kita, T. Tiba and M. Sugimura. (1984) Histological observations on the regression of corpus luteum graviditatis in the Japanese serow (*Capricornis crispus*). *Res. Bull. Fac. Agr. Gifu Univ. 49*, 273–82 (in Japanese with English summary)

Kanomata, K. and M. Izawa. (1982) Some observations on estrus in female Japanese serows *Capricornis crispus* in captivity. *J. Japan. Assoc. Zool. Gard. Aq. 7*, 53–61 (in Japanese)

Kita, I., M. Sugimura, Y. Suzuki and T. Tiba. (1983a) Reproduction of wild Japanese serows based on the morphology of ovaries and fetuses. *Proc. Vth World Conf. Anim. Product.*, Tokyo 2, 243–4

Kita, I., M. Sugimura, Y. Suzuki and T. Tiba. (1983b) Reproduction of female Japanese serows, *Capricornis crispus*, based on pregnancy and macroscopical ovarian findings. *Rep. Bull. Fac. Agr. Gifu Univ. 48*, 137–46 (in Japanese with English summary)

Kiuchi, M., F. Kudo, M. Yoshida, M. Miyasaka, M. Hoshino, K. Yamazaki, S. Kato and C. Umezu. (1979) Social organization and habitat use of the Japanese serow in Asahi mountain ranges. *Rep. Natl. Conserv. Soc. Japan 56*, 19–72 (in Japanese)

Kloek, J. (1961) The smell of some steroid sex hormones and their metabolites. Reflections and experiments concerning the significance of smell for the mutual relation of the sexes. *Psychiat. Neurol. Neurochir. 64*, 309–44

Kodera, S., Y. Suzuki and M. Sugimura. (1982) Postnatal development and histology of the infraorbital glands in the Japanese serow, *Capricornis crispus. Japan J. Vet. Sci. 44*, 839–43

Komori, A. (1975) Survey on the breeding of Japanese serows, *Capricornis crispus*, in captivity. *J. Japan Assoc. Zool. Gard. Aq. 7*, 53–61 (in Japanese)

Miura, S. and K. Yasui. (1985) Validity of tooth eruption-wear patterns as age criteria in Japanese serow, *Capricornis crispus. J. Mammal. Soc. Japan 10*, 169

Müller-Schwarze, D. (1969) Complexity and relative specificity in a mammalian pheromone. *Nature (London) 223*, 525–6

Müller-Schwarze, D., C. Müller-Schwarze, A.G. Singer and R.M. Silverstein. (1974) Mammalian pheromone: identification of active component in the subauricular scent of the male pronghorn. *Science 183*, 860–2

Mykytowycz, R. (1965) Further observations on the territorial function and histology of the submandibular cutaneous (chin) gland in the rabbit, *Oryctolagus cuniculus* (L.). *Anim. Behav. 13*, 400–12

Niimura, S., K. Ishida and M. Yamaguchi. (1979) Effects of gonadoectomy and injections of steroid hormones on the abdominal sebaceous glands of Mongolian gerbils. *Res. Bull. Fac. Agr. Niigata Univ. 31*, 119–23

Pocock, R.I. (1910) On the specialized cutaneous glands of ruminants. *Proc. Zool. Soc. London* 840–986

Sasada, H., T. Sugiyama, K. Yamashita and J. Masaki. (1982) Identification of specific odor components in mature male goat during the breeding season. *Japan. J. Zootech. Sci. 54* (6), 401–8

Sato, H., N. Manabe, S. Watanabe and T. Ishibashi. (1981) Histological pattern and histochemistry of the anal gland in the Japanese weasel, *Mustela itatsi. Zool. Mag. (Japan) 90*, 15–20 (in Japanese with English abstract)

Schaffer, J. (1940) *Die Hautdrüsenorgane der Saugetiere.* Urban & Schwarzenberg, Berlin and Vienna, pp. 248–80

Stoddart, D.M. (1973) Preliminary characterization of the caudal organ secretion of *Apodemus flavicolis. Nature (London) 246*, 501–3

Sugano, M., N. Tsuchimoto, M. Sugimura and Y. Suzuki. (1982) Morphometrical study on the skeleton of Japanese serow. I. Vertebral column and appendicular skeleton. *Res. Bull. Fac. Agr. Gifu Univ. 46*, 205–14 (in Japanese with English summary)

Sugimura, M., Y. Suzuki, S. Kamiya and T. Fujita. (1981) Reproduction and prenatal growth in the wild Japanese serow, *Capricornis crispus. Japan. J. Vet. Sci. 43*, 553–5

Sugimura, M., Y. Suzuki, I. Kita, Y. Ide, S. Kodera and M. Yoshizawa. (1983) Prenatal development of Japanese serows, *Capricornis crispus,* and reproduction in females. *J. Mammal. 64* (2), 302–4

Thiessen, D.D. and F.E. Regnier. (1974) Identification of a ventral scent-marking pheromone in the male Mongolian gerbil *Meriones unguiculatus. Science 184*, 83–4

Uno, K., M. Sugimura, Y. Suzuki and Y. Atoji. (1984) Morphological study on vagina, vestibule and external genitalia of Japanese serows. *Res. Bull. Fac. Agr. Gifu Univ. 49*, 183–95 (in Japanese with English summary)

Vandenbergh, J.G. (1983) Chemical signals and reproductive behavior. In *Pheromones and reproduction in mammals.* Academic Press, London, pp. 3–37

Yokohata, Y., M. Sugimura, Y. Suzuki, T. Nakamura and Y. Atoji. (1985) Histology and lipid analysis of sebaceous gland portion of infraorbital gland in female Japanese serows. *Res. Bull. Fac. Agr. Gifu Univ. 50*, 185–91 (in Japanese with English summary)

Yokoyama, H., Y. Atoji, Y. Suzuki and M. Sugimura. (1985) Prenatal development of infraorbital glands in Japanese serows, *Capricornis crispus. Res. Bull. Fac. Agr. Gifu Univ. 50*, 193–201 (in Japanese with English summary)

22

Brachial and lumbosacral plexuses and brains of the Japanese serow

Yasuro Atoji, Yoshitaka Suzuki and Makoto Sugimura
Department of Veterinary Anatomy, Faculty of Agriculture, Gifu University, Gifu
501-11, Japan

INTRODUCTION

This study was carried out to elucidate the morphology of the brachial and lumbosacral plexuses and the brain of the Japanese serow, a wild ruminant, and to compare the findings with those obtained from domestic and wild ruminants.

MATERIALS AND METHODS

Sixty-six Japanese serows, *Capricornis crispus*, were used. They were killed from January to February 1984, and from December 1984 to March 1985 in Japan. For the observation of plexuses, the specimens (16 males and ten females for the brachial plexus; 15 males and 15 females for the lumbo-sacral plexus) were dissected in the fresh state without fixation. Brains (five males and five females) were fixed with 10 per cent formalin. Six goat brains were used for comparison.

RESULTS

Brachial plexus

The brachial plexus was formed by the ventral rami of the sixth (C_6), seventh (C_7) and eighth (C_8) cervical nerves and the first (T_1) thoracic nerve (98.1 per cent) (Figure 22.1). These rami passed in between the ventral and middle scalenic muscles. In one case, T_1 received a small branch of the second (T_2) thoracic nerve (1.9 per cent). C_8 was the largest, and C_7 and T_1 were nearly equal in size. C_6 was thin, but the longest. In the case where T_2 was found, it was considerably smaller than C_6. C_6 and C_7 ran caudolaterally, C_8 coursed laterally, and T_1 travelled a short distance craniolaterally in the axillary space.

Figure 22.1: Brachial plexus on left side consists of C_6 to T_1. A, A. axillaris; M, N. medianus; R, N. radialis; S, N. suprascapularis

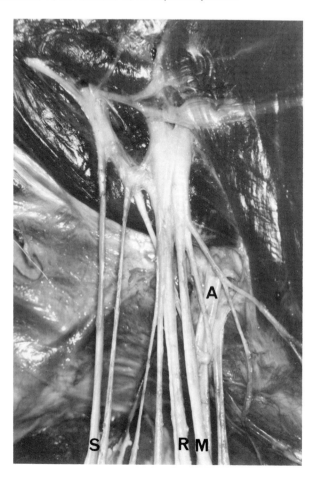

C_6 and C_7 united to form the cranial trunk, while C_8 joined with $T_1(+T_2)$ to make up the caudal trunk. The caudal trunk was very much thicker than the cranial one, but its length was quite short. Cranial and caudal trunks ran laterally and split into dorsal and ventral divisions, respectively (Figure 22.2). According to the manner of branching of the divisions, they were classified into one type of dorsal division and four types of ventral division.

Pattern of the dorsal divisions

The cranial dorsal division gave off a thick suprascapular nerve from the front margin of its base; next, sent one or two thin subscapular nerves from the middle region; and thereafter formed the dorsal cord with a thick

Figure 22.2: Schematic diagram of a typical brachial plexus. 1, N. suprascapularis; 2, Nn. subscapulares; 3, N. thoracodorsalis; 4, N. axillaris; 5, N. radialis; 6, N. musculocutaneus; 7. N. medianus;8, N. ulnaris; 9, N. pectorales craniales; 10, N. pectorales caudales; 11, N. thoracolateralis; 12, Branch to thoracic trunk; 3, N. thoracoventralis; 14, N.phrenicus; 15, N thoracicus longus; 16, N. dorsalis scapulae

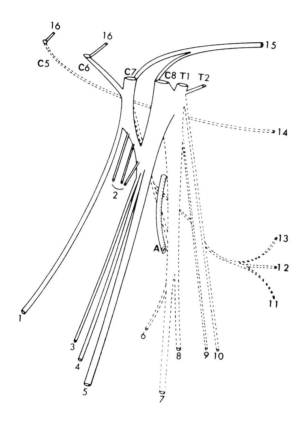

dorsal division arising from the caudal trunk. The short dorsal cord gave off the thoracodorsal nerve from the lateral side and also sometimes a thin subscapular nerve. The axillary nerve arose from the great part of the cord started as the radial nerve, which was the largest among the branches emerging from the brachial plexus.

Pattern of the ventral divisions

The ventral divisions were subject to variation. There were four types, as follows (Figure 22.3).

Type I. The thin cranial and thick caudal ventral divisions united to form the ventral cord. As a result, a loop (the ansa axillaris) was formed between

Figure 22.3: Variation in union of the diversions and divergence of the nerves. Explanation of types I–IV is given in the text. M, N. medianus; MC1, Ramus muscularis proximalis; MC2, Ramus muscularis distalis, U, N. ulnaris

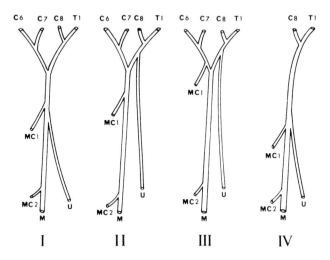

these ventral divisions. The cord gave off a proximal muscular ramus of the musculocutaneous nerve, and the median and ulnar nerves. A distal muscular ramus emerged from the distal half of the median nerve. This type was found in 35 of 52 cases (67.3 per cent).

Type II.   The caudal ventral division from C_8 and T_1 bifurcated to form the ulnar nerve and a branch to the median nerve. The cranial ventral division joined with the branch to the median nerve and formed the median nerve. The proximal and distal muscular rami arose from the median nerve. This type was encountered in ten cases (19.2 per cent).

Type III.   The cranial ventral division bifurcated to form the proximal muscular ramus and a branch to the median nerve. The caudal division forked into the ulnar nerve and a branch to the median nerve as in Type II. Thereafter two branches united to form the long and thick median nerve. The distal muscular ramus appeared from the distal half of the median nerve. This type occurred in six cases (11.6 per cent).

Type IV.   Evidence was that the cranial ventral division from C_6 to C_7 did not appear. The caudal ventral division itself formed the ventral cord which gave off the proximal muscular ramus, and the median and ulnar nerves. Since the cranial ventral division did not join the median nerve, the ansa axillaris was not formed. This type appeared in one case (1.9 per cent).

Types found in individual animals were basically the same on both sides. However, Type IV was seen on the right side while the opposite side showed Type I.

Lumbosacral plexus

The number of lumbar vertebrae was not constant. Twenty-eight of 30 had six lumbar vertebrae and the remaining had five. The spinal nerve roots in the plexus were composed of the third (L_3) lumbar to fifth (S_5) sacral nerves. Almost all of L_1 and L_2 travelled independently and laterally along the abdominal wall, but they rarely connected with each other (1.7 per cent) and did not join the plexus. The segmental origins of each nerve in the lumbosacral plexus clearly varied in individuals. The origin of each nerve is shown in Figure 22.4.

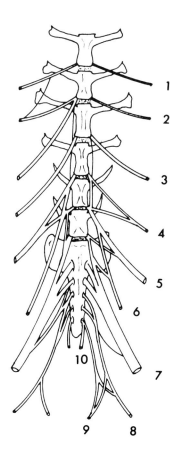

Figure 22.4: Lumbosacral plexus in six-lumbar-vertebrae type (ventral view).
1, N. iliohypogastricus;
2, N. ilioinguinalis;
3, N. genitofemoralis;
4, N. cutaneus femoris lateralis;
5, N. femoralis;
6, N. obturatorius;
7, N. ischiadicus; 8, N. dorsalis penis; (N. dorsalis clitoridis);
9, Distal branch of N. pedundus;
10, Nn. rectales caudales

Examples showing a conspicuous difference in segmental origins of the femoral and ischiatic nerves among specimens of the six-lumbar-vertebrae type are designated in Figure 22.5 as prefixed, median fixed and postfixed plexuses. In the prefixed plexus, the femoral and ischiatic nerves arose from more cranial lumbar nerves than did the nerves in the other cases. The femoral nerve was composed of large L_4 and small branches of L_3 and L_5. The ischiatic nerve consisted mainly of L_6 and S_1, and received small fibres from L_5 and S_2. In the postfixed plexus, two nerves originated from more caudal spinal segments than the nerves in the other cases. The femoral nerve sprang chiefly from L_5. The ischiatic nerve received the great majority of S_1 and S_2, and small contributions from L_6 and S_3. In the median fixed plexus, two nerves had an intermediate position. Although respective plexuses varied, they could be roughly classified into three fixed

Figure 22.5: Left views of three lumbosacral plexuses in six-lumbar-vertebrae type; prefixed (top), median fixed (middle and postfixed (bottom) plexuses. Note segmental difference of the femoral and ischiatic nerves. 1, N. iliohypogastricus; 2, N. ilioinguinalis; 3, N. genitofemoris; 4, N. cutaneus femoris lateralis; 5, N. femoralis; 6, N. obturatorius; 7, N. gluteus cranialis; 8, N. gluteus caudalis; 9, N. ischiadicus; 10 N. dorsalis penis or N. dorsalis clitoridis; 11, distal branch of N. pudendus; 12, Nn. rectales caudalés

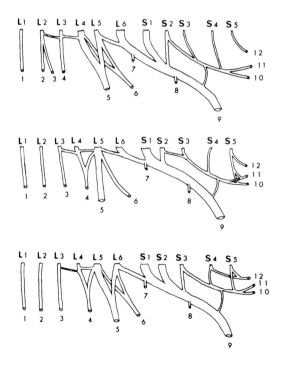

types regardless of small differences in individuals. The frequency of pre-fixed, median fixed and postfixed plexuses was 14:36:6 in 56 cases. This was calculated to have an approximate ratio of 2:6:1. Two different origins of the femoral nerve in the five-lumbar-vertebrae types are shown in Figure 22.6. Half of the specimens examined revealed a symmetry of both sides; the remaining showed only a small difference.

Figure 22.6: Left views of two lumbosacral plexuses in five-lumbar-vertebrae type. Main spinal root of the femoral nerve emerges from L_4 (A) and L_5 (B). Abbreviations: see Figure 22.5

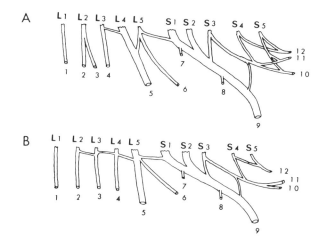

Brain

Viewed dorsally the cerebrum was pear-shaped; the cerebellum, which was hardly covered by the cerebrum, was diamond-shaped (Figure 22.7). On the ventral surface, the piriform lobe and pons were well developed (Figure 22.8). The trapezoid body and olive were less developed.

The average values of measurements were as follows; 141 g in weight (whole brain), 134 ml in volume (whole brain), 7.7 cm in length (cerebrum), 6.3 cm in width (cerebrum) and 4.6 cm in height (cerebrum).

The sulci of the cerebrum were fewer than in the goat (Figures 22.7 and 22.9). The lateral rhinal sulcus extended rostrocaudally on the lateral surface of the cerebrum at a height of one-fourth. The sylvian fissure was found perpendicularly at the middle of the cerebrum. The cruciate sulcus was not easily seen from the dorsal view. It ran caudally and continued to the splenial sulcus on the medial surface. The splenial sulcus was not parallel to the corpus callosum, but ran obliquely towards the dorsal surface and appeared at the caudal surface. The cingulate gyrus occupied a large

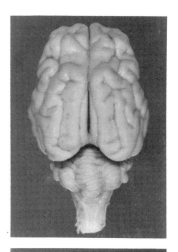

Figure 22.7: Brain (dorsal view)

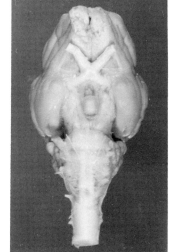

Figure 22.8: Brain (ventral view)

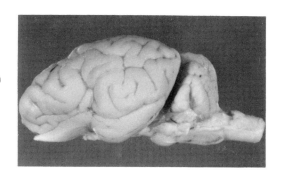

Figure 22.9: Brain (lateral view)

portion of the medial surface. Its area was approximately one-and-a-half times as large as that of the goat. The suprasylvian sulcus showed a distinct groove, ran on the lateral surface of the cerebrum above the sylvian gyrus, and communicated with the coronal sulcus. In most specimens, the marginal gyrus extended rostrocaudally on the entire dorsal surface of the hemisphere. In the caudal one-third of the cerebrum, the marginal gyrus extended laterally from the longitudinal fissure. The caudal region of the cingulate gyrus was exposed on the dorsal surface (Figures 22.7 and 22.10). The patterns of the sulci and gyri of the cerebrum were variable in individuals.

Figure 22.10: Brain (median view)

DISCUSSION

Brachial plexus

In the ox the brachial plexus is derived from C_6-T_1 ($+T_2$: 64 per cent) (Reimers 1925) or C_6-T_2 (Magilton *et al.* 1968). Hara (1966) dissected an adult ox and found C_5 joining the brachial plexus via the phrenic nerve on one side. He also reported the frequency (62.5 per cent) of the union of T_2 into T_1. In goat and sheep it was formed by C_6-T_1 (Reimers 1925, Magilton *et al.* 1968). In wild ruminants there are two reports dealing with different animals. Paterson (1896) observed C_6-T_1 in the gnu. Wakuri *et al.* (1970) reported that the giraffe's plexus was made up of C_6-T_1. These findings indicate that C_6-T_1 always constitutes the brachial plexus in ruminants and, in addition, T_2 frequently joins it.

In the Japanese serow the brachial plexus consists of C_6-T_1 ($+T_2$: 1.9 per cent). This is largely in agreement with other ruminants. There is, however, quite a difference compared with the ox regarding the frequency of T_2 occurrence in the brachial plexus. The Japanese serow showed a very low value (1.9 per cent), whereas that of the ox is very high (at least >62.5 per cent) (Reimers 1925, Hara 1966). This indicates that the Japanese serow probably manifests a high degree of conformity in the composition of the brachial plexus, while the occurrence of T_2 possibly shows a very rare variation. Thus, it is likely that the roots of the brachial plexus of the Japanese serow are similar to those of the goat or sheep among domestic ruminants.

Based on the divergence of the musculocutaneous, median and ulnar nerves, the ventral divisions of the serow brachial plexus have been classified into four types. Type I (67.3 per cent) was frequent in the present study. Type I appears to be standard in the Japanese serow. Type IV was found in one case and probably represents a rare variation. Applying the present classification to other results obtained from ruminants, Type I is found in the ox, goat, sheep (Reimers 1925) and giraffe (Wakuri *et al.* 1970). Type II is consistent with results from the ox (Hara 1966). In the report given by Magilton *et al.* (1968), the ox and sheep are Type I. In the goat 90 per cent of cases have been shown to be Type I. In the remainder the musculocutaneous nerve is separated from the median nerve and does not join it. The latter has not been observed in the present study. No report in the literature seems to belong to Type III or IV.

On the other hand, concerning the dorsal divisions, the Japanese serow only exhibited one type. This result is consistent with the giraffe (Wakuri *et al.* 1970). In observations of the ox, goat and sheep performed by Reimers (1925), it was evident that the dorsal division of C_7 bifurcated into cranial and caudal branches. The cranial branch united with the dorsal division of C_6 to form the axillary nerve. The radial nerve was made up of the caudal branch and dorsal divisions from C_8 to T_1. Magilton *et al.* (1968) also reported similar findings in the goat and sheep, and they stressed that C_6 in the ox did not join the axillary nerve. These findings are quite different from those of other reports including the present one on the Japanese serow.

Lumbosacral plexus

The lumbosacral plexus consists of the ventral branches of the last four lumbar nerves and five sacral nerves in domestic ruminants (Schreiber 1956, Linzell 1959, Ghoshal and Getty 1970, 1971). The iliohypogastric and ilioinguinal nerves arising from L_1 and L_2 independently run parallel. In this study there was a connection of L_1 and L_2 in two individual cases (Figure 22.4). However, these nerves do not join the plexus.

It is difficult to point out a standard type of lumbosacral plexus unlike the situation that applies to the brachial plexus, since the plexus is exceedingly variable in individuals. Restricted to the femoral and ischiatic nerves, the median fixed plexus may be regarded as a standard type of the Japanese serow. This is in good agreement with a variation in the goat described by Linzell (1959), but none of these three types was found in the sheep, goat and cattle reported by other investigators (Schreiber 1956, Ghoshal and Getty 1970, 1971). In these cases, however, there was little difference in the extent of the segmental origins of each nerve between the Japanese serow and domestic ruminants. This indicates that the lumbo-sacral plexus of the Japanese serow appears to be similar to that of domestic ruminants.

A slender nerve anastomosis between the ischiatic and pudendal nerves illustrated in Figures 22.4 to 22.6 passes through the lesser ischiatic foramen in all specimens. This anastomosis corresponds to the caudal cutaneous femoral nerve in domestic ruminants as reported by other investigators (Schreiber 1956, Larson and Kitchell 1958, Habel 1966, Ghoshal and Getty 1970, 1971). They state that the caudal cutaneous femoral nerve sends out a branch from the ischiatic nerve dorsolateral to the broad sacrotuberal ligament, travels through the lesser ischiatic foramen and then soon joins the pudendal nerve. In the cow it further divides into medial and lateral branches (Habel 1966). In the present report it is described only as an anastomosis instead of the caudal cutaneous femoral nerve, because the form is a loop or an ansa, and after joining it is impossible to detect its course and destination with the naked eye.

Sugano *et al.* (1982) reported that the number of lumbar vertebrae in each of 73 Japanese serows was six. In this study it is of interest that there were two specimens out of 30 with five lumbar vertebrae; all of the remaining had six. In the five-lumbar-vertebrae type, each nerve derived from the plexus was not significantly different from that of the six-lumbar-vertebrae type, since the segmental origins of each nerve in the five lumbar vertebrae were variations of the six-lumbar-vertebrae type. It is, however, obscure as to which lumbar segment is missing. Ogata (1984) investigated clinically the relation between the decrease in the number of lumbar verte-brae and the functional abnormality of the hind limb in the cat. He found that 37 of 112 animals had fewer than the normal number and concluded that the variability of lumbar vertebrae was not directly related to dysfunc-tion of the hind limb.

Brain

There are two main differences between the Japanese serow and the goat.

The sulci of the Japanese serow are fewer than in the goat, so each gyrus is broad and the gyrus pattern is simple. For example, whereas the ansate sulcus which divides the postcruciate and marginal sulci is evident in the goat, its groove in the Japanese serow is so shallow and short that the marginal sulcus appears to run continuously to the rostral edge of the frontal lobe. Furthermore, the suprasylvian sulcus is continued to the coronal sulcus without an interruption of the gyrus.

The cingulate gyrus in general underlies the marginal gyrus and can only be seen from the median view. In the Japanese serow the caudal part of the cingulate gyrus is not covered with the marginal gyrus and is observed from a dorsal view. The area of the cingulate gyrus is larger than in the goat.

In conclusion, the nerve plexuses of the Japanese serow are similar to those of ruminants, but the brain surface is less complex.

REFERENCES

Ghoshal, N.G. and R. Getty. (1970) The lumbosacral plexus (plexus lumbosacralis) of the goat (*Capra hircus*). *Iowa State J. Sci. 45*, 269

Ghoshal, N.G. and R. Getty. (1971) The lumbosacral plexus (plexus lumbosacralis) of the sheep (*Ovis aries*). *New Zealand Vet. J. 19*, 85

Habel, R.E. (1966) The topographic anatomy of the muscle, nerves and arteries of the bovine female perineum. *Am. J. Anat. 119*, 79

Hara, A. (1966) The comparative anatomical study on the plexus brachialis. *Igakuno Kenkyu (Acta Medica) 36*, 59

Larson, L.L. and R.L. Kitchell. (1958) Neuronal mechanism in sexual behavior. II Gross neuroanatomical and correlative neurophysiological studies of the external genitalia of the bull and the ram. *Am. J. Vet. Res. 19*, 853

Linzell, J.L. (1959) The innervation of the mammary glands in the sheep and goat with some observations on the lumbosacral autonomic nerves. *Quart. J. Exp. Physiol. 44*, 160

Magilton, J.H., R. Getty and N.G. Ghoshal. (1968) A comparative morphological study of the brachial plexus of domestic animals (goat, sheep, ox, pig and horse). *Iowa State J. Sci. 42*, 245

Ogata, M. (1984) Relation between vertebral variation and functional abnormality in the cat. *J. Jap. Small Anim. Vet. Assoc. Suppl.* 9 (in Japanese)

Reimers, H. (1925) Der plexus brachialis der Haussaugetiere. Ein vergleichend-anatomische Studie. *Z. Anat. Entw-geschicht. 76*, 653

Schreiber, J. (1956) Die anatomischen Grundlagen der Leitungsanasthesie beim Rind. *Wiener Tierartl. Mschr. 43*, 673

Sugano, M., M. Tsuchimoto, M. Sugimura and Y. Suzuki. (1982) Morphometrical study on the skeleton of Japanese serows. I. Vertebral column and appendicular skeleton. *Res. Bull. Fac. Agr. Gifu Univ. 46*, 205

Wakuri, H., Y. Kano and H. Hori. (1970) Anatomical study on the brachial plexus of a giraffe (*Giraffa australis*). *J. Vet. Med. 521*, 627

23

What can serow horns tell us?

Shingo Miura

Department of Biology, Hyogo College of Medicine, Mukogawa 1-1, Nishinomiya 663, Japan

INTRODUCTION

Japanese serow (*Capricornis crispus*), as well as other caprids, have true horns that are never shed and grow continuously throughout their lifespan. As with the horns of goral and mountain goat, serow horns are short, bent slightly backwards, sharply pointed, blackish, and conical. Horns have been used not only in taxonomic research, but also in ecological and demographic studies. Horn growth is a complex set of metabolic events which are known to be influenced by age, sex, nutritional status, environmental conditions, population quality, and genetic factors. Thus, the development of the horn may reflect these effects in its shape, size and growth rate. In addition, intermittent growth of the horn during winter causes the formation of distinct annual rings. Horn annulation makes it possible to trace past metabolic changes of individuals. Age, sex, habitat condition and reproductive history in females can be derived from horn measurements. Detailed results of the first and last items have been published elsewhere (Miura 1985 and in preparation).

MATERIALS AND METHODS

The samples of this study consisted of three groups collected from 1977 to 1986. For the first group, the horns from over 3500 serow were collected during November to March in Gifu and Nagano prefectures in the central part of Honshu, Japan. For the second sample, the horns from 53 serows that had died natural deaths during different seasons were collected from a wide range of localities in Honshu and Shikoku. Finally, the horns from 25 known-age serow (five females with known reproductive records) were used to verify the reliability of the estimates.

Horns were weighed and their maximum length and basal diameter were measured. The first horn growth increment was measured from the most distal tip to the first annual ring which formed during the second winter of life (at about $1\frac{1}{2}$ years old). The rest of the increments were measured between successive rings to the nearest 0.5 mm along the median line of the back side of the horn using vernier callipers.

RESULTS AND DISCUSSION

Age determination

Figure 23.1 shows a median sagittal section of a horn which was made by means of a strip of contact print. Consistent and distinct growth layers form in the keratinised zone. These layers are upside down 'Vs', which indicates that horns grow as a series of superimposed cones, and that as new keratin is added the horn is pushed upwards. These thin, fairly regular layers indicate discontinuous keratin deposits. By examining known-age samples and samples collected in different seasons, it had been confirmed that the first layer is formed during the first winter of life and one is formed each winter thereafter. However, the first layer is rather indistinct, probably because of the rapid growth that takes place in the early stages of the animal's life.

Furthermore, it is notable that, with the exception of the first layer, these growth layers are related to the rings that form deep grooves circling the horn (Figure 23.2). Thus, the ages of serow are easily determined by the distinct annual rings without sectioning the horn or tooth (Miura 1985). This result is consistent with findings for mountain goats, Himalayan tahr, and chamois (Brandborg 1955, Caughley 1965, Knaus and Schröder 1983, Bargagli and Lovari 1984). The horns of lambs were easily distinguished from those of yearlings and older serow by their length (less than 80 mm). The annual rings were distinct for almost all horns. Even when the front of the horns had been worn down by intensive rubbing, the rings could still be identified on the back of the horn. Ages can be also determined by the eruption and replacement patterns of the teeth up to about 30 months (Miura and Yasui 1985).

Sex differentiation

The horns of Japanese serow show little sexual dimorphism in length or weight (Miura 1986), but some minor differences can be detected in shape, ring formation, and growth pattern. However, each character in isolation should be regarded only as a very rough and fallible guide to sex identification. A combination of these characters is desirable for ascertaining sex.

Shape

Male horns are thick and bulky, whereas female horns tend to be more

Figure 23.1: A median sagittal section of a horn, made by means of a strip of contact print. Consistent and distinct growth layers can be seen in the keratinised sheath

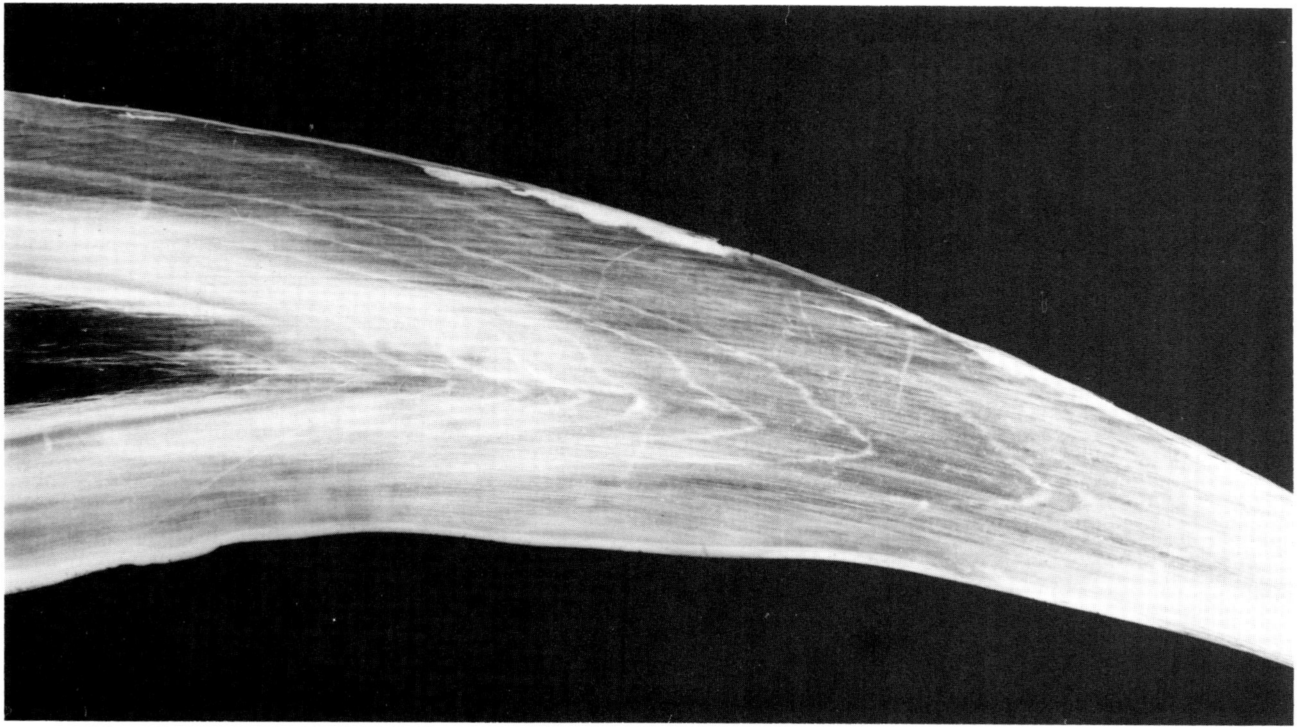

Figure 23.2: Difference in horn shape and annual ring formation of male and female Japanese serow

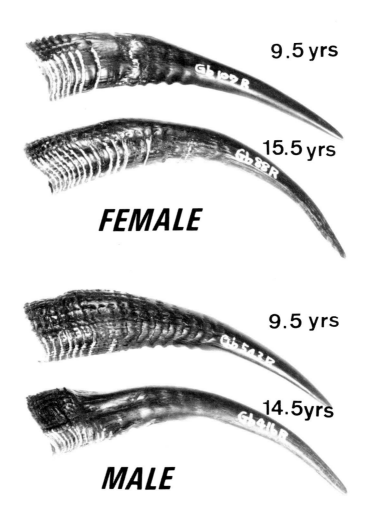

slender (Figure 23.2). The mean of the basal diameter of the horns of males 3.5 years old and older is significantly greater than the diameter of the horns of females (31.82 mm for males and 30.61 mm for females; $t = 7.66$, df = 552, $P < 0.001$), though a considerable overlap exists.

Additionally, both sexes frequently rubbed their horns on tree trunks (Kishimoto 1981) and wore down the frontal surface of the horn. The rate of wearing down, however, seems to be different between the sexes. Intensive wearing down, which resulted in changes in the shape as shown in Figure 23.2, was not observed in female horns.

Ring formation

In general, the grooves of annual rings of female horns were more distinct and deeper. Furthermore, obscure corrugations consisting of very narrow grooves which formed irregularly between true annual rings tended to occur in male horns.

Growth pattern

Detailed examinations indicated that the growth rate of horns of younger serow was slightly greater in males than in females; the first and second horn growth increments were significantly larger (Miura 1986). The growth rates of male horns after the fourth year were smaller in proportion to the first and second growth increments, and tended to become constant earlier than in females (Figure 23.2). Moreover, as described later, the annual growth increments of male horns tended to decrease steadily as age increased, whereas the growth increments of female horns were very irregular, possibly due to the effects of reproduction.

Habitat evaluation

The best measure of the factors influencing horn development is the first growth increment because it is the longest, generally accounting for about two-thirds of the total horn length in adult serow. Most serow are born from late May to early June (Sugimura *et al.* 1983) and their horns start to erupt by the age of about 5 months (Miura 1985). The size of the first growth increment which grows from the first autumn until the second winter of life is influenced by maternal weight and condition, birth date, and environmental conditions. It is probable that the first two factors play an important role in the early growth of horn until the first winter of life, but the proportion of early growth to the total length of the first growth increment is relatively low. The size of the first growth increment therefore depends mainly on environmental conditions, especially the severity of the first winter and early spring range conditions. Bunnell (1978) indicated that the spring range quality is the most important factor influencing horn growth in Dall sheep and thus horn growth can aid in the assessment of range conditions.

To examine the relationship between horn growth and environmental variables, I compared the length of the first growth increment in yearling males with the severity of the previous winter. Meteorological records indicated that the winter of 1980–81 was severe (mean snow depth 48.0 cm), and the 1981–82 winter was more moderate (snow depth

9.6 cm). Differences in mean growth between the two years were found in the horns of yearlings. The horns of both sexes were significantly longer during the second winter than during the first ($t=10.91$, df$=54$, $P<0.001$ for males, $t=3.32$, df$=50$, $P<0.001$ for females). These data suggest that the first growth increment can be used as a sensitive measure of environmental conditions during the first winter and the second spring of life.

Furthermore, serow seldom break or split the tips of their horns. Although the horns were often worn down by rubbing behaviour, the worn-down area was concentrated on the frontal surface of the horn. Thus, the first growth increment was preserved in a state of perfection for almost all horns. These observations lead to the conclusion that range conditions in the first winter and second spring of life can be assessed by the measurement of horns regardless of age.

Reproductive history in females

In general, the size of the yearly horn growth increments versus age for males tended to decrease steadily as they became older, whereas the growth patterns for females showed frequent significant reductions in the size of the growth increments after the second year of life. It was assumed that these smaller growth increments might have been caused by reproduction, which is energetically costly and results in less nutrition being available for the horns. To elucidate the relationship between horn growth and reproduction, I compared the formation of smaller growth increments with the results of ovary histological examinations made by Sugimura *et al.* (1984) and Kita *et al.* (Chapter 28, this volume). A comparison of the horns and reproductive tracts of 132 females showed that the number of smaller growth increments observed in each horn was approximately equivalent to the number of elastoid bodies (which are thought to be the retrograde corpora lutea of pregnancy) in a pair of ovaries (Kita *et al.* 1983, Sugimura *et al.* 1984). Female Japanese serow generally give birth to singletons (Sugimura *et al.* 1983). Thus, horn growth measurements can be used to trace the reproductive history of female serow.

Why is there a close relationship between horn growth and reproduction? Estimates of energy requirements for some ungulates have indicated that costs for females were highest in the spring and early summer during the last third of the gestation period and early lactation (see Oftedal 1985). This is the case for serow, and these seasons coincide completely with the peak of horn growth (Miura 1985). It would seem plausible, therefore, that the smaller horn growth increments would correspond to the low nutritional level which is associated indirectly with reproduction. Of course, other factors may also influence horn growth. For example, the nutritional environment, especially the spring range quality (Bunnell 1978), could cause the smaller growth increment. However, as this factor influences the horn growth of both sexes, it is possible to distinguish the smaller growth increments caused by this factor from those caused by reproduction by chronologically comparing horn growth patterns between sexes of the same age. Secondly, contingent diseases that would decrease an animal's nutritional level could cause smaller growth increments. However, no data for this possibility are available.

This estimate based on horn growth measurement is very useful for many phases of research because it is a simple guide to both age and parturition that does not require any specialised equipment and is applicable to live animals (Miura, in preparation).

REFERENCES

Bargagli, R. and S. Lovari. (1984) Age correlation between horn segment and tooth cementum annuli in the chamois (*Rupicapra rupicapra ornata*) (Artiodactyla, Bovidae). *Saügetierkundl. Mitt. 31*, 179–83

Brandborg, S.M. (1955) Life history and management of the mountain goat in Idaho. *Idaho Dept. Fish Game Wildl. Bull. 2*, 1–142

Bunnell, F.L. (1978) Horn growth and population quality in Dall sheep. *J. Wildl. Mgmt 42*, 764–75

Caughley, G. (1965) Horn rings and tooth eruption as criteria of age in the Himalayan tahr, *Hemitragus jemlahicus. N. J. J. Sci. 8*, 333–51

Kishimoto, R. (1981) Behavior and spatial organization of the Japanese serow (*Capricornis crispus*). Unpublished MS thesis, Osaka City University, Osaka, 59 pp.

Kita, I., M. Sugimura, Y. Suzuki and T. Tiba. (1983) Reproduction of female Japanese serows, *Capricornis crispus*, based on pregnancy and the macroscopical findings of the ovary. *Res. Bull. Fac. Agr. Gifu Univ. 48*, 137–46 (in Japanese with English abstract)

Knaus, W. and W. Schröder. (1983) *Das Gamswild*, 3rd edn. Paul Parey, Berlin and Hamburg

Miura, S. (1985) Horn and cementum annulation as age criteria in Japanese serow. *J. Wildl. Mgmt 49*, 152–6

Miura, S. (1986) Body and horn growth patterns in the Japanese serow, *Capricornis crispus. J. Mammal. Soc. Japan, 11*, 1–13

Miura, S. and K. Yasui. (1985) Validity of tooth eruption-wear patterns as age criteria in the Japanese serow, *Capricornis crispus. J. Mammal. Soc. Japan, 10*, 169–78

Oftedal, O.T. (1985) Pregnancy and lactation. In R.J. Hadson and R.G. White (eds) *Bioenergetics of wild herbivores*. CRC Press, Boca Raton, FLA, pp. 215–38

Sugimura, M., I. Kita, Y. Ide, S. Kodera and M. Yoshizawa. (1983) Parental development of Japanese serow, *Capricornis crispus*, and reproduction in females. *J. Mammal. 64*, 302–4

Sugimura, M., I. Kita, Y. Suzuki, Y. Atoji and T. Tiba. (1984) Histological studies on two types of retrograde corpora lutea in the ovary of Japanese serow, *Capricornis crispus. Zool. Anz. Jena 213*, 1–11

Part Six:
Diseases of the Rupicaprini

24

Clostridial infections in chamois (*Rupicapra rupicapra*) in captivity

H. Wiesner

Münchener Tierpark, Munich, Federal Republic of Germany

INTRODUCTION

Clostridial infections are quite common in various species of domestic and wild animals all over the world. Domestic animals, e.g. cattle, horses, goats and pigs are affected and, especially in sheep, heavy losses may occur as a result (Behrens 1979). Clostridial infections in wild and zoo animals are described for buffalo, deer (B. Röken, pers. comm.), elephant (Nelson 1978, cited in Davis *et al.* 1981), bear (Williamson *et al.* 1978 cited in Davis *et al.* 1981), gnu (Wiesner 1976), saiga antelope (W. Zimmermann pers. comm.), mink and birds (Mayr *et al.* 1984). Furthermore, the Caprinae tribe, which includes *Ibex* spp. (Chenitir 1968, Göltenboth and Klös 1974), tahrs (*Hemitragus jemlahicus*) and chamois (Wiesner 1976), seems to be highly perceptible to this anaerobic disease-causing group of organisms.

CLASSIFICATION

Infections by *Clostridium* spp. are classified according to their epidemiology and clinical importance (Mayr *et al.* 1984). Three groups exist:

(1) Epidemic clostridiosis (emphysema gangrenosum, septicaemia gangrenosa, toxaemia), caused by *Cl. chauvoei, Cl. septicum, Cl. novyi, Cl. perfringens* type A.
(2) Enterotoxaemia, caused by *Cl. perfringens* types A–E.
(3) Clostridiosis of wounds (septicaemia gangrenosa, malignant oedema), caused by *Cl. septicum, Cl. novyi* and *Cl. perfringens* type A and mixed infections with types A–E.

Of these three groups of infection, enterotoxaemia is the most common in wild animals kept under human management conditions.

ENTEROTOXAEMIA

Cl. perfringens is widely spread out in the soil and in the intestinal tract of animals. It measures $4-8 \times 0.1-1$ μm, and appears as a relatively large, anaerobic, oval-spore-forming, immobile, rod-shaped, Gram-positive organism with the ability to produce potent endotoxins. There are five types of endotoxin (A–E) but only types A, B and C are alone or together responsible for enterotoxaemia in mammals. The pathogenic types are probably acquired by ingestion of contaminated food. The clinical symptoms in most species are a peracute diarrhoea, listlessness, loss of appetite and sudden death. In the post-mortem findings the major lesion in all species is a haemorrhagic enteritis with ulcerations of the mucosa.

Our experiences

In the years 1972–1977 we had five deaths by enterotoxaemia in our flourishing chamois colony. The losses occurred in 6- to 13-year-old animals and not in kids. The clinical symptoms were listlessness, loss of appetite, stupor, dysentery, and fetid profuse diarrhoea tinged with blood within a few hours or overnight. All the different therapeutic efforts, such as high dosages of antibiotics, permanent infusions, hyperimmuneserum, corticoids, etc. were in vain. The animals died within a few hours. Coccidiosis with or without infestation by strongylids, a sudden change of food, feeding of frozen corn silage, a sudden change in the weather, stress caused by transporting, or social stress in unsuitable enclosures probably enhanced the predisposition to the outbreak of this polyfactorial disease in our stock. In the two cases in which *Cl. perfringens* could be isolated, a classification of the toxin type was not performed. In regard to the clinical symptoms and the peracute process of the disease in every case, a mixed infection with type A, B and/or C appears probable. In all post-mortems, acute haemorrhagic typhlitis, colitis and abomaso-duodeno-jejunitis, petechial bleedings of the heart and an hyperaemia of the parenchyma were diagnosed.

SEPTICAEMIA GANGRENOSA

At the same time we lost another adult female through an infection by *Cl. septicum*. The anamnesis was a fight between two females in a high range position. The animal was found dead overnight with bloody foam in the nostrils, but without any signs of wounds or other symptoms the day before. The autopsy revealed a generalised gaseous malignant oedema of the subcutis and an emphysematous phlegmon of the muscles. Touching the skin of the animal caused a crackling noise, and the swollen body seemed to have doubled in size. In the bacteriological samples *Cl. septicum* could be isolated. Apart from a non-gaseous malignant oedema in a female alpine ibex, which was probably caused by *Cl. novyi*, no other types of clostridiosis has occurred in our chamois or in other specimens of the Caprinae tribe in our zoo.

PROPHYLAXIS

As a prophylactic measure we now apply twice a year a polyvalent clostridium vaccine (Covexin 8® Burroughs Wellcome) by the blow-pipe system. In addition the enclosure of the chamois has been enlarged to 4240 m^2 for 10−15 animals with a well structured surface to avoid injuries during fights. Throughout the year, we provide the chamois, a selective browser, with leaves, shoots or branches. Since taking these measures no more clostridiosis has occurred in our zoo, although we stopped the vaccination programme 9 years ago.

DISCUSSION

Although there are only few reports on clostridial infections in chamois, the mortality of 20.68 per cent in 13 years in our stock reveals a high disposition of this species to this polyfactorial disease. Comparing the biology of other members of the Caprinae tribe with that of the chamois, we can assume a similar disposition. In stocks with problems of enterotoxaemia (*Cl. perfringens*), vaccination with a polyvalent vaccine is recommended. Wound infection with *Clostridium* (*Cl. septicum, Cl. novyi, Cl. perfringens*) seems to play a minor part, but nevertheless it should be considered in surgical treatment of these species. Therefore, when using a distance immobilisation method in these animals, only systems with low impact (cold gas systems) are suitable. After darting Caprinae, the application of antibiotics and of a polyvalent vaccine is recommended.

ACKNOWLEDGEMENT

We thank Prof. Dr V. Sandersleben, of the Veterinary Pathology Institute of the University of Munich, for carrying out the post-mortems.

REFERENCES

Behrens, H. (1979) *Lehrbuch der Schafkrankheiten.* Paul Parey, Berlin and Hamburg

Chenitir, H. (1968) Untersuchungen über *Clostridium perfringens* Infektionen bei Zootieren und über die möglichen Infektionsquellen. *Vet. Med. Diss. F. U. Berlin*

Göltenboth, R. and H.G. Klös. (1974) Zu einigen Erkrankungen und Todesfällen bei Elefanten des Zoologischen Gartens Berlin. Verhandlungsber. *XVI Int. Symp. Erkrank. Zootiere* 175–9

Mayr, A., G. Eissner and B. Mayr-Bibraek. (1984) *Handbuch der Schutzimpfungen in der Tiermedizin* Paul Parey, Berlin and Hamburg

Wiesner, H. (1976) Zur Bestandsprophylaxe bei Hochgebirgstieren in Gefangenschaft. Verhandlungsber. *XVIII Int. Symp. Erkrank. Zootiere* 89–93

25

Pathological studies on Japanese serow (*Capricornis crispus*)

Yoshitaka Suzuki, Makoto Sugimura and Yasuro Atoji
Department of Veterinary Anatomy, Faculty of Agriculture, Gifu University, Gifu
501-11 Japan

INTRODUCTION

The Japanese serow, *Capricornis crispus*, is the only wild bovine ruminant in Japan and is strictly conserved by law, so systemic pathological and epidemiological investigations have never been performed before. However, during the period from 1979 to 1985, a large number of serows caught in Gifu Prefecture were brought to our laboratory with the permission of the Agency of Cultural Affairs. Based on a necropsy of about 2000 Japanese serows, the main pathological findings can be summarised as follows: (1) parasitological lesions, (2) infectious diseases, (3) neoplasms, and (4) other.

PARASITOLOGICAL LESIONS

In 1981, parasitological investigations were carried out on 39 serows by Dr Yagi (Department of Parasitology, Faculty of Veterinary Medicine, Hokkaido University). Nine species of helminths were recognised, and six of the nine were specific to serows (Table 25.1). Among these helmiths, the most frequent lesions encountered at necropsy were pulmonary nodules caused by *Protostrongylus shiozawai* and subcutaneous lesions due to *Onchocerca* spp. 1 and 2.

PATHOLOGICAL STUDY ON LUNGWORM DISEASE IN WILD JAPANESE SEROW (SUZUKI *et al.* 1981)

Materials in the present study consisted of 119 cases including fawns, yearlings and adults. Pulmonary lesions were found in over 70 per cent of the cases examined.

Macroscopically, pulmonary lesions were mostly distributed bilaterally in the caudal part of the posterior lobes. They were somewhat hard in

Table 25.1: Helminth fauna in Japanese serows in Gifu Prefecture

Parasite	Invasion site	% Incidence (male/female)
Trematoda		
Ogmocotyle capricorni		
Machida, 1970	Small intestine	86.8 (33/38)
Dicrocoelium dendriticum		
Rudolphi, 1819	Gall bladder and bile duct	29.4 (5/17)
Cestoda		
Moniezia monardi Fuhrmann, 1933	Small intestine	31.6 (12/38)
Nematoda		
Okapinema japonica Machida, 1970	Abomasum	100 (38/38)
Skrjabinema sp.	Caecum and colon	59.5 (22/37)
Trichuris discolor Von Linstow, 1906	Caecum	2.6 (1/38)
Protostrongylus shiozawai		
Ohbayashi et Ueno, 1974	Lung	100 (38/38)
Onchocerca sp. 1	Subcutis	71.8 (28/39)
Onchocerca sp. 2	Carpal and tarsal joints	66.7 (26/39)

Source: Yagi, 1982

consistency, yellowish-white in colour, as large as a soybean to a thumb tip in size, and had rather sharp margins.

Microscopically, the lesions could be characterised as follows.

(1) Living and dead parasites including adults, larvae and eggs were noted in the cavity of bronchioles, alveolar ducts and alveoli (Figure 25.1).

(2) Cellular reactions consisted of eosinophils, neutrophils, lymphocytes, plasma cells, macrophages and multinucleated giant cells (Figure 25.2).

(3) Parasitic minute granulomas were evident (Figure 25.3). Moreover, a remarkable finding was that of numerous globule leucocytes in the bronchiolar epithelium with affected lobuli (Figure 25.4). The cells were characterised by conspicuous eosinophilic granules or droplets in the cytoplasm. Ultrastructurally, typical granules consisted of homogeneous, dense material and were bounded with a membrane (Yoshizawa *et al.* 1982).

Ohbayashi and Ueno (1974) examined two Japanese serows from Nagano Prefecture in Japan described *Protostrongylus shiozawai* as a new species. Thereafter, the parasites were also found in serows from Tohoku Prefecture in Japan (Chihaya *et al.* 1976, Kato 1979). Parasitologically, the lungworms in the present study were identical to those described in the previous report (Ohbayashi and Ueno 1974), and pathological changes were also similar to those described in a previous report (Shiozawa *et al.* 1975). Therefore, these findings suggest that lungworm diseases were not just enzootic but epidemic among wild Japanese serows.

Figure 25.1: Numerous larvae and eggs (E) in the alveoli. H–E stain × 280

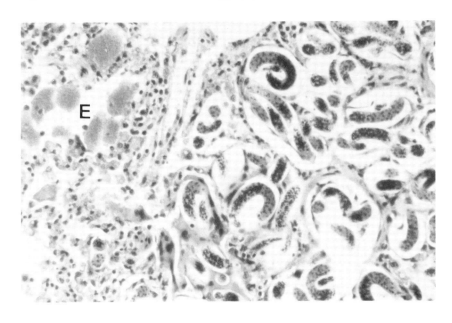

Figure 25.2: Cellular infiltration and multinucleated giant cells enclosing dead larvae. H–E stain. × 250

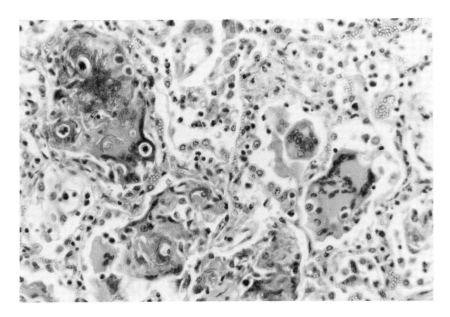

Figure 25.3: Parasitic minute granulomas. H–E stain. ×220

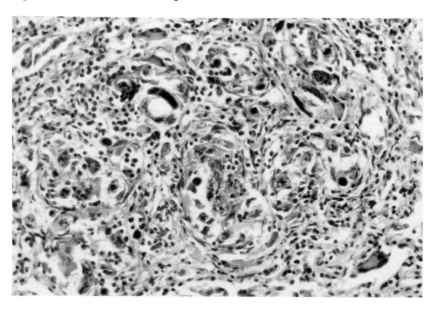

Figure 25.4: Numerous globule leucocytes in bronchial epithelium. Epon thick section. Toluidine Blue stain. × 450

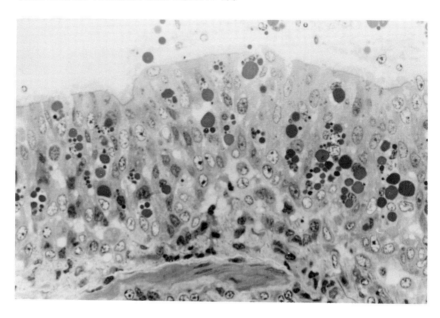

ONCHOCERCIASIS IN WILD JAPANESE SEROW (SUZUKI *et al.* 1982)

Materials used in the present study consisted of 278 cases. Upon necropsy, two types of lesion were noticed in the subcutis. One was of yellowish-white nodules sporadically distributed anywhere on the body; the other was of a fibrous bursa formation restricted to the carpal and tarsal regions. The lesions were found in 60 per cent and 78 per cent of the cases examined, respectively.

The former lesions due to *Onchocerca* sp. 1 were most frequently encountered in the thoracic area and pelvic limbs, and were irregular in shape, measuring about 0.5 to 10 cm (Figure 25.5). Histopathology revealed that these lesions represented parasitic granulomas (Figure 25.6).

Whereas the latter lesions, the fibrous bursa, contained yellowish coarse materials intermingled with numerous parasites, which were also present in the thick fibrous wall of the bursa (Figures 25.7 and 25.8), the inner surface of the bursa had an epithelium-like lining with papillary projections (Figure 25.9). Microfilariae were occasionally found near mature nematodes. The lesion was significantly prominent in the forelimbs, with a tendency to increasing severity with age (Table 25.2).

Two types of *Onchocerca* species found in the present study were different from those of other species of animals (Muller 1979).

Figure 25.5: Degenerated *Onchocerca* sp. 1 (arrow) in the subcutis of the elbow area

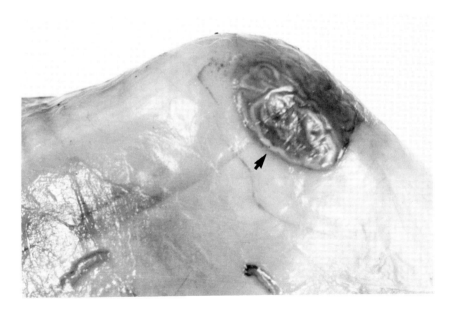

Figure 25.6: Subcutaneous parasitic nodule including *Onchocerca* sp. 1.
H–E stain × 95

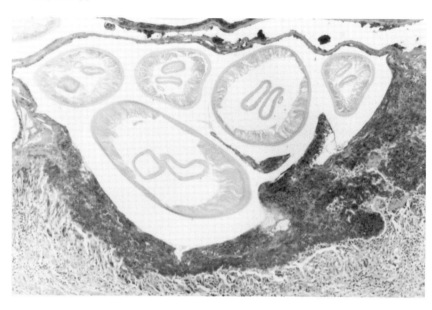

Figure 25.7: Longitudinal section through the carpal joint. Well developed fibrous bursa (arrow) in the subcutis

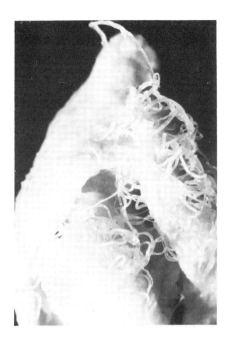

Figure 25.8: Cut surface of a fibrous bursa with numerous parasites (*Onchocerca* sp. 2) in the bursal lumen and wall

Figure 25.9: Bursa covered with an epithilium-like lining and parasites within granulomatous tunnels. H–E × 60

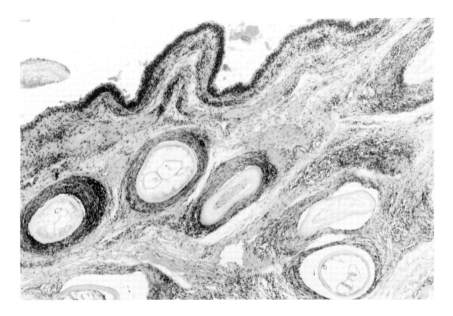

Table 25.2: Occurrence of lesion and age of animals

Age group[a]	Nodule (Onchocerca sp. 1)	Fibrous bursa formation (Onchocerca sp. 2)
0	6/44 (14%)	2/44 (4.5%)
1	15/26 (58%)	17/26 (65%)
2-I	28/46 (61%)	40/46 (86%)
2-II	29/49 (59%)	47/49 (96%)
2-III	33/44 (75%)	43/44 (97%)
2-IV	34/44 (77%)	43/44 (97%)
2-V	21/25 (84%)	25/25 (100%)

[a]Age-class as judged by teeth eruption and wear of incisors (Miura and Yasui 1979, Sugimura et al. 1981)

INFECTIOUS DISEASES

During the 6 years the necropsies were performed the most remarkable fact was an outbreak of parapox infection in the period from December 1984 to March 1985.

Materials subjected to the present investigation consisted of 402 cases including fawns, yearlings and adults. Macroscopically, papular or nodular lesions were found on the mucosa and/or integument in the naso-oral, external genital, auricular and udder parts as well as in the oesophagus and rumen (Figures 25.10 to 25.12). They developed sporadically or multi-focally. The lesions ranged in size from that of a rice grain to a thumb tip, and were occasionally covered with scabs.

Histopathologically, the early lesions were characterised by focal acanthosis of the mucosa and epidermis (Figure 25.13). The mucosa was thickened in most cases, and occasionally twice the normal thickness with long slender rete pegs. Cytological changes were most evident in the prickle cells of the stratum spinosum, namely reticular and ballooning degeneration. The vacuolated cells frequently contained spherical eosinophilic or basophilic cytoplasmic inclusion bodies of various sizes (Figure 25.14). They were homogeneous or finely granular in appearance and were slightly positive in Feulgen reaction.

Electron microscopically, the cytoplasm of the prickle cells of the upper layer was oedematous and contained free viral particles and a granular matrix with mature and immature viral particles. In the negatively stained specimens, observations revealed cylindrical viral particles measuring 360 by 200 nm with the fine criss-cross pattern typical of the parapox virus group (Figure 25.15). In addition to these epithelial changes, inflammatory cell infiltration, hyperaemia, haemorrhagy, capillarisation and oedema were often observed in the propria or dermis.

Figure 25.10: Multiple papular lesions on the lips and muzzle

Figure 25.11: Nodular lesions with scabs on the udder

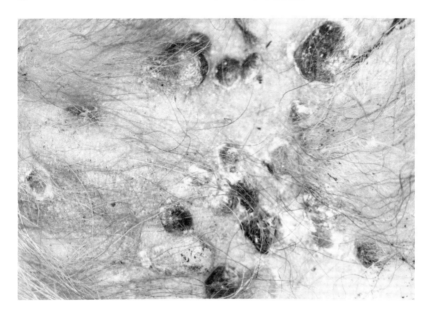

Figure 25.12: Hard palate severely affected with parapox infection

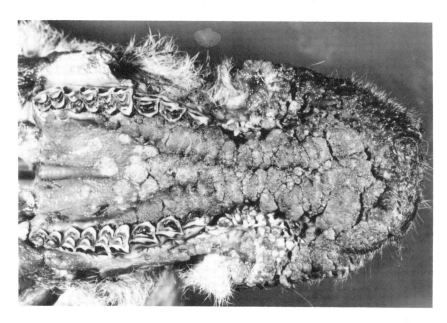

Figure 25.13: Marked acanthosis and vacuolar degeneration in the mucosa of the hard palate. H–E stain. × 55

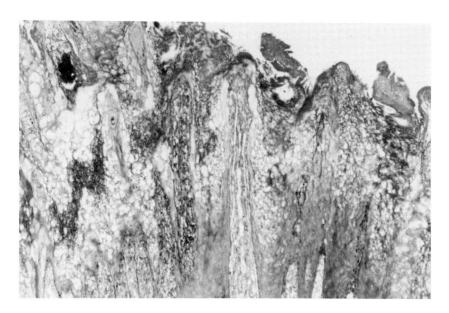

Figure 25.14: Reticular degeneration of prickle cells containing cytoplasmic spherical inclusion bodies (arrows). H–E stain. × 300

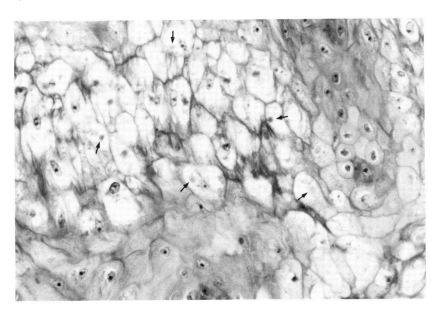

Figure 25.15: Electron microscopy of prickle cell containing viroplasm (V) and mature virus particles (MP). × 12000. Insert. A parapox virus particle showing criss-cross pattern × 125000. (Dr Minamoto's original, Gifu University)

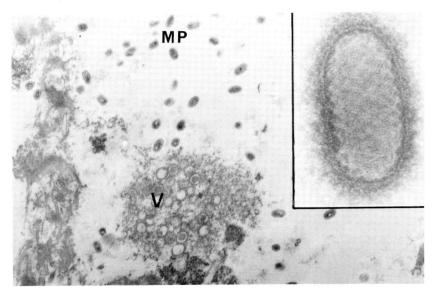

In advanced lesions, microabscess formation and hyperkeratosis were occasionally observed. The pathological findings described above were closely similar to those on Japanese serows in Tohoku Prefecture which had been reported as cases of papular stomatitis (Kato *et al.* 1980), or contagious papular dermatitis (Okada *et al.* 1984a, b).

As shown in Table 25.3, the disease was widely distributed in Gifu Prefecture, and the lesions were found in 155 of the 402 cases examined (39 per cent). Taking into consideration the distribution pattern of pathological changes and the fluctuation of the disease's occurrence, various factors, such as the serows' breeding season and the possibility of fomites infection by parapox virus were considered to play an important role in the widespread parapox infection at this time.

Table 25.3: Occurrence of parapox infection

Region	Dec./1984	Jan./1985	Feb./1985	March/1985	Total
Kamiyahagi	10/15 (67%)	13/18 (72%)	1/2 (50%)	Not captured	24/35 (69%)
Nakatsugawa	11/17 (65%)	3/4 (75%)	6/15 (40%)	0/2 (0%)	20/38 (53%)
Kawaue	1/7 (14%)	2/10 (20%)	1/4 (25%)	3/8 (38%)	7/29 (24%)
Tsukechi	4/12 (33%)	3/12 (25%)	1/8 (13%)	1/3 (33%)	9/35 (26%)
Kashimo	2/5 (40%)	7/17 (41%)	2/7 (29%)	2/2 (100%)	13/31 (42%)
Gero	4/6 (67%)	8/17 (47%)	0/5 (0%)	1/1 (100%)	13/29 (45%)
Hagiwara	6/13 (46%)	7/16 (44%)	1/9 (11%)	0/2 (0%)	14/40 (35%)
Osaka	8/31 (26%)	14/29 (48%)	6/13 (46%)	2/5 (40%)	30/78 (38%)
Kukuno	Not captured	4/13 (30%)	0/4 (0%)	Not captured	4/17 (24%)
Takane	10/16 (63%)	8/25 (32%)	2/20 (10%)	1/9 (11%)	21/70 (30%)
Total	56/122 (46%)	69/161 (43%)	20/87 (23%)	10/32 (31%)	155/402 (39%)

NEOPLASMS

Table 25.4 shows the incidence of neoplasm for the 6 years from 1979 to 1984. Tumour lesions were found in only 5 of 1951 cases examined; 4 of the 5 cases were found in the female genital system (Figures 25.16 and 25.17). The other was diagnosed as liver cell adenoma with focal haematopoiesis. All of them were benign tumours and appeared in aged serows. Therefore, the incidence of tumour lesions in wild Japanese serows was very low compared with those of domestic animals (Moulton 1978, Cotchin 1984).

The results obtained in the present study may be useful for a survey of wild animal oncology, about which little has yet been clarified.

Table 25.4: Occurrence of neoplasms

Year	Number of cases examined	Number of positive cases	Age[a]	Sex	Diagnosis
1979	119	0	—	—	—
1980	293	2	2-IV (9.5 y)	Female	Cystoadenoma of the ovary
			2-V (16.5 y)	Female	Fibroma of the vagina
1981	395	1	2-IV (12.5 y)	Male	Liver cell adenoma with haematopoiesis
1982	370	2	2-V (unexamined)	Female	Cystoadenoma of the vagina
			2-V (unexamined)	Female	Adenoma of the vagina
1983	372	0			
1984	402	0			

[a] Age-class by teeth eruption and wear of incisors (Miura and Yasui 1979, Sugimura *et al.* 1981)

Figure 25.16: Cut surface of cystoadenoma in the ovary

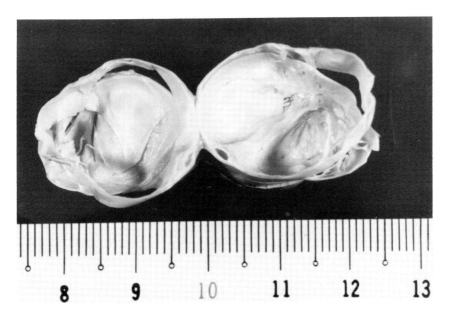

Figure 25.17: Cystoadenoma in the vagina

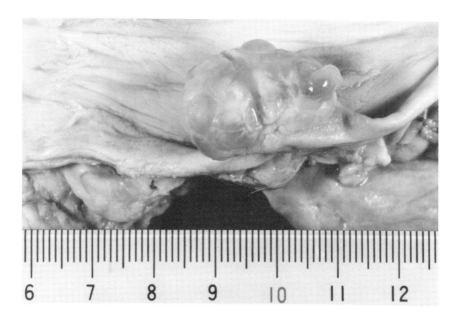

OTHER

Sporadic lesions observed at necropsy included the following:

(1) repaired lesion of bone fracture;
(2) gallstone;
(3) trichobezoar in the abomasum;
(4) cutaneous horn on the dorso-lumbar region (Figure 25.18);
(5) cryptorchism.

On the basis of the pathological findings described above, we can conclude that environmental conditions for serows in Gifu Prefecture seem to be generally favourable as no serow has suffered from severe lesions that eventually proved fatal.

Figure 25.18: Cutaneous horn in the dorsolumbar region

REFERENCES AND FURTHER READING

Chihaya, Y., K. Ohshima, S. Miura and S. Numakunai. (1976) Pathological study on cutaneous papillomatosis in a Japanese serow (*Capricornis crispus*). *Japan. J. Vet. Sci. 38*, 327–38

Cotchin, E. (1984) Veterinary oncology: a survey. *J. Pathol. 142*, 101–27

Davis, J.W., L.H. Karstadt and D.O. Trainer (1981) *Infectious diseases of wild animals.* Iowa State University Press, Ames, Iowa

Kato, H. (1979) Dictyocaulosis in Japanese serows *Capricornis crispus. J. Japan. Zool. Aq. 21*, 39–41 (in Japanese)

Kato, H., K. Sato, Y. Ishikawa, S. Takahashi, Y. Gonai and K. Yatsu. (1980) Papular stomatitis with dictyocaulosis in Japanese serow, *Capricornis crispus*, in captivity. *J. Japan. Zool. Aq. 22*, 46–50 (in Japanese)

Miura, S. and K. Yasui. (1979) Preliminary report on a study of age determination of Japanese serow by cementum layers. *Rep. Conserv. Soc. Japan 56*, 105–13

Moulton, J. (1978) *Tumors in domestic animals.* Univ. of California Press, London

Muller, R. (1979) Identification of *Onchocerca. Brit. Soc. Parasitol. Symp. 17*, 175–206

Ohbayashi, M. and H. Ueno. (1974) A new lungworm, *Protostrongylus* (*Davtianostrongylus*) *shiozawai* n. sp., from the Japanese serow, *Capricornis crispus* (Temminck). *Japan. J. Vet. Res. 22*, 111–15

Okada, H.M., K. Okada, S. Numakunai and K. Ohshima. (1984a) Histopathologic studies on mucosal and cutaneous lesions in contagious papular dermatitis of Japanese serows (*Capricornis crispus*), *Japan. J. Vet. Sci. 46*, 257–64

Okada, H.M., K. Okada, S. Numakunai and K. Ohshima. (1984b) Electron microscopy on mucosal and cutaneous lesions in contagious papular dermatitis of Japanese serows (*Capricornis crispus*). *Japan. J. Vet. Sci. 46*, 297–302

Shiozawa, M., M. Isoda, S. Aoki, H. Ueno and H. Chiba. (1975) Observation on the histopathology of verminous pneumonia of Japanese serow. *Bull. Nippon Vet. Zootech. Coll. 24*, 76–86 (in Japanese)

Sugimura, M., Y. Suzuki, S. Kamiya and T. Fujita. (1981) Reproduction and prenatal growth in the wild Japanese serow, *Capricornis crispus. Japan. J. Vet. Sci. 43*, 553–5

Suzuki, Y., M. Sugimura, M. Kamiya, T. Fujita and M. Yoshizawa. (1981) Pathological study on lungworm disease in the wild Japanese serow, *Capricornis crispus. Japan. J. Vet. Sci. 43*, 281–5

Suzuki, Y., M. Sugimura, Y. Atoji, N. Minamoto and T. Kinjo. (1986) Widespread parapox infection in wild Japanese serows, *Capricornis crispus. Japan J. Vet. Sci. 48*, 1279–82

Suzuki, Y., M. Sugimura, K. Yagi, M. Ohbayashi and C. Shoho. (1982) Onchocerciasis in wild Japanese serows, *Capricornis crispus. Japan. J. Vet. Sci. 44*, 823–5

Yagi, K. (1982) Helminth fauna of the Japanese serow, *Capricornis crispus crispus* Temminck, in Gifu prefecture. *Japan. J. Vet. Res. 30*, 40

Yoshizawa, M., Y. Suzuki and M. Sugimura. (1982) Morphology of globule leucocytes in the lung of Japanese serows. *Res. Bull. Fac. Agr. Gifu Univ. 46*, 223–30. (In Japanese with English summary.)

26

Serological survey for selected microbial pathogens in Japanese serow (*Capricornis crispus*) in Gifu Prefecture, Japan

Toshio Kinjo and Nobuyuki Minamoto
Department of Veterinary Public Health, Faculty of Agriculture, Gifu University,
Gifu 501-11, Japan

INTRODUCTION

The importance of wild animals has been recognised in the epidemiology of a wide variety of infectious diseases. Many wild animal species can act as reservoirs of the diseases for other wild or domestic animals and even humans. The assessment of the situation of infectious diseases in wild animals is therefore very important in wildlife management as well as for the protection of domestic animals and humans. However, studies along these lines have not been systematically carried out in Japan. We had an opportunity to collect blood samples from wild Japanese serows (*Capricornis crispus*), the only wild bovine ruminant in Japan. We attempted to determine by means of a serological survey whether certain microbial pathogens which can affect cattle and other animal species are present in the serow population. The possibility that serows could serve as potential sources of the infections for domestic animals and humans is also discussed here.

MATERIALS AND METHODS

Serum samples

Serum samples tested were obtained from a total of 766 serows comprising 393 males and 373 females which were killed in Gifu Prefecture, Japan, during the winter months of 1981 to 1984. The serow age was determined by Dr Sugimura and his staff according to eruption of teeth and wear of incisors (Sugimura *et al.* 1981).

Microbial pathogens and serological tests

The prevalence of antibodies to the following eight pathogens was investigated by the indicated test procedures.

(1) *Toxoplasma gondii.* The latex agglutination test was used with commercial antigen (Eiken Chemical Co., Tokyo) (Kobayashi *et al.* 1977). Titres higher than 1:32 were referred to as positive.
(2) *Brucella abortus.* Both the agglutination and complement fixation (CF) tests recommended by the Brucella Committee in Japan were conducted using antigens provided by the Institute of Animal Health, Japan.
(3) *Leptospira* species. The latex agglutination test with commercial antigens (Denka-Seiken Co., Tokyo) was used. The test was conducted by the indicated procedure with a slight modification. Antibodies to five serovars including *L. icterohaemorrhagiae, L. autumnalis, L. hebdomadis, L. australis* and *L. canicola* were examined. Serum showing agglutination with any one of the five antigens at a serum dilution of 1:10 was considered as positive. Though there were cases that reacted with more than two serovars, no further tests to determine corresponding serovars were performed.
(4) *Chlamydia psittaci.* The CF test with commercial antigen (Denka-Seiken Co., Tokyo) was used, and a titre of 1:10 or more was referred to as positive.
(5) Rotavirus. The immune adherence haemagglutination test (Inoue *et al.* 1981) with tissue culture antigen of the Lincoln strain of bovine origin was used. A titre of 1:8 or more was taken as positive.
(6) Akabane virus. Detection of antibodies to this virus was done by the haemagglutination inhibition (HI) test using commercial antigen (Kyoto-Biken Lab., Kyoto). The test was performed according to the manufacturer's directions. A titre of 1:10 or more was recorded as positive.
(7) Japanese encephalitis virus. The HI test using the JaGAr 01 strain was employed (Nakamura and Nakamura 1983).
(8) Ibaraki virus. Antibodies to this agent were tested by the HI test using commercial antigen (Kyoto-Biken Lab., Kyoto), according to the manufacturer's indicated procedure.

Statistical analysis

Differences in prevalence were tested for significance by chi-square analysis.

RESULTS

Toxoplasma gondii

As shown in Table 26.1, a total of 765 serows were examined, and 41 of them (5.4 per cent), were seropositive. The rate of reactors ranged from 2.6 to 7.8 per cent, and the statistical difference ($P < 0.05$) was observed between the rates in 1983 and 1984. Serows having higher antibody titres of 1:128 or more were detected.

Next, we analysed antibody prevalence by serow age and sex (Table 26.2). Seropositive rates varied from 3.2 to 8.5 per cent, but no statistical differences were present. Similarly, no difference was obtained between the positive rates of males and females.

Table 26.1: Prevalence of antibodies[a] to *Toxoplasma gondii* in Japanese serows

Date of collec-tion	No. of sera tested	No. of sero-reactors	Rate of reactors	Antibody titres						
				<16	16	32	64	128	256	512
1981	363	21	5.8	326	16	13	5	3	0	0
1982	51	4	7.8	41	6	4	0	0	0	0
1983	196	5	2.6 } *	188	3	4	0	0	1	0
1984	155	11	7.1 }	127	17	5	2	3	0	1
Total	765	41	5.4	682	42	26	7	6	1	1

[a]Detected by latex agglutination test using commercial antigen
[b]Titres of $\geqslant 32$ are taken as positive
*Significantly different ($P < 0.05$)

Table 26.2: Prevalence of antibodies to *Toxoplasma gondii* in Japanese serows by age and sex

Age				Sex			
Age groups[a] (years)	No. tested	No. positive	% Positive	Sex	No. tested	No. positive	% Positive
≤ 1	95	3	3.2	Male	393	22	5.6
1.5~2.5	118	10	8.5	Female	372	19	5.1
3~5	264	13	4.9	Total	765	41	5.4
6~9	144	9	6.3				
$\geqslant 10$	144	6	4.2				
Total	765	41	5.4				

[a]Age groups based on dental eruption and wear

Brucella abortus

Of 718 serow samples tested, all were negative for antibodies to *Brucella abortus*.

Leptospira species

As shown in Table 26.3, 42 out of a total of 404 (10.4 per cent) were seroreactors for at least one of the five serovars tested. No statistical differences in antibody prevalence were observed for the years studied. Although seropositive rates seemed to increase with age, there were no statistical differences among them. However, a 5 per cent level of significant difference was found between the 13.5 per cent of males and the 7.1 per cent of females (Table 26.4).

Table 26.3: Prevalence of leptospirosis antibodies in Japanese serows by the latez agglutination test

Date of collection	No. of sera tested	No. of seroreactors[a]	Rate of reactors
1981	105	15	14.3
1982	51	5	9.8
1983	141	10	7.1
1984	107	12	11.2
Total	404	42	10.4

[a]No. of positive sera against one or more of five serovars tested; *L. icterohaemorrhagiae, L. autumnalis, L. hebdomadis, L. australis* and *L. canicola*

Table 26.4: Prevalence of antibodies to *Leptospira* spp. in Japanese serows by age and sex

Age				Sex			
Age groups[a] (years)	No. tested	No. positive	% Positive	Sex	No. tested	No. positive	% Positive
≤ 1	57	4	7.0	Male	207	28	13.5 }*
1.5~2.5	69	6	8.7	Female	197	14	7.1
3~5	126	15	11.9	Total	404	42	10.4
6~9	72	10	13.9				
≥ 10	80	7	8.8				
Total	404	42	10.4				

*Significantly different ($P < 0.05$)

Chlamydia psittaci

In all, 36 out of 335 samples tested were seropositive, and the seropositive rate was 10.7 per cent (Table 26.5). The positive rates persisted at almost similar levels throughout the four years studied. Antibody titres were relatively low, e.g. 1:16 or less. There were no statistical differences in positive rates by age and sex.

Table 26.5: Prevalence of antibodies[a] to *Chlamydia psittaci* in Japanese serows

Date of collection	No. of sera tested	No. of seroreactors[b]	Rate of reactors	No. of sera giving reciprocal titres of:			
				< 8	8	16	32
1981	158	17	10.8	141	12	5	0
1982	39	4	10.3	35	3	1	0
1983	71	9	12.7	62	8	1	0
1984	67	6	9.0	61	3	3	0
Total	335	36	10.7	299	26	10	0

[a]Detected by complement fixation test using commercial antigen
[b]Titres of $\geq$ 8 are taken as positive

Rotavirus

Table 26.6 shows the antibody prevalence for rotavirus. Overall, 95 out of 370 tested (25.7 per cent) were seroreactors. The positive rate varied from year to year; the minimum rate of 7.7 per cent was attained in 1982, and the maximum rate of 40.4 per cent in 1983. As indicated in the footnote to Table 26.6, there exist 1 or 5 per cent levels of significant difference between the rates of four out of six combinations. Serows with high antibody titres such as 1:64 or more were also detected in 1981 and 1983 in parallel with the high rate of reactors.

As shown in Table 26.7, from the distribution data of seropositivity by age, serows were divided into two groups, the less than 1-year-old group and the more than $1^{1}/_{2}$-year-old group. There was a 5 per cent level of significant difference between positive rates of the two groups. In contrast, no sex difference was observed in the positive rates.

Akabane virus

Out of 519 examined, 30 serows (5.8 per cent), were seropositive (Table 26.8). Serum antibody titres were relatively low, e.g. 1:40 or less. From the

303

Table 26.6: Prevalence of antibodies[a] to rotavirus in Japanese serows

Date of collection	No. of sera tested	No. of seroreactors[b]	Rate of reactors	No. of sera giving reciprocal titres of:							
				<8	8	16	32	64	128	256	512
1981	120	29	24.2	91	12	6	5	3	2	0	1
1982	39	3	7.7	36	1	2	0	0	0	0	0
1983	104	42	40.4	62	4	18	12	4	3	1	0
1984	107	21	19.6	86	16	3	2	0	0	0	0
Total	370	95	25.7	275	33	29	19	7	5	1	1

[a]Detected by immune adherence haemagglutination test using tissue culture antigen of the Lincoln strain of bovine origin
[b]Titres of ≥8 are taken as positive

	1981	1982	1983	1984
1982	*			
1983	*	**		
1984	—	—	**	

$* \; P < 0.05 \quad **P < 0.01$

Table 26.7: Prevalence of antibodies to rotavirus in Japanese serows by age and sex

Age				Sex			
Age groups (years)	No. tested	No. positive	% Positive	Sex	No. tested	No. positive	% Positive
≤ 1	49	5	10.2	Male	186	49	26.3
1.5~2.5	60	15	25.0	Female	184	46	25.3
3~5	106	25	23.6	Total	370	95	25.7
6~9	78	22	28.2				
≥ 10	77	28	36.4				
Total	370	95	25.7				

*Significantly different ($P < 0.05$)

Table 26.8: Prevalence of antibodies[a] to Akabane virus in Japanese serows

Date of collection	No. of sera tested	No. of seroreactors[b]	Rate reactors	No. of sera giving reciprocal titres of:			
				< 10	10	20	40
1981	142	8	5.6	134	6	2	0
1982	92	4	4.3	88	3	0	1
1983	178	8	4.5	170	2	6	0
1984	107	10	9.3	97	7	1	2
Total	519	30	5.8	489	18	9	3

[a]Detected by haemagglutination inhibition test using commercial antigen
[b]Titres of ≥ 10 are taken as positive

seropositive rates of each age group, serows were divided into two groups, those under 3 years of age with lower seropositive rates, and those over 3 years of age with higher rates (Table 26.9). There was a 5 per cent level of significant difference between seropositive rates of these two groups. No sex difference in antibody prevalence was observed.

Japanese encephalitis virus and Ibaraki virus

The antibodies to Japanese encephalitis virus were checked in 208 serows for two years and antibodies to Ibaraki virus in 337 serows for three years. As indicated in Table 26.10, no antibody to each of the two viruses was detected.

Table 26.9: Prevalence of antibodies of Akabane virus in Japanese serows by age and sex

Age				Sex			
Age groups (years)	No. tested	No. positive	% Positive	Sex	No. tested	No. positive	% Positive
≤ 1	68	2	2.9 ⎫	Male	271	15	5.5
1.5~2.5	84	1	1.2 ⎬ ⎫	Female	248	15	6.0
3~5	150	8	5.3 ⎫ ⎬ *	Total	519	30	5.8
6~9	104	10	9.5 ⎬				
≥ 10	113	9	8.0 ⎭				
Total	519	30	5.8				

*Significantly different ($P < 0.05$)

Table 26.10: Serological survey for Japanese encephalitis virus and Ibaraki virus in Japanese serows

Date of collection	Japanese encephalitis virus[a]		Ibaraki virus[b]	
	No. tested	No. positive	No. tested	No. positive
1981	142	0	142	0
1982	62	0	92	0
1983	0	—	103	0
1984	0	—	0	—
Total	208	0	337	0

[a] Tested by haemagglutination inhibition (HI) test using JaGAr 0-1 strain
[b] Tested by HI test using commercial antigen

DISCUSSION

With the continuing expansion of human activities, wildlife habitats are constantly shrinking and overlapping with that of domestic animals and humans. Thus, contact of humans and domestic animals with many species of wild animals is increasing. From this perspective one cannot overlook the important role of wild animals in the epidemiology of infectious diseases affecting humans and domestic animals.

We have been investigating serologically the prevalence of zoonoses in wild animals since 1980, when serum samples of wild Japanese serows were obtained. The serow, the only wild bovine ruminant in Japan, has been protected as a special natural monument. But in recent years, conversely, to prevent forest damage by overbreeding serows, the Agency of Cultural Affairs of the Japanese Government has permitted the killing of serows within limited numbers and locations. Therefore, we attempted to determine by serological survey using these valuable serum samples whether certain microbial infections were present that would affect the serow population and could serve as potential sources of infection for domestic animals and humans. Data obtained were analysed by serow age and sex.

Toxoplasma gondii can infect most species of animals including man and cause abortion, stillbirths, encephalitis, pneumonia and neonatal mortality in the affected animals. The cat and other Felidae are considered the definitive host of the parasite and act as sources of infection for other animals. The herbivores may be infected by grazing areas accidentally contaminated with toxoplasma oocysts excreted from infected felids (Costa *et al.* 1977). The seropositive rate of 5.4 per cent in serows obtained in the present study does not seem to be so high, but of the seropositive cases eight individuals show antibody titres of 1:128 or more, suggesting recent infection at the time of capture. Because there was no difference in antibody prevalence by serow age and sex, it was assumed that infected felids might be present in or near the serow habitat, and that the chances of contacting serows shedding oocysts might be ongoing. The seropositivity of cattle kept in the adjacent areas is obscure. However, even in the case of swine, the seropositive rate is only 2 to 3 per cent. At any rate, toxoplasmosis may certainly be one of the infectious diseases occurring among the serow population.

The disease of cattle caused by infection with *Brucella abortus* is characterised by abortion and a subsequent high rate of infertility. In Japan, severe storms of abortions by this bacterium occurred in pregnant cows from 1955 to 1965. After that they decreased rapidly thanks to an eradication programme, and no outbreak has been reported since 1980. It was very interesting to determine whether this disease occurred among the serow population as well, but no serological evidence for the presence of this disease was found among 718 serows tested. The serow, like other ruminants, is thought to be susceptible to this bacterium, so its habitat may well be free of any reservoirs of it.

Leptospirosis is known to cause jaundice, haemoglobinuria, etc. in cattle and other animal species. Although all parasitic serovars are potential pathogens to all species of domestic animals, we examined antibodies to only five serovars (*L. icterohaemorrhagiae, L. autumnalis, L. hebdomadis, L. australis* and *L. canicola*), which are known to be the main causative serovars of human leptospirosis in Japan. The diagnostic antigens of these five serovars are available commercially. The corresponding serovar should be determined carefully. We considered a serow as seropositive for leptospirosis when serum showed agglutination with any one of the five antigens at a serum dilution of 1:10. Many of the positive sera reacted similarly with two or more antigens; of them, sera reacting with *L. icterohaemorrhagiae* was the most common, followed by those with *L. autumnalis* and *L. canicola.* From the present study, the seropositive rate of leptospirosis was 10.4 per cent. In Japan, leptospirosis is widespread among many animal species including cattle, horses, swine, dogs, cats and rats (Yanagawa and Takashima 1974). A recent survey run by the Japanese Veterinary Medical Association (1984) showed that 13.8 per cent (63 out of 457) of dogs in Japan were seropositive. Although no serological survey on humans and domestic animals living near areas where serows were captured was carried out, it can be said that reservoirs of leptospirosis exist in the serow's habitat. The percentage of seropositive males was significantly higher than that of females. Similar data were obtained in dogs (Ryu *et al.* 1975, Torten, 1979). The reason for the above phenomenon is not certain, but it is probably attributable to the male habit of sniffing females and also of wandering widely in areas where infected rodents or contaminated inanimate vehicles such as soil and water are present.

Chlamydia psittaci can affect not only psittacine birds but also a wide variety of mammals including humans. The present study showed a seropositive rate of 10.7 per cent without any difference by serow age, sex or years studied. From their serological survey of bovine chlamydiosis in neighbouring areas, Omori *et al.* (1960) reported a positive rate of 34.7 per cent (35/101). Recently, Fukushi *et al.* (1985) confirmed a positive rate of 30.2 per cent (316/1048) by CF test. These and our own data together suggest that the chlamydial reservoirs are widely distributed and continuously maintained in the habitats of cattle and serows.

Rotavirus is a common cause of gastroenteritis in the young of many mammalian and avian species. In the present study, the immune adherence haemagglutination (IAHA) test was employed with a self-prepared tissue-culture antigen of the Lincoln strain of bovine origin (Sugiyama *et al.* 1984). In prior tests, we had confirmed that IAHA titres are closely correlated with neutralising antibody titres. Overall, 25.7 per cent of serows were seropositive. This positive rate was the highest among those of the eight microbial pathogens tested in this study. The positive rate varied from year to year; higher rates of 40.4 per cent and 24.2 per cent were attained in 1983 and 1981, respectively. Serows with high antibody titres such as 1:64 or more were detected in 1983 and 1981. These data suggested that rotavirus infection might have occurred among the serow population in these years.

It is generally assumed that calf-rotavirus infection occurs during the winter months, but we could not observe any port-mortem finding suggesting diarrhoea, though serows were killed in winter periods. There is a possibility that rotavirus infection in the serow population may occur during summer to autumn. The positive rate of the 1-year-old group was significantly ($P<0.05$) lower than those of older groups for reasons that are unclear. Woode and Bridger (1975) and Woode *et al.* (1975) showed experimentally that passively derived maternal antibody in the blood did not protect the calf against oral challenge of rotavirus. They also demonstrated that the antibody content of the colostrum showed a rapid decline within two days of calving. These findings could explain field outbreaks of diarrhoea in calves less than 5 weeks old (Inaba 1978). Thus the antibody prevalence of the <1-year-old group was presumed to be on a level with that of the older group. A higher rate of seropositivity in the older group might be partially due to recurrent infections by persistently infected animals around serow populations (McNulty 1978). In Japan, rotavirus infection has been shown by serological tests to be widespread among cattle (Sato *et al.* 1981), so it could very well spread widely among future serow populations. Even at present, this infection may be one of the important diseases prevalent among serows.

Akabane virus, which belongs to the Simbu group of Bunya viruses, is known as an agent causing epizootic abortion and congenital arthrogryposis-hydranencephaly in cattle, sheep and goats. The disease occurred among cattle in Japan from 1972 and 1974 (Inaba *et al.* 1975, Kurogi *et al.* 1975), and then decreased rapidly through vaccination. In the present study, a relatively low seropositive rate (5.8 per cent) was obtained by HI test, and seropositivity of serows over 3 years of age was significantly higher ($P<0.05$) than in those under 3 years of age. Akabane disease is thought to be transmitted by haematophagous arthropods (Oya *et al.* 1961, Doherty *et al.* 1972), so such arthropod vectors and reservoirs may exist in the habitat of serows, and mild outbreaks may occur sporadically. Judging from the clinical cases of bovine Akabane disease, the possibility of an outbreak in serow populations is undeniable.

As to the Japanese encephalitis virus, we presumed that examinations would reveal seropositive cases, but no serological evidence of the disease was found in 208 serows tested. It was shown that Japanese encephalitis is not prevalent among the serow population as in cattle.

Finally, the Ibaraki virus is known to cause epizootis disease of cattle resembling bluetongue, affecting the oesophageal and laryngo-pharyngeal musculature. In Japan, Ibaraki disease occurred in cattle from 1951 to 1960 (Omori *et al.* 1969), but no outbreak was subsequently reported because of vaccination measures. However, in 1982, after an interval of 22 years, this disease occurred in three cattle in Kyushu in southern Japan (Uchikoshi *et al.* 1983). In the present study, antibodies to this virus were checked in 337 serows for three years, but no antibody was detected. Reservoirs of this virus probably do not exist in serow habitats.

As discussed here, we checked antibodies to the eight pathogens, and demonstrated antibody prevalence to these pathogens except for *Brucella abortus*, Japanese encephalitis virus and Ibaraki virus. The results suggest that infectious diseases caused by at least these five pathogens have occurred among the wild Japanese serow population in Japan. In particular, rotavirus infection might be one of the important diseases prevalent among the serow population. Although the territory inhabited by Japanese serows seems to be quite removed from those of domestic animals or humans, the serological evidence is clear that direct or indirect contacts exist between them.

Since it is almost impossible to vaccinate wild animals, in conserving and controlling Japanese serows as a special natural monument one must also take into consideration the possible spread of disease from domestic animals and humans. Conversely, reservoirs of certain diseases may be maintained for a long time in mountainous regions even after the diseases have been eradicated in urban areas. Thus, these reservoirs could act as continuing sources of infection for domestic animals and also humans.

This is the first serological survey report of several pathogens of large wild animals in Japan, and it suggests the need for more examinations to clarify the situation of infectious diseases prevalent in wild Japanese serow populations. The authors are now pursuing further studies in this regard.

ACKNOWLEDGEMENTS

We thank M. Sugimura, Professor at Gifu University, and his staff, for providing us with the serow blood samples. This work was supported in part by a Grant-in-Aid for Co-operative Research (58362001; Chief researcher: M. Sugimura) from the Ministry of Education, Science and Culture, Japan.

REFERENCES

Costa, A.J., F.G. Araujo, J.O. Costa, J.D. Lima and E. Nascimento. (1977) Experimental infection of bovines with oocysts of *Toxoplasma gondii. J. Parasitol. 63*, 212

Doherty, R.L., J.G. Carley, H.A. Standfast, A.L. Dyce and W.A. Snowdon. (1972) Virus strains isolated from arthropods during an epizootic of bovine ephemeral fever in Queensland. *Aust. Vet. J. 48*, 81

Fukushi, H., H. Ogawa, T. Morikoshi, Y. Okuda, S. Shimakura and K. Hirai. (1985) Chlamydial complement fixing antibodies in cows, horses and pigs from 1980 to 1983. *Res. Bull. Fac. Agr. Gifu Univ. 50*, 259

Inaba, Y. (1978) Neonatal calf diarrhea — with special reference to rotavirus infection (review). *J. Japan. Vet. Med. Assoc. 31*, 127 (in Japanese)

Inaba, Y., H. Kurogi and T. Omori. (1975) Akabane disease: epizootic abortion, premature birth, stillbirth and congenital arthrogryposis-Hydranencephaly in cattle, sheep and goats caused by Akabane virus. *Aust. Vet. J. 51*, 584

Inoue, S., S. Matsuno and R. Kono. (1981) Difference in antibody reactivity between complement fixation and immune adherence hemagglutination tests with virus antigens. *J. Clin. Microbiol. 14*, 241

310

Japanese Veterinary Medical Association (1984) Serologic survey for zoonoses. *J. Japan. Vet. Med. Assoc. 37*, 546 (in Japanese)

Kobayashi, A., N. Hirai, Y. Suzuki, H. Nishikawa and N. Watanabe. (1977) Evaluation of a commercial toxoplasma latex agglutination test. *Japan. J. Parasitol. 26*, 175 (in Japanese)

Kurogi, H., Y. Inaba, Y. Goto, H. Takahashi, K. Sato, T. Omori and M. Matsumoto. (1975) Serologic evidence for etiologic role of Akabane virus in epizootic abortion–arthrogryposis–hydranencephaly in cattle in Japan, 1972–1974. *Arch. Virol. 47*, 71

McNulty, M.S. (1978) Rotaviruses. *J. Gen. Virol. 40*, 1

Nakamura, Y. and H. Nakamura. (1983) Haemagglutination inhibition test for Japanese encephalitis virus. *Nisseiken-Tayori 29*, 29 (in Japanese)

Omori, T., Y. Inaba, T. Morimoto, Y. Tanaka, R. Ishitani, H. Kurogi, K. Munakata, K. Matsuda and M. Matsumoto. (1969) Ibaraki virus, an agent of epizootic disease of cattle resembling bluetongue I. *Japan. J. Microbiol. 13*, 139

Omori, T., S. Ishi and M. Matsumoto. (1960) Miyagawanellosis of cattle in Japan. *Am. J. Vet. Res. 21*, 564

Oya, A., T. Okuno, T. Ogata, I. Kobayashi and T. Matsuyama. (1961) Akabane, a new arbovirus isolated in Japan. *Japan. J. Med. Sci. Biol. 14*, 101

Ryu, E., M. Bessho, K. Iyoda and N. Shimohara. (1975) Investigation on leptospiral agglutination of dogs in the Shikoku region. *J. Japan. Vet. Med. Assoc. 28*, 369 (in Japanese with English summary)

Sato, K., Y. Inaba, T. Shinozaki and M. Matsumoto. (1981) Neutralizing antibody to bovine rota virus in various animal species. *Vet. Microbiol. 6*, 259

Sugimura, M., Y. Suzuki, S. Kamiya and T. Fujita. (1981) Reproduction and prenatal growth in the wild Japanese serow, *Capricornis crispus. Japan. J. Vet. Sci. 43*, 553

Sugiyama, M., N. Minamoto, T. Kinjo and A. Hashimoto. (1984) A serological survey on rotavirus infection in dogs by immune adherence haemagglutination test. *Japan. J. Vet. Sci. 46*, 767

Torten, M. (1979) Leptospirosis. In J.H. Steel (ed.) *CRC Handbook Series on zoonoses.* CRC Press, FLA, pp. 363–421

Uchikoshi, N., N. Enaga, H. Shimodaira and F. Minamikawa. (1983) An outbreak of Ibaraki disease in Kyusyu, Japan in 1982. *J. Japan. Vet. Med. Assoc. 36*, 648 (in Japanese with English summary)

Woode, G.N. and J.C. Bridger. (1975) Viral enteritis of calves. *Vet. Rec. 25*, 85

Woode, G.N., J. Jones and J. Bridger. (1975) Levels of colostral antibodies against neonatal calf diarrhoea virus. *Vet. Rec. 23*, 148

Yanagawa, R. and I. Takashima. (1974) Animal leptospirosis. *J. Japan. Vet. Med. Assoc. 27*, 211 (in Japanese)

27

Haematological and biochemical findings on Japanese serow

Hiroshi Hori[1] and Hiroshi Takeuchi[2]

[1]Kanazawa Natural Park Zoo, Yokohama; and [3]Laboratory of Veterinary Physiology, College of Agriculture & Veterinary Medicine, Nihon University, Tokyo

INTRODUCTION

In Kanazawa Natural Park Zoo, Yokohama, Japan, we are now keeping four male Japanese serows and two females, some of which were kept in Nogeyama Zoological Garden from 1979 till May 1984. In Yokohama Zoological Garden we have often carried out haematological and biochemical examination of animals in order to control the health of the animals. Unlike domestic animals and pets, animals kept in the zoo perform the same behaviours as wild ones. The blood condition of the animals was checked to determine their previous infections. In Kanazawa Natural Park Zoo, we examined the blood of ten Japanese serows, five of which are now kept in Kanazawa; the other five were caught in November 1985 in the town of Nakanojo in Gunma Prefecture, using RabAΣ.

RESULTS

Three male Japanese serows and two females, which have been kept in good condition, and two wild males and three wild female serows were examined. One of the females was born in Kanazawa Natural Park Zoo on 15 June 1984 (Table 27.1). Until now the duration of keeping serows at Yokohama Zoo has ranged between 6 years and 5 months.

The serows in captivity were fed lucerne hay (400 g), pellets for grass eaters (280 g), high-protein feed (130 g), hay cubes (130 g), carrots (280 g), calcium (8 g) and valinase (3 g) per day. In spring and summer, the serows were given Italian ryegrass silage (1000 g) instead of lucerne hay (200 g). We could not, however find out what the wild serows had been eating.

Blood samples were collected from the jugular vein by using a normal 10 ml vacuum blood sample tube (Vacutainer SST) and a heparinised 2 ml vacuum blood sample tube. The tubes were centrifuged at 2600 r.p.m. for 10

Table 27.1: Breeding history of Japanese serow at Kanazawa Natural Park Zoo

Ind. No.	Hause Name	Sex	Age (years)	Remark
No. 1	Pao	♂	4	Capt.
No. 2	Kotarou	♂	12	Capt.
No. 3	Teppei	♂	3	Capt.
No. 4	—	♂	6	Wild
No. 5	—	♂	16	Wild
No. 6	Kater	♀	3	Capt.
No. 7	Azusa	♀	1.4	Zoo born
No. 8	—	♀	14	Wild
No. 9	—	♀	7	Wild
No. 10	—	♀	3	Wild

min. A Unopette 5851 for counting RBC and a Unopette 5856 for counting WBC were employed to analyse the whole blood.*

The number of RBC of every male serow in captivity and of male wild serow was not less than $10^7/\mu l$. The number of RBC of every kept and wild female individual was also not less than $10^7/\mu l$. On the other hand, the number of WBC of Nos 11 and 12 was $5 \times 10^3/\mu l$, and that of No. 10 $3 \times 10^3/\mu l$.

It is suggested that the difference between the former and the latter might be caused by the ages of animals. The serum concentrations of T.p, Alb. and iron were higher in the kept male individuals than in the wild ones. In contrast, G.O.T., G.P.T. and β-LP levels were higher in the wild male individuals than in the kept ones. The concentration of iron was higher in the blood of kept female ones, and G.O.T. and G.P.T. levels were higher in the wild ones.

There was a significant difference between the results of the haematological and biochemical analyses of all the male and female Japanese serows (Table 27.2). However, differences between WBC, Glob., A/G, Z.T.T., L.D.H. and iron levels of all the kept serows and those of all the wild ones were significant ($P< 0.05$) (Table 27.3).

Significant differences between T.p, Glob. and G.P.T. concentrations of kept males (2.5 ± 0.7 g/dl, 10 ± 1, Karmen) and those of wild males (8.2 ± 0.8 g/dl, 33 ± 8 Karmen) were also noted.

Glob., A/G, Z.T.T. and G.O.T. levels of kept females and those of wild females were significantly different ($P<0.05$) (Table 27.4). On the other hand, there was no difference between the haematological and biochemical findings on kept males and females. We found significant differences

---

*For explanation of abbreviations see Table 27.2

Table 27.2: Results of blood examination between male and female

			Male			Female	
		n	Mean	SD	n	Mean	SD
RBC	$10^4/mm^3$	8	1212 ±	298	5	1185 ±	165
Hematocrit	%	8	41.5 ±	10.4	5	31.5 ±	8.5
Hemoglobin	g/dl	6	14.4 ±	4.3	5	12.8 ±	2.5
WBC	$/mm^3$	8	3275 ±	923	5	4505 ±	1259
T.p	g/dl	8	8.5 ±	2.6	5	7.4 ±	1.0
Alb.	g/dl	6	4.1 ±	0.5	5	4.2 ±	1.4
Glob.	g/dl	6	4.4 ±	3.0	5	3.3 ±	1.8
A/G		6	1.4 ±	1.0	5	2.0 ±	1.8
Z.T.T.	Kunkel	6	5.5 ±	3.3	5	6.0 ±	3.5
T.T.T.	Maclagan	6	0.6 ±	0.4	5	0.7 ±	0.7
G.O.T.	Karmen	6	214 ±	191	5	239 ±	37
G.P.T.	Karmen	6	18 ±	12	4	107 ±	152
A.L.P.	K.A.U.	6	75.4 ±	84.1	5	9.8 ±	4.3
L.D.H.	W.U.	6	1450 ±	1281	5	2221 ±	1434
L.A.P.	G.R.U.	6	162 ±	36	5	155 ±	36
Glu.	mg/dl	6	158 ±	115	5	87 ±	62
B.U.N.	mg/dl	5	28.5 ±	12.1	4	22.3 ±	18.3
T-Bil	mg/dl	6	1.3 ±	0.9	5	0.9 ±	0.6
cho-E	mg/dl	6	50.8 ±	10.4	4	46.8 ±	28.6
Fe	μ g/dl	6	175 ±	74	5	118 ±	100
Ca	mg/dl	6	12.9 ±	3.0	5	13.6 ±	5.6
U.A.	mg/dl	6	1.4 ±	0.5	5	1.1 ±	1.3
Crea.	mg/dl	6	1.6 ±	0.9	5	2.0 ±	1.6
I.P.	mg/dl	6	4.3 ±	1.4	5	6.6 ±	3.0
T.G.	mg/dl	6	45 ±	25	5	116 ±	74
HDL cho-E	mg/dl	5	39.4 ±	15.8	3	50.2 ±	9.6
Amy.	Somogyi	5	96 ±	47	4	140 ±	155
β-LP.	mg/dl	5	44.0 ±	39.6	5	94.5 ±	52.0
K	mEq/l	5	5.1 ±	2.7	4	6.7 ±	4.8
Mg	mg/dl	5	2.0 ±	0.3	4	3.7 ±	1.3
A.C.P.	K.A.U.	3	1.1 ±	0.4	—	—	—
γ-G.T.P.	IU/l	4	70.8 ±	47.8	3	28.0 ±	45.9
α-H.B.D.	R.U.	4	418 ±	106	3	749 ±	245
ch E	ΔpH	5	0.1 ±	0.01	4	0.09 ±	0.04
C.P.K.	IU/l	5	535 ±	435	3	320 ±	485

* $P < 0.05$

Notes

RBC	: erythrocyte or red blood cell		Mbl.	: megakaryoblast
WBC	: leukocyte or white blood cell		T.p	: total protein
Baso.	: basophil		Alb.	: albumin
Eos.	: eosinophil		Glob.	: globulin
Myelo.	: myeloblast		A/G	: albumin-globulin ratio
Bands.	: band neutrophil		Z.T.T.	: zinc turbidity test
Sag.	: segmenter neutrophil		T.T.T.	: thymol turbidity test
Lympho.	: lymphocyte		G.O.T.	: glutamic oxaloacetic transaminase
Mono.	: monocyte		G.P.T.	: glutamic pyruvic transaminase
Plasma.	: plasma cell		A.L.P.	: alkali phosphatase

L.A.P.	: leucine aminopeptidase
Glu.	: glucose
B.U.N.	: blood urea nitrogen
T-Bil.	: total bilirubin
cho-E	: total cholesterol
Fe	: Iron
Ca	: calcium
U.A.	: uric acid
Crea.	: creatinine
I.P.	: inorganic phosphorus
T.G.	: triglyceride

HDL cho-E	: high density lipoprotein cholesterol
Amy.	: amylase
β-LP	: β-lipoprotein
K	: kalium or potassium
Mg	: magnesium
A.C.P.	: acid phosphatase
γ-G.T.P.	: γ-glutamyl transpeptidase
α-H.B.D.	: α-hydroxybutyric dehydrogenase
ch E	: cholinesterase
C.P.K.	: creatine phosphokinase

Table 27.3: Results of blood examination of animals in captivity and in the wild

		in Captivity			in Wild		
		n	Mean	SD	n	Mean	SD
RBC	$10^4/mm^3$	8	1330 ±	193	5	997 ±	182
Hematocrit	%	8	42.4 ±	9.3	5	26.1 ±	14.8
Hemoglobin	g/dl	6	13.6 ±	1.4	5	13.8 ±	5.4
WBC	/mm³	8	3109 ±	795*	5	4770 ±	1006*
T.p	g/dl	8	7.2 ±	1.2	5	9.5 ±	2.7
Alb.	g/dl	6	4.6 ±	0.9	5	3.5 ±	0.6
Glob.	g/dl	6	2.1 ±	0.8*	5	6.0 ±	2.1*
A/G		6	2.6 ±	1.3*	5	0.6 ±	0.2*
Z.T.T.	Kunkel	6	3.5 ±	1.6*	5	8.4 ±	2.5*
T.T.T.	Maclagan	6	0.4 ±	0.2	5	0.8 ±	0.7
G.O.T.	Karmen	6	104 ±	19	5	371 ±	126
G.P.T.	Karmen	6	11 ±	2	4	117 ±	145
A.L.P.	K.A.U.	6	79.1 ±	80.4	4	4.6 ±	5.0
L.D.H.	W.U.	6	857 ±	333*	5	2932 ±	1200*
L.A.P.	G.R.U.	6	155 ±	41	5	164 ±	29
Glu.	mg/dl	6	150 ±	88	5	77 ±	109
B.U.N.	mg/dl	6	27.7 ±	12.7	3	21.7 ±	20.0
T-Bil	mg/dl	6	1.1 ±	1.0	5	1.2 ±	0.5
cho-E	mg/dl	5	48.8 ±	13.2	5	49.6 ±	24.2
Fe	μ g/dl	6	219 ±	41*	5	66 ±	33*
Ca	mg/dl	6	13.1 ±	3.3	5	13.3 ±	5.3
U.A.	mg/dl	6	1.6 ±	0.9	5	1.4 ±	1.2
Crea.	mg/dl	6	1.2 ±	0.5	5	2.4 ±	1.5
I.P.	mg/dl	6	5.7 ±	2.2	5	4.9 ±	3.0
T.G.	mg/dl	6	59 ±	26	5	98 ±	88
HDL cho-E	mg/dl	4	38.8 ±	17.8	4	48.2 ±	9.6
Amy.	Somogyi	4	99 ±	49	5	129 ±	138
β-LP	mg/dl	4	22.8 ±	22.2	5	85.4 ±	62.0
K	mEq/l	4	4.4 ±	1.0	5	7.0 ±	4.6
Mg	mg/dl	4	2.3 ±	0.5	5	3.1 ±	1.6
A.C.P.	K.A.U.	4	1.1 ±	0.3	—	—	—
γ-G.T.P.	IU/l	4	70.8 ±	47.8	3	28.0 ±	45.9
α-H.B.D.	R.U.	4	402 ±	87	3	770 ±	208
ch E	Δ pH	4	0.09 ±	0.02	5	0.10 ±	0.03
C.P.K.	IU/l	4	191 ±	196	4	718 ±	464

* $P < 0.05$

For explanation of abbreviations see Table 27.2

Table 27.4: Results of blood examination of female animals in captivity and in the wild

		in Captivity			in Wild		
		n	Mean	SD	n	Mean	SD
RBC	10⁴/mm³	2	1320 ±	88	3	1096 ±	143
Hematocrit	%	2	39.0 ±	4.2	3	26.5 ±	6.5
Hemoglobin	g/dl	2	14.9 ±	1.0	3	11.5 ±	2.2
WBC	/mm³	2	3213 ±	230	3	5367 ±	599
T.p	g/dl	2	7.0 ±	1.1	3	7.7 ±	1.0
Alb.	g/dl	2	5.6 ±	1.1	3	3.2 ±	0.2
Glob.	g/dl	2	1.4 ±	0*	3	4.5 ±	0.9*
A/G		2	4.0 ±	0.8*	3	0.7 ±	0.1*
Z.T.T.	Kunkel	2	2.4 ±	1.5*	3	8.5 ±	0.8*
T.T.T.	Maclagan	2	0.4 ±	0.1	3	0.9 ±	0.9
G.O.T.	Karmen	2	94 ±	7*	3	335 ±	53*
G.P.T.	Karmen	2	13 ±	4	2	200 ±	186
A.L.P.	K.A.U.	2	12.8 ±	0.2	3	7.9 ±	4.6
L.D.H.	W.U.	2	945 ±	13	3	3072 ±	1182
L.A.P.	G.R.U.	2	136 ±	49	3	168 ±	28
Glu.	mg/dl	2	109 ±	62	3	73 ±	70
B.U.N.	mg/dl	2	24.2 ±	14.0	2	20.3 ±	28.1
T-Bil	mg/dl	2	0.6 ±	0.3	3	0.8 ±	0.4
cho-E	mg/dl	—	—	—	3	50.7 ±	33.7
Fe	μ g/dl	2	220 ±	58	3	51 ±	34
Ca	mg/dl	2	13.6 ±	5.0	3	13.6 ±	7.0
U.A.	mg/dl	2	0.5 ±	0.3	3	1.5 ±	1.6
Crea.	mg/dl	2	1.2 ±	0.8	3	2.5 ±	1.9
I.P.	mg/dl	2	7.1 ±	3.9	3	6.3 ±	3.3
T.G.	mg/dl	2	76 ±	11	3	142 ±	91
HDL cho-E	mg/dl	—	—	—	2	49.0 ±	13.2
Amy.	Somogyi	—	—	—	3	177 ±	166
β-LP	mg/dl	—	—	—	3	108 ±	54
K	mEq/l	—	—	—	3	7.2 ±	5.6
Mg	mg/dl	—	—	—	3	3.0 ±	2.7
A.C.P.	K.A.U.	—	—	—	—	—	—
γ-G.T.P.	IU/l	—	—	—	2	41.5 ±	55.9
α-H.B.D.	R.U.	—	—	—	2	890 ±	20
ch E	Δ pH	—	—	—	3	0.10 ±	0.04
C.P.K.	IU/l	—	—	—	2	469 ±	581

* P<0.05

For explanation of abbreviations see Table 27.2

between the T.p (12.2±1.6 g/dl) and Glob. (8.2±0.8 g/dl) concentrations of wild males and the T.p (7.7±1.0 g/dl) and Glob. (4.5±0.9 g/dl) levels of wild females.

CONCLUSION

The kept Japanese serows are fed a much higher protein feed than the wild ones bearing in mind their reduced amounts of exercise, so at first it was supposed that the concentrations of T.p, cho-E and T.G. (triglyceride) would be higher in the blood of the kept ones than in that of the wild ones. However, the differences between the haematological and biochemical states of kept ones and wild ones were not so remarkable. As a result, there would seem to be no problem in our method of feeding of the Japanese serows to keep them healthy.

Part Seven:
Endocrinology and
Reproduction of *Capricornis*

28

Reproduction of female Japanese serow based on the morphology of ovaries and fetuses

Isao Kita,[1] Makoto Sugimura,[2] Yoshitaka Suzuki,[2] Toshiro Tiba[1] and Shingo Miura[3]

[1]Department of Theriogenology, and [2]Department of Veterinary Anatomy, Faculty of Agriculture, Gifu University, Gifu 501-11, Japan; and [3]Department of Biology, Hyogo College of Medicine, Mukogawa 1-1, Nishinomiya 663, Japan

INTRODUCTION

The Japanese serow (*Capricornis crispus*) is the only wild bovine ruminant preserved as a special natural monument in Japan. Until the present authors reported some data on the reproduction of males (Tiba *et al.* 1981a,b) and females (Sugimura *et al.* 1981, 1983, 1984, Kita *et al.* 1983a,b, 1986), there was little information in the literature regarding the reproduction of the serow (Asdell 1964). In the present study, the data previously reported and further information on the reproduction of female serows are presented based on morphological studies of the ovary and the fetus.

GROWTH OF FETUSES

The relationship between the growth of fetuses and the season when their mothers were captured offered abundant information regarding the reproduction of the animals. Two-hundred and sixty-one embryos and fetuses were obtained from 259 pregnant females including two cases with twins. Their mothers were captured from December to March, 1979 to 1983, in Gifu Prefecture in Central Japan. After determining the body weight, crown–rump length and sex, embryos and fetuses were preserved in 10 per cent formalin solution. The external features of embryos and fetuses were observed macroscopically or, if necessary, with the aid of a stereomicroscope.

Growth curve of serow fetuses

The crown–rump lengths of 261 embryos and fetuses ranged between 1.0 and 34.0 cm. Each length was plotted against the date of capture of the mother serows (Figure 28.1). By referring to the growth curve of domestic ruminant embryos (Evans and Sack 1973), the entire prenatal growth curve of serows was approximated. On the basis of this approximation, the length at birth was estimated to be about 48 to 50 cm, based on a gestation period of 210 to 220 days in captive serows (Ito 1971, Komori 1975).

Fetal length−weight relationship

Embryos and fetuses from 1.0 to 34.0 cm in length varied from 0.12 to 1300 g in weight. There was an obvious linear relationship between the length and weight when both variables were plotted on a logarithmic scale. Consequently, a straight line was fitted to log-transformed data, giving an equation of $\log W = 2.76 \log L - 1.12$, where W = fetal weight in grams and L = fetal length in centimetres.

Figure 28.1: Prenatal development of Japanese serow

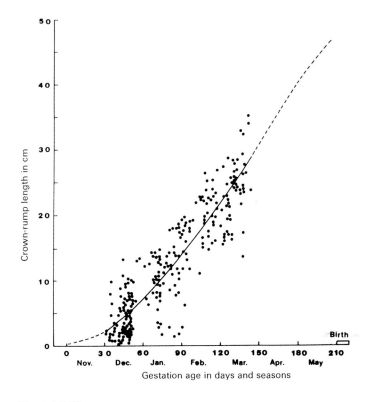

Source: Kita *et al.* 1983b

Extrapolating from the above equation, the suggested birth weight was calculated as 3313 to 3708 g from the suggested length at birth.

Sex ratio of serow fetuses

The sex of serow fetuses more than 2.7 cm in length is determined by the differentiation of external genital organs. The sex ratio was almost equal, female 1.03 to male 1.

Litter size

Out of 259 pregnant females, 257 begot a single offspring and only two females had twins. It is assumed that serows usually have a single offspring.

BREEDING SEASON AND PARTURITION SEASON

If it is assumed that all fetuses grow at a similar rate, most of the scatter shown in Figure 28.1 is due to variations in the conception date. Therefore, the breeding season is calculated to be from September to January, with peak conception from late October to early November (Figure 28.2). Because the gestation period is 215 days on average, the season of parturition is also assumed to extend from April to July, with a peak from late May to early June (Figure 28.3).

HISTOLOGY OF RETROGRADE CORPORA LUTEA OF OESTRUS AND PREGNANCY

Some basic data on the reproduction of Japanese serow were also obtained from histological studies of serow ovaries (Sugimura *et al.* 1984, Kita *et al.* 1983a,b, 1986). Ovaries were obtained from 189 female Japanese serows captured from December to March, 1979 to 1983. Some 77 females among them were pregnant. The age estimation was made on the basis of the tooth eruption wear patterns (Miura and Yasui 1985). In some animals, age was determined based on the annual layers of the cementum of the teeth and horn rings (Miura 1985).

Ovaries were fixed in 10 per cent formalin solution, and then serially sliced at 1.5–2.0 mm thickness with a razor. The size of the largest follicles, corpora lutea and retrograde corpora lutea observed on the ovarian slices was recorded. Ovarian slices were embedded in paraffin, and sectioned at 5–7 μm thickness, semi-serially at intervals of 100 μm in most cases.

Figure 28.2: Half-monthly distribution of conceptions calculated from fetal length

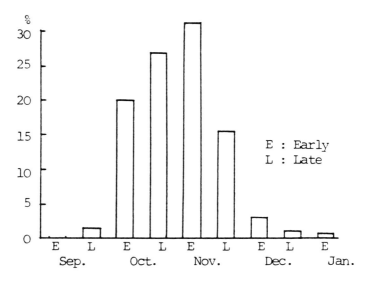

Source: Kita *et al.* 1983b

Figure 28.3: Half-monthly distribution of birth calculated from fetal length

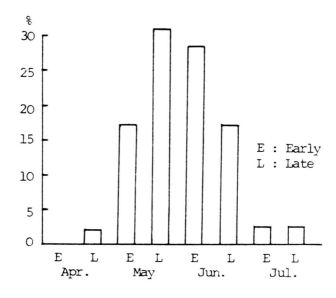

Source: Kita *et al.* 1983b

Sections were stained with Haematoxylin−Eosin, Weigert's resorcin−fuchsin for elastic fibres, or Masson's trichrome.

Corpora lutea and retrograde corpora lutea

The corpus luteum of oestrus consisted of the fibrous capsule, the trabeculae and the parenchyma. In the parenchyma there were numerous lutein cells, a rich capillary network and a small amount of arterioles without elastic fibres or lamina elastica interna and venules. In the capsule, small arteries and veins were observed.

The corpus luteum of pregnancy in the ovaries of the serows which conceived fetuses less than 15 cm in crown−rump length were histologically similar to the corpus luteum of oestrus. In the ovaries of the serows having fetuses more than 17 cm in crown−rump length, however the corpora began to decrease in size. Histologically, many arterioles and degenerated lutein cells were observed in the parenchyma. It was a notable finding that elastic fibres increased in the tunica media of the arterioles and fibrous capsule (Ito *et al.* 1984, Sugimura *et al.* 1984).

Two types of retrograde corpora lutea, hyaline and elastoid bodies, were distinguished. Hyaline bodies were observed in the ovaries of more than $1^1/_2$-year-old serows, and elastoid bodies in the ovaries of more than $3^1/_2$-year-old serows (Kita *et al.* 1983a).

The hyaline body showed the hyalinisation of parenchyma accompanied by degeneration of lutein cells and rapid degeneration of blood vessels without elastic fibres (Figure 28.4). The elastoid body had numerous coiled elastoid arterioles and degenerated lutein cells in the parenchyma, and was surrounded by an elastic capsule (Figure 28.5).

Based on the continuity of histology, the authors suggested that the corpora lutea of oestrus and of pregnancy change into hyaline and elastoid bodies, respectively (Sugimura *et al.* 1984). This suggestion was supported by the fact that, although pregnancies of serows were first recorded at the age of 30 months, elastoid bodies were found only in the ovaries of more than $3^1/_2$-year-old serows, whereas hyaline bodies were found also in the ovaries of $1^1/_2$ and $2^1/_2$-year-old serows. There were a few bodies with structures slightly different from those of the typical elastoid body. They had many degenerated lutein cells and a few degenerated arterioles in the parenchyma. The authors propose to call them 'pseudo-elastoid bodies'. Probably, these bodies are formed in connection with an interrupted pregnancy (Kita *et al.* 1986) (Figure 28.6). The origin and the fate of the hyaline and elastoid bodies are illustrated in Figure 28.7.

Figure 28.4: A hyaline body. A small hyaline body consists of a group of degenerated lutein cells and a hyaline zone. Almost no blood vessels are detected and the appearance of elastic fibres is observed. Weigert, × 240

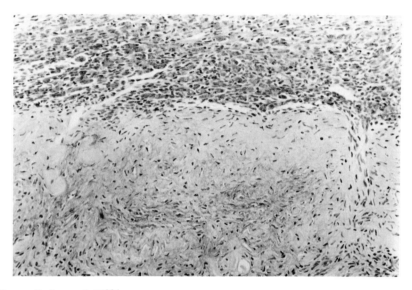

Source: Sugimura *et al.* 1984

Figure 28.5: An elastoid body (medium-sized). Elastosis is observed in coiled arterioles and capsule. Weigert, × 120

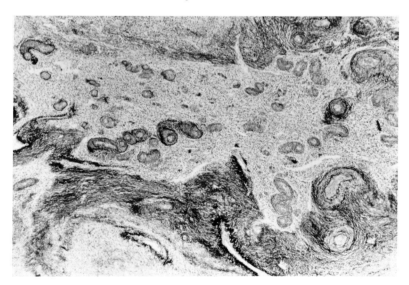

Source: Sugimara *et al.* 1984

326

Figure 28.6: A pseudo-elastoid body. Many degenerated lutein cells and a few degenerated arteries showing elastosis are observed. Weigert, × 500

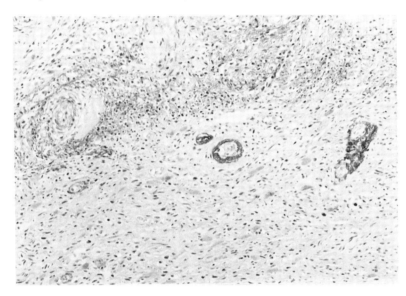

Source: Kita *et al.* 1986

Frequency of ovulation during one breeding season

The frequency of ovulation may be suggested from the total number of corpora lutea and hyaline bodies, because serows usually ovulate one ovum. From the results obtained, the ovulation was one to six times during one breeding season. About 70 per cent of serows probably conceived at the first or second ovulation, because out of 58 pregnant serows 13 females had one corpus and 26 females two corpora. It has been suggested that hyaline bodies disappear within about 5 months, since the maximum number of hyaline bodies encountered in a pair of ovaries was six, of which the smallest body was disappearing in the fibrous parenchyma of the ovarian cortex (Sugimura *et al.* 1984), and the oestrous cycle in captive serows ranges from 20 to 21 days (Ito 1971).

Sexual maturity and reproductive life

On the basis of ovulations, as determined by the presence of corpus luteum or hyaline body (Figure 28.8), about half of the female serows seemed to become sexually mature at around 2½ years old; at 4½ years old the

327

Figure 28.7: Structural changes of follicles and corpora lutea in Japanese serows

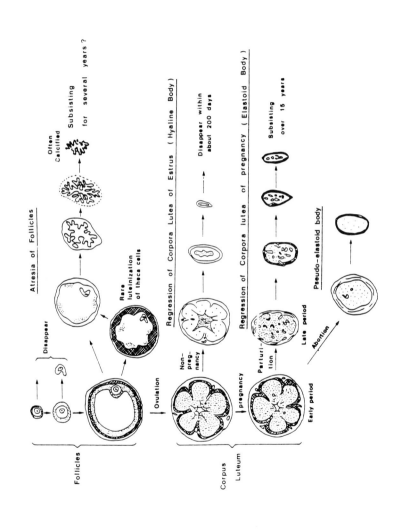

Figure 28.8: Relationship between age and frequency of ovulation or pregnancy

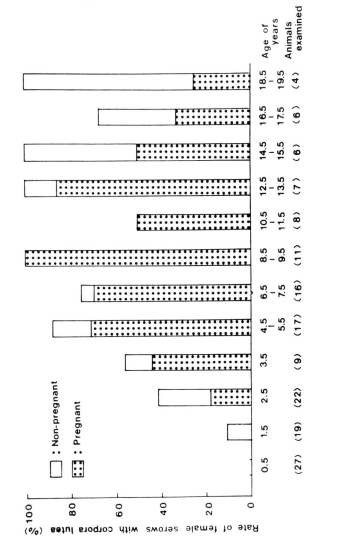

majority were sexually mature. It was also noted that one-tenth of 1½-year-old animals had experienced ovulation. Reproductive life in the serow seemed to last until about 19½ years of age. However, it is interesting that the pregnancy rate in 2½-year-old and more than 14½-year-old serows decreased in spite of a high rate of ovulations.

Frequency of past parturition or past pregnancy

Serows usually ovulate one ovum and deliver a single offspring (Kita *et al.* 1983b). Therefore, the total number of elastoid bodies in one pair of ovaries of one animal seemed to indicate the frequency of past parturitions. From calculations of the bodies in paired ovaries, the past parturitions of one animal were suggested to be 15 at maximum. Therefore, the elastoid bodies may not disappear over 15 years and probably remain for life.

As shown in Figure 28.9, a high correlation was recognised between the age and the number of past pregnancies estimated by the method described

Figure 28.9: Relationship between age and total number of elastoid and pseudo-elastoid bodies

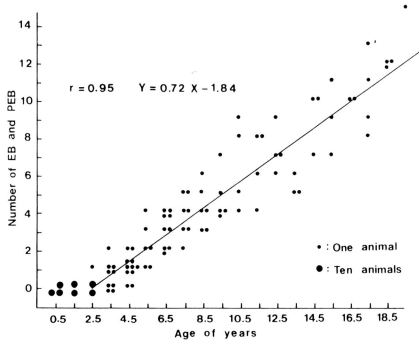

Source: Kita *et al.* 1986

330

above ($r = 0.95$**; $Y = 0.72\,X - 1.84$, where $X =$ age and $Y =$ number of past pregnancies). These results suggest that adult serows deliver about twice every three years. The approximate age of serows can also be estimated with the above equation.

REFERENCES

Asdell, S.A.L. (1964) *Patterns of mammalian reproduction.* Cornell University Press, Ithaca and New York, pp. 621

Evans, H.E. and W.O. Sack. (1973) Prenatal development of domestic and laboratory mammals. *Anat. Histol. Embryol. 2,* 11–45

Ito, T. (1971) On the oestrous cycle and gestation period of the Japanese serow, *Capricornis crispus. J. Mammal. Soc. Jap.,* 5, 104–8 (in Japanese with English summary)

Ito, Y., I. Kita, T. Tiba and M. Sugimura. (1984) Histological observations on the regression of corpus luteum graviditatis in the Japanese serow (*Capricornis crispus*). *Res. Bull. Fac. Agr. Gifu Univ.* (*49*) 273–282 (in Japanese with English summary)

Kita, I., M. Sugimura, Y. Suzuki and T. Tiba. (1983a) Reproduction of female Japanese serows, *Capricornis crispus,* based on pregnancy and macroscopical ovarian findings. *Res. Bull. Fac. Agr. Gifu Univ.* (*48*), 137–46 (in Japanese with English summary)

Kita, I., M. Sugimura, Y. Suzuki and T. Tiba. (1983b) Reproduction of wild Japanese serows based on the morphology of ovaries and fetuses. *Proc. Vth World Conf. Anim. Product., Tokyo 2,* 243-4

Kita, I., T. Tiba, M. Sugimura, Y. Suzuki and S. Miura. (1986) Frequency of past parturition estimated by retrograde corpora lutea of pregnancy, elastoid bodies, in Japanese serow ovary. *Zool. Anz. Jena* in press

Komori, A. (1975) Survey on the breeding of Japanese serow, *Capricornis crispus,* in captivity. *J. Japan. Assoc. Zool. Gard. Aq. 17,* 53–61 (in Japanese)

Miura, S. (1985) Horn and cementum annulation as age criteria in Japanese serow. *J. Wildl. Mgmt 49,* 152–6

Miura, S. and K. Yasui. (1985) Validity of tooth eruption-wear patterns as age criteria in the Japanese serow, *Capricornis crispus. J. Mammal. Soc. Japan 10,* 169–78

Sugimura, M., Y. Suzuki, S. Kamiya and T. Fujita. (1981) Reproduction and prenatal growth in the wild Japanese serow, *Capricornis crispus. Japan. J. Vet. Sci. 43,* 553–5

Sugimura, M., Y. Suzuki, I. Kita, Y. Ide, S. Kodera and M. Yoshizawa. (1983) Prenatal development of Japanese serows, *Capricornis crispus,* and reproduction in females. *J. Mammal. 64,* 302–4

Sugimura, M., I. Kita, S. Suzuki, Y. Atoji and T. Tiba. (1984) Histological studies on two types of retrograde corpora lutea in the ovary of Japanese serows, *Capricornis crispus. Zool. Anz. Jena 213,* 1–11

Tiba, T., M. Sugimura and Y. Suzuki. (1981a) Kinetik der Spermatogenese bei der Wollhaargemse (*Capricornis crispus*). I. Geschlechtreife und Jahreszeitliche Schwankung. *Zool. Anz. Jena, 207,* 16–24

Tiba, T., M. Sugimura and Y. Suzuki. (1981b) Kinetik der Spermatogenese bei der Wollhaargemse (*Capricornis crispus*). II. Samenepithelzyklus und Samenepithelwelle. *Zool. Anz. Jena 207* 25–34

29

Seasonal changes in male reproductive functions of Japanese serow

Toshiro Tiba,[1] Mikio Sato,[1]* Tadahiro Hirano,[1†] Isao Kita,[1] Makoto
Sugimura[2] and Yoshitaka Suzuki[2]
[1]Department of Theriogenology, and [2]Department of Veterinary Anatomy, Faculty
of Agriculture, Gifu University, Gifu 501-11, Japan

INTRODUCTION

Since 1979 the authors have conducted basic studies on the reproductive
physiology of male Japanese serows, and have published two reports on
spermatogenesis in 1981 (Tiba *et al.* 1981a,b). From these previous
studies, it was strongly suggested that reproductive activities in the male are
subject to seasonal fluctuation; that is, in fully adult males over $2^1/_2$ or 3
years of age, spermatogenesis showed a decline from December to March
of every year. It was also clarified in the previous studies that sexual
maturation in the male begins within 6 or 7 months after birth. Further
studies have been performed on the endocrine activity of the testis and on
secretory activity in the seminal vesicle. This chapter considers seasonal
changes in testicular and serum testosterone levels and the fructose content
of the seminal vesicle.

MATERIALS AND METHODS

From 1982 to 1985, materials were collected during the authorised season
for capture of the animals; that is, from December to March. A total of 558
animals were examined. Animals shot in their habitats were transported to
Gifu University for examination. Most materials were taken from the
animals transported, but for the purpose of precise histological observations on spermatogenesis, fresh testicular materials were removed at the
site of capture within a few hours after the animal's death.

---

*Present address: Handa Health Centre of Aichi Prefecture, Handa, Aichi 475, Japan
†Present address: Institute for Safety Research, Ono Pharmaceutical Company Limited,
Mikuni, Fukui 913, Japan

Size of testis.   The size of testis was measured in a total of 504 animals during three consecutive capturing seasons. The product of three dimensions of each testis was obtained, and the products for paired testes were summed.

Spermatogenic activity.   The seminiferous tubule diameter was measured in 31 animals, and the number of pachytene primary spermatocytes per cross-section of seminiferous tubule was counted in fresh testicular materials from 13 animals.

Testosterone concentration in testicular tissue.   Determination of the concentration was carried out in 207 animals for three consecutive seasons by means of radioimmunoassay.

Testosterone concentration in serum.   Serum was obtained from the coagulated blood in the heart of 106 animals, and the concentration of testosterone per millilitre of serum was determined by radioimmunoassay.

Weight of seminal vesicles.   The seminal vesicles were collected from 402 animals in three seasons. After obtaining the wet weight, the seminal vesicles were stored at $-20°C$ for determination of the fructose concentration.

Fructose concentration in the seminal vesicle.   The fructose concentration of the seminal vesicles of 174 animals was determined by the method of Lindner and Mann (1960).

Body weight.   As a supplementary method for determining the end of sexual maturation, the body weight of 545 animals captured in three consecutive seasons was obtained.

Classification of age groups.   The age estimation that underlies this study was based on the tooth-eruption wear patterns. The animals were classified into six age groups; that is, group 0 (7 to 10 months old), group 1 (19 to 22 months), group 2-I° (31 to 34 months), group 2-I · II (4.6 ± 1.8 years), group 2-III (7.4 ± 2.7 years) and group 2-IV · V (12.7 ± 3.9 years). Animals of groups 0 to 2-I° were immature ones in which the second dentition was not completed; those of groups 2-I · II to 2-IV · V are adults in which all teeth are permanent.

RESULTS

Size of testis

The means and standard deviations shown in Figure 29.1 were calculated from measurements made over three capturing seasons. Before these means were obtained, however, a possible significant difference in the mean value for each single season was investigated. No significant difference was found among the three seasons. There was also no significant difference in size between paired testes.

It is very clear from the figure that the testes grow rapidly between age groups 0 and 2-I°, that is, from 7 months to about $2^{1}/_{2}$ years of age. The testes continue to grow slowly in age groups 2-I° to 2-I · II, and thereafter the testis size remains relatively constant. In adults, a decrease in size occurs every year from December to March. A mean value taken from all adults each month shows a highly significant difference ($P<0.01$).

Testosterone concentration in testicular tissue

The first appearance of testicular testosterone is demonstrable in fawns 7 to 10 months old. The concentration increased rapidly between the age groups 0 and 2-I · II. In adults the concentration varied widely. In Table 29.1, the means and standard errors calculated from all the adults belonging to 2-I · II to 2-IV · V are given for each month. The decrease in mean value between December and January is highly significant ($P<0.01$), but the increase between January and March is not significant.

Table 29.1: Testosterone concentration and fructose concentration (means $\pm$ SE) in male adult Japanese serows in December to March. ( ) = Number of animals examined

	December	January	February	March
Testosterone in testicular tissue (ng/g)	496.26±57.46 (39)	223.40±25.12 (46)	266.96±71.27 (27)	305.55±40.57 (18)
Testosterone in serum (ng/g)	7.17±0.89 (22)	3.20±0.36 (24)	3.19±0.54 (18)	3.08±3.31 (11)
Frutose in seminal vesicle (mg/g)	1.09±0.03 (59)	0.81±0.09 (38)	0.65±0.07 (28)	0.58±0.08 (20)

Testosterone concentration in serum

The concentration increase with age in immature animals was not so clear as with testicular testosterone. The mean value in all adults in December

Figure 29.1: Seasonal changes in combined size of paired testes (M ± SD, XII 1982-III 1985)

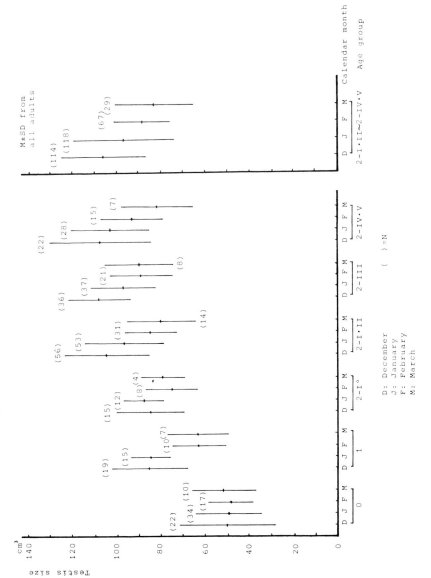

was significantly higher than in any other month ($P<0.01$, Table 29.1).

It was statistically clarified that there is a highly significant correlation between the concentration in serum and that in testicular tissue ($r = 0.814**$).

Correlation between spermatogenic activity and testosterone level

A highly significant correlation was shown between the seminiferous tubule diameter and the testosterone concentration in testicular tissue ($r = 0.532**$). On the other hand, in the fresh testicular materials obtained shortly after the animals' death, a highly significant correlation was found between the seminiferous tubule diameter and the number of primary spermatocytes at the pachytene stage per cross-section of the tubule ($r = 0.820**$).

Weight of seminal vesicle

A rapid increase in weight was found between age groups 0 and 2-I°. Seasonal changes in adults were not clear. The mean weight (0.73 g) of all adults in December was not significantly different from any other month.

Fructose concentration in the seminal vesicles

Owing to the tiny size of the seminal vesicles of young animals, it was extremely difficult to obtain the minimum volume of material necessary for this chemical assay. For this reason, insufficient data were acquired from young animals, but the values obtained seemed lower than in adults. A decrease in concentration from December to March was demonstrated clearly in adults. The mean concentration of all adults in December was significantly higher than in any other month ($P<0.01$, Table 29.1).

Correlation between fructose concentration in seminal vesicle and testosterone concentration in serum

A statistical analysis of the correlation was conducted with the same animals in which both substances were evaluated. A highly significant correlation coefficient was found between the two ($r = 0.753**$).

Body weight

It was clarified that the growth of male animals ends at $2^1/_2$ to 3 years of age, and thereafter the body weight is maintained at a relatively constant level. However, it was demonstrated statistically that there is a tendency for a decrease in weight to occur from December to February or March in most adult animals.

DISCUSSION

It can be concluded from the results that testicular functions in this animal are subject to seasonal fluctuations. This was indicated by the previous studies on spermatogenesis, and is now clearly demonstrated by the findings that the testosterone levels in testicular tissue as well as in serum are significantly higher in December than in any other month. Moreover, the demonstrated positive correlation between the testicular testosterone concentration and the seminiferous tubule diameter shows that seasonal fluctuations in this animal are, just as in other species, dependent upon the secretory functions of testosterone from the testis. It has also been clearly demonstrated in this study that the contents of fructose in the seminal vesicle are seasonally changeable, depending upon the testosterone level. Based on this knowledge, it may now be assumed that the secretory functions not only of the seminal vesicles but also of other accessory sex glands fluctuate seasonally under the control of testosterone.

According to fieldwork on sexual behaviour in males (Akasaka 1978, 1979, Akasaka and Maruyama 1979), the most frequent mating occurred in October and November. Out materials, therefore, must have been obtained in the period of lower reproductive activity, and we have never clarified a complete series of seasonal fluctuations in reproductive function in this animal. In any case, there is no doubt that our results are indicative of cyclically changeable reproductive functions in male Japanese serows.

REFERENCES

Akasaka, T. (1978) Social organization of Japanese serows in Kasabori. *Rep. Nat. Conserv. Soc. Japan. No. 55*, 225 (in Japanese)
Akasaka, T. (1979) Social organization of Japanese serows in Nibetsu, Akita Pref. — especially on the social units. *Rep. Nat. Conserv. Soc. Japan. No. 56*, 5 (in Japanese)
Akasaka, T and N. Maruyama. (1977) Social organization and habitat use of Japanese serow in Kasabori. *J. Mammal. Soc. Japan 7*, 87
Lindner, H.R. and T. Mann. (1960) Relationship between the content of androgenic

steroids in the testes and the secretory activity of the seminal vesicles in the bull. *J. Endocrinol. 21*, 341

Tiba, T., M. Sugimura and Y. Suzuki. (1981a) Kinetik der Spermatogenese bei der Wollhaargemse (*Capricornis crispus*) I. Geschlechtsreife und jahreszeitliche Schwankung. *Zool. Anz. 207*, 16

Tiba, T., M. Sugimura and Y. Suzuki. (1981b) Kinetik der Spermatogenese bei der Wollhaargemse (*Capricornis crispus*) II. Samenepithelzyklus und Samenepithelwelle. *Zool. Anz. 207*, 25

30

Steroid hormone synthesis and secretion by adrenals of wild Japanese serow, *Capricornis crispus*

Takao Nakamura, Yoshitaka Suzuki and Makoto Sugimura
Faculty of Agriculture, Gifu University, Gifu 501-11, Japan

INTRODUCTION

The adrenal corticoid production pattern from endogenous precursors had been reported in cattle (Beyer and Samuels 1956), sheep (Ford and Engel 1974), rat (Nakamura and Tamaoki 1964) and chicken (Nakamura *et al.* 1978). In these species, little or no 17α-hydroxylase activity was found in the adrenal tissue, and the major corticosteroids were corticosterone and aldosterone. By contrast 17α-hydroxylase was present in the adrenal gland of the higher mammals such as humans (Cameron *et al.* 1969, Oshima *et al.* 1969, Whitehouse and Vinson 1969) and the monkey (Kittinger and Beamer 1969). However, the pathway of corticoidogenesis in the adrenal of Japanese serows has not yet been thoroughly investigated. The present experiment was performed to establish the metabolic pathways of steroids in the Japanese serow *in vitro*, and to demonstrate intracellular distribution of the enzymatic activities related to transformation of the steroids. The results obtained from serow adrenals are discussed in comparison with the metabolic pathways in other mammals and in other species.

MATERIALS AND METHODS

After female Japanese serows had been decapitated, the adrenal glands were isolated. Pooled adrenals were weighed and homogenised with a loose-fitting glass−Teflon homogeniser in an ice-cold 0.33 M sucrose solution buffered with 0.05 M Tris-HC1 buffer (pH 7.4) and 0.005 M MgC1$_2$. After centrifugation of the homogenates at $800 \times g$ for 20 min, the supernatant fluid was centrifuged at $10000 \times g$ for 20 min, and the precipitate thus obtained was used as the mitochondrial fraction. The subsequent supernatant fluid was again centrifuged at $105000 \times g$ for 60 min. The precipitate was resuspended in 0.33 M sucrose solution and used as the microsomal fraction, and the supernatant was used as the cytosol fraction.

[4-^{14}C]-labelled pregnenolene (55.7 mCi/mmol), progesterone (29.3 mCi/mmol), and 11-deoxycorticosterone (54.3 mCi/mmol) were purchased from Daiichi Pure Chemical Co., Tokyo, and New England Nuclear Corp., Boston, Massachusetts. NAD$^+$ and NADPH were obtained from Boehringer, Mannheim. Three millilitres of the adrenal homogenate and each of the subcellular fractions were mixed with 2 ml of 0.33 M sucrose solution containing NAD$^+$ or NADPH as cofactor and then transferred into the incubation flasks, in which the radioactive steroid precursor had been added with 3 drops of propylene glycol. The final volume of the incubation mixture was 5 ml per flask with 120 µmol of the cofactor. The whole homogenates and subcellular fractions of the Japanese serows were incubated with 4-^{14}C-labelled pregnenolone (10 µg, 9.82 × 10^4 dpm), progesterone (10 µg, 9.73 × 10^4 dpm) and 11-deoxycorticosterone (10 µg, 9.91 × 10^4 dpm) in the presence of NAD$^+$ or NADPH for 60 min. The protein concentrations of the incubation mixture added to the incubation flasks were 20.5 ± 1.41 mg (mean ± SE) per flask as whole homogenate, 19.3 ± 1.24 mg per flask as mitochondrial fraction, 21.2 ± 1.34 mg per flask as microsomal fraction, and 18.9 ± 1.12 mg per flask as cytosol fraction. Incubation was carried out with shaking for 60 min in O$_2$ gas at 37 ± 0.5°C. Immediately after incubation, 15 ml of methylene dichloride was added to each flask. The procedures for isolation and detection of the products were the same as described by Nakamura and Tanabe (1972). Radioactive products were chromatographed on thin layer in benzene–acetone (4:1 by volume) and chloroform–methanol (94:6 by volume) systems, and their mobilities on a thin layer chromatogram were compared with those of the standard preparations identically treated by the above-stated procedures. Finally, their identities were confirmed by repeated crystallisation with various solvent systems after the addition of a corresponding authentic preparation. The criterion for the constancy of specific activity of crystals was based on a coefficient of variation of less than 5 per cent at least in the last three consecutive crystallisations, also taking into account the specific activity of the solid in the mother liquor as shown in Table 30.1. Half of the eluate was transferred to a counting vial and the radioactivity was measured with a liquid scintillation spectrometer. The yield of metabolites was calculated and expressed as a percentage of the substrate after correcting for protein concentration of the incubated tissue preparation. Blood samples were collected with a heparinised syringe from heart or venous vessels of male and female Japanese serows. The blood samples were centrifuged for the separation of plasma, which was then frozen and kept at −20°C until used for assay. One-tenth of a millilitre of plasma was used for the determination of corticosterone and cortisol. The radioimmunoassay for corticosterone and cortisol were carried out by our assay methods as described previously (Tanabe et al. 1983).

Table 30.1: Recrystallisation of metabolites formed from each substrate to constant specific activity (dpm/mg)

Substrate precursor	Metabolite	1st	Crystal 2nd	3rd	Solid in mother liquor
Pregnenolone	Progesterone	314	320	331	352
Progesterone	11-Deoxycorticosterone	579	584	562	589
	Corticosterone	438	425	453	451
11-Deoxycorticosterone	Corticosterone	512	540	537	551
	18-Hydroxycorticosterone	215	232	218	245
	Aldosterone	243	231	219	245

The solvent systems used for 1st, 2nd and 3rd crystallisation were acetone–n-heptane, ethylacetate-n-heptane and $CHCl_3$-methanol, respectively

RESULTS AND DISCUSSION

[4-^{14}C]-labelled pregnenolone, progesterone and 11-deoxycorticosterone were incubated with the adrenal tissue homogenates and the subcellular fractions of 2-year-old Japanese serows in the presence of NAD$^+$ or NADPH for 60 min. The metabolites detected and their yields are shown in Tables 30.2–30.4. When [4-^{14}C]pregnenolone was incubated with the whole homogenates and the subcellular fractions of the serow adrenals in the presence of NAD$^+$, the major metabolite was progesterone and the minor one was 11-deoxycorticosterone. But it was not converted into 17α-hydroxyprogesterone. The enzymatic activities in these reactions were higher in the microsomal fraction than in the other fractions. [4-^{14}C]-Progesterone in the presence of NADPH was mainly converted into 11-deoxycorticosterone. Corticosterone was detected in smaller amounts as the metabolite from progesterone; however, progesterone was not converted to 11-deoxycortisol and cortisol. The 21-hydroxylase activity was high in the microsomal fraction, but low in the mitochondrial and cytosol fractions. After [4-^{14}C]11-deoxycorticosterone was incubated with the whole homogenates and the subcellular fractions of the serow adrenals in the presence of NADPH, the major metabolite was corticosterone, but other minor metabolites such as 18-hydroxycorticosterone and aldosterone were also obtained. The 11β-hydroxylase related to the corticosterone formation was higher in the mitochondrial fraction than in the other fractions. The intracellular distribution of the adrenal enzyme activities related to corticoidogenesis is illustrated in Figure 30.1. The activity of 3β-hydroxysteroid dehydrogenase associated with Δ^5-Δ^4 isomerase is expressed as the total conversions to progesterone and 11-deoxycorticosterone from pregnenolone; 21-hydroxylase is the conversion to 11-deoxycorticosterone and

Table 30.2: The conversion of [4-^{14}C] pregnenolone to corticosteroids by the 60-min incubation of adrenal tissue of the Japanese serow

Metabolite	Radioactivity (%) in enzymatic material[a]			
	Whole homogenate	Mitochondrial fraction	Microsomal fraction	Cytosol fraction
Residual substrate	58.2 ± 7.60	77.3 ± 9.31	42.7 ± 6.73	79.4 ± 11.7
Progesterone	26.4 ± 4.16	9.7 ± 1.20	35.9 ± 5.24	7.2 ± 0.93
11-Deoxycorticosterone	3.3 ± 0.45	1.4 ± 0.24	7.1 ± 0.91	1.2 ± 0.15
17α-Hydroxyprogesterone	ND	ND	ND	ND
Unidentified substances	3.1 ± 0.40	2.9 ± 0.32	4.2 ± 0.62	2.9 ± 0.33
Total recovery	91.0 ± 6.28	91.3 ± 7.53	89.9 ± 6.67	90.7 ± 6.92

[a]All values are the average of five incubation flasks ± SEM corrected for the concentration of protein. The protein concentrations of the incubation mixture added to incubation flasks were 20.5 ± 1.41 mg per flask as whole homogenate, 19.3 ± 1.24 mg per flask as mitochondrial fraction, 21.2 ± 1.34 mg per flask as microsomal fraction, and 18.9 ± 1.12 mg per flask as cytosol fraction. ND, Not determined

Table 30.3: The conversion of [4-^{14}C] progesterone to corticosteroids by the 60-min incubation of adrenal tissue of the Japanese serow

Metabolite	Radioactivity (%) in enzymatic material[a]			
	Whole homogenate	Mitochondrial fraction	Microsomal fraction	Cytosal fraction
Residual substrate	64.6 ± 7.50	74.9 ± 9.32	52.4 ± 7.46	75.5 ± 9.73
11-Deoxycorticosterone	19.1 ± 2.35	5.7 ± 0.71	28.5 ± 3.73	4.6 ± 0.63
Corticosterone	3.9 ± 0.41	3.5 ± 0.40	2.4 ± 0.33	1.7 ± 0.20
17α-Hydroxyprogesterone	ND	ND	ND	ND
11-Deoxycortisol	ND	ND	ND	ND
Cortisol	ND	ND	ND	ND
Unidentified substances	4.7 ± 0.67	6.1 ± 1.04	6.2 ± 0.70	7.5 ± 0.93
Total recovery	92.3 ± 6.51	90.2 ± 6.12	89.5 ± 5.82	89.3 ± 5.71

[a]All values are the average of five incubation flasks ± SEM corrected for the concentration of protein. The protein concentrations of the incubation mixture added to incubation flasks were as in Table 30.2. ND, Not determined

Table 30.4: The conversion of [4-^{14}C] deoxycorticosterone to corticosteroids by the 60-min incubation of adrenal tissue of the Japanese serow

Metabolite	Radioactivity (%) in enzymatic material[a]			
	Whole homogenate	Mitochondrial fraction	Microsomal fraction	Cytosal fraction
Residual substrate	68.3 ± 10.1	60.2 ± 7.92	79.8 ± 8.28	80.1 ± 9.63
Corticosterone	11.4 ± 1.82	17.4 ± 2.72	3.2 ± 0.43	1.7 ± 0.24
18-Hydroxycorticosterone	4.1 ± 0.67	6.9 ± 0.85	1.0 ± 0.11	0.5 ± 0.07
Aldosterone	3.7 ± 0.48	3.1 ± 0.47	0.7 ± 0.01	0.2 ± 0.01
Cortisol	ND	ND	ND	ND
Cortisone	ND	ND	ND	ND
Unidentified substances	3.2 ± 0.54	4.3 ± 0.34	4.9 ± 0.52	5.7 ± 0.62
Total recovery	90.7 ± 5.83	91.9 ± 7.42	89.6 ± 9.65	88.2 ± 9.53

[a]All values are the average of five incubation flasks ± SEM corrected for the concentration of protein. The protein concentrations of the incubation mixture added to incubation flasks were as in Table 30.2. ND, Not determined

342

Figure 30.1: Intracellular distribution of adrenal enzyme activities related to corticoidogenesis. Mt, Ms and Cy indicate the mitochondria, microsome and cytosol fractions, respectively. (Average of five flasks corrected for the concentration of protein)

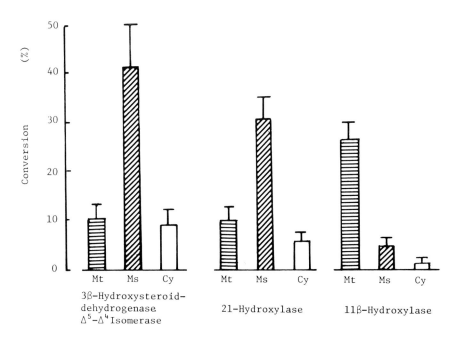

corticosterone from progesterone; and that of 11β-hydroxylase is the conversion to corticosterone, 18-hydroxycorticosterone and aldosterone from 11-deoxycorticosterone. The *in vitro* studies clearly demonstrated that the microsome of the serow adrenal tissue was the major site of the conversion of pregnenolone to progesterone in the presence of NAD⁺, and progesterone to 11-deoxycorticosterone in the presence of NADPH. Furthermore, the mitochondria were the major site of the conversion of 11-deoxycorticosterone to corticosterone in the presence of NADPH. These results agree with those reported on the adrenal gland of cattle (Beyer and Samuels 1956) and on that of the rat (Nakamura and Tamaoki 1964). 17α-Hydroxylase, which converts progesterone to 17α-hydroxyprogesterone, the precursor of 11-deoxycortisol and cortisol, was not detected in the adrenal gland of the Japanese serows. By contrast 17α-hydroxylase was present in the adrenal gland of the higher mammals such as monkey (Kittinger and Beamer 1969) and humans (Cameron *et al.* 1969, Oshima *et al.* 1969, Whitehouse and Vinson 1969). The plasma concentrations of corticosterone and cortisol of the male and female Japanese serows are given in Table 30.5. Corticosterone concentrations of plasma were signi-

343

Table 30.5: Plasma corticosterone and cortisol concentrations of male and female Japanese serow ($n = 20 \pm$ SEM)

Steroid (ng/ml)	Male	Female
Corticosterone	3.62 ± 0.64	2.77 ± 0.56
Cortisol	0.23 ± 0.03	0.21 ± 0.02

Figure 30.2: Site of steroidogenesis and related enzymes on the pathway of corticosterone biosynthesis

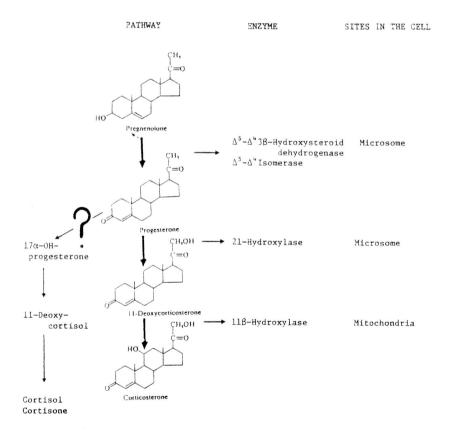

ficantly higher in the male than the female. Plasma cortisol levels were much less than the corticosterone levels. No significant sex difference was observed in plasma cortisol concentrations. In reptilian species, although corticosterone is the main adrenal corticoid, the presence of small amounts of cortisol in the plasma of the grass snake, *Natrix natrix* (Phillips and Chester Jones 1957), and *in vitro* production of small amounts of cortisol have been reported in the turtle, *Pseudemys scripta* (Sandor *et al.* 1964), and in the cobra snake, *Naja naja* (Gottfried *et al.* 1967). Bush (1953) has shown that the quantities of steroidal hormones in adrenal venous blood vary among different species of mammals. The noticeable species differences are in the relative quantities of corticosterone and cortisol. Noting the lack of any obvious relationship between the ratios and the dietary or environmental habits of the species, the suggestion is that the differences in adrenocortical secretions are genetically determined. These results demonstrate that the main pathway for steroidogenesis in the adrenal gland of the Japanese serows is proposed to be: pregnenolone → progesterone → 11-deoxycorticosterone → corticosterone and aldosterone, as illustrated in Figure 30.2.

REFERENCES

Beyer, K.F. and L.T. Samuels. (1956) Distribution of steroid-3β-ol-dehydrogenase in cellular structures of the adrenal. *J. Biol. Chem. 219*, 69

Bush, I.E. (1953) Species differences in adrenocortical secretion. *J. Endocrinol. 9*, 95

Cameron, E.H.D., T. Jones, D. Jones, A.B.M. Anderson and K. Griffiths. (1969) Further studies on the relationship between C_{19}- and C_{21}-steroid synthesis in the human adrenal gland. *J. Endocrinol. 45*, 215

Ford, H.C. and L.L. Engel. (1974) Purification and properties of the Δ^5-3β-hydroxy-steroid dehydrogenase−isomerase system of sheep adrenal cortical microsomes. *J. Biol. Chem. 249*, 1363

Gottfried, H., D.P. Huang, B. Lofts, J.G. Phillips and W.H. Tam. (1967) *In vitro* production of steroids by the adrenal and testicular tissues of the cobra. *Gen. Comp. Endocrinol. 8*, 18

Kittinger, G.W. and N.B. Beamer. (1969) *In vitro* corticoidogenesis in primates: a comparative study. *Gen. Comp. Endocrinol. 13*, 236

Nakamura, Y. and B. Tamaoki. (1964) Intracellular distribution and properties of steroid 11β-hydroxylase and steroid 18-hydroxylase in rat adrenal. *Biochim. Biophys. Acta 85*, 350

Nakamura, T. and Y. Tanabe. (1972) *In vitro* steroidogenesis by testes of the chicken (*Gallus domesticus*). *Gen. Comp. Endocrinol. 19*, 432

Nakamura, T., Y. Tanabe and H. Hirano. (1978) Evidence of *in vitro* formation of cortisol by the adrenal gland of embryonic and young chickens. *Gen. Comp. Endocrinol. 35*, 302

Oshima, H., T. Sarada, K. Ochiai and B. Tamaoki. (1969) A comparison study of steroid biosynthesis *in vitro* in clear cell adenoma and its adjacent tissue of human adrenal gland. *Endocrinol. Japon. 16* (1), 47

Phillips, G.G. and I. Chester Jones. (1957) The identity of adrenocortical secretion in lower vertebrates. *J. Endocrinol. 16,* iii

Sandor, T., J. Lamoureux and A. Lanthier. (1964) Adrenocortical function in reptiles. The *in vitro* biosynthesis of adrenal cortical steroids by adrenal slices of two coon North American turtles, the slider turtle and the painted turtle. *Steroids 4,* 213

Tanabe, Y., T. Yano and T. Nakamura. (1983) Steroid hormone synthesis and secretion by testes, ovary, and adrenals of embryonic and postembryonic ducks. *Gen. Comp. Endocrinol. 49,* 144

Whitehouse, B.J. and G.P. Vinson. (1969) Pathway for cortisol biosynthesis in the foetal adrenal cortex. *Nature (London) 221,* 1051

Part Eight:
Nutritional Status of
Capricornis

31

The quantity of food taken by raised Japanese serow

Sanji Chiba

Ohmachi Alpine Museum, Ohmachi City, Nagano Prefecture 398, Japan

INTRODUCTION

A report on the amount of food taken by feral serows has been made in 1976 by Haneda *et al.*, but the amount per day or in each season was not reported. Accordingly we have chosen raised serows as the subjects of a study to investigate the amount per day, its seasonal variation and its difference according to age.

MATERIALS AND METHODS

As the subjects of the study five serows were chosen. Three of them were captured as lambs and were raised; of the remainder, one had been protected when young and the other was born and raised at Ohmachi Alpine Zoo (Table 31.1). The amount of food taken by two of them (OMC-4, OMC-6) was investigated throughout one year and the others were excluded except on one occasion.

The main food was the leaves of trees, and as a supplement fruits, edible roots and an artificial diet were given. As such subsidiary food was fed in restricted quantities, it was almost all consumed each time. The average amount of food per day was investigated as follows: the ones studied for 12 months were given food twice a day on 3 days in succession each month; the others were given food once a day for 7 days. In each case food was taken away after 3 hours. Dry weight is the amount of food dried in a drying oven (80°C, 24 h).

RESULTS

Annual quantity of food

The annual quantity of food taken by the two in a day is indicated in Figure

Table 31.1: Average quantity of food and excrement per day

Month	Food (w.wt)	Food (d.wt)	Average hydrous percentage of food	Excrement (w.wt)	Excrement (d.wt)	Average hydrous percentage of excrement
OMC-4						
June	2390.0	736.7	69.2	326.7	167.6	48.0
July	2376.7	816.2	65.7	726.7	313.3	56.9
Aug.	2268.3	776.4	65.8	718.3	277.0	61.4
Sept.	2386.7	709.7	70.3	826.7	273.3	66.9
Oct.	2363.3	991.1	58.1	600.0	364.3	39.3
Nov.	1579.7	581.6	63.2	337.7	177.6	47.4
Dec.	1880.0	726.6	61.4	557.0	265.6	52.3
Jan.	1594.3	579.0	63.7	467.0	201.3	56.9
Feb.	1787.0	657.2	63.2	486.7	202.6	58.4
March	1554.7	556.6	64.2	370.3	183.6	50.4
April	1371.3	464.6	66.1	268.0	160.0	40.3
May	1380.0	482.0	65.1	540.7	223.0	58.8
OMC-6						
June	1725.0	643.1	62.7	202.3	117.0	46.9
July	2055.0	741.8	63.9	405.0	165.6	59.1
Aug.	1831.7	635.6	65.3	351.7	141.0	59.9
Sept.	2016.7	688.1	65.9	526.7	196.6	62.7
Oct.	2178.7	879.9	59.6	576.7	253.3	56.1
Nov.	1052.7	451.8	57.1	345.3	170.3	50.7
Dec.	1340.0	559.8	58.2	668.0	306.0	54.2
Jan.	1070.0	396.1	63.0	461.7	177.3	61.6
Feb.	1120.0	458.2	59.1	470.7	189.6	59.7
March	1050.0	410.1	60.9	415.0	154.3	62.8
April	1130.0	359.7	68.2	268.0	170.0	36.6
May	1160.0	369.0	68.2	329.7	166.0	49.7

31.1. The peak of the amount (wet weight) by OMC-4 was 2390 g a day in June and the lowest was 1371.3 g in April. From the viewpoint of dry weight, the true peak was in October (991.1 g) and the lowest was 464.6 g in April. Thus the peak differs between wet and dry weight, but such a difference results from the percentage of water contained in the different kinds of food given in each month (in October the food had the lowest percentage of water). OMC-6 ate most in October: 2178.7 g (w.wt), and least in March: 1050 g (w.wt). This was also the case with OMC-4: 879.9 g in October, 359.7 g in April (dry weight).

The amount of food taken by raised serows was different in each case, and they ate most in autumn. This tendency to eat much in autumn is notable for hibernating animals, but though serows never hibernate, they took much food. This is because they need a high intake of energy in order to store up subcutaneous fat and to grow thick fur for the coming winter.

The increase of food resulted in a gain in weight. With a decrease in the amount of food the body weight lessened from winter to spring because of

Figure 31.1: Quantity of food per day in each month

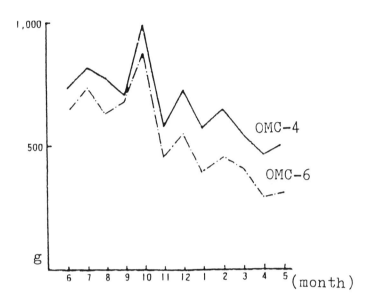

the shortage of nutritious feeding plants damaged by snow. Presumably serows have adapted to these constitutional conditions over a long period of time.

Thus, the accumulated fat in autumn makes up for malnutrition in winter, and there is a consequent gradual decrease in weight. A decrease in weight is seen in summer as well. This is not because serows eat less than they do in winter, because they took more or nearly as much food in August and June. The weight decrease was chiefly because they consumed much energy in respiration for thermoregulation in response to temperatures of 30°C at 760 m above sea level. In this connection the annual quantity of feed taken by OMC-4 was 697.8 kg (w.wt), 245.9 kg (d.wt), and it gained 1700 g in a year. (OMC-6 consumed 540.34 kg (w.wt), 200.8 kg (d.wt) and gained 2250 g in a year.)

The quantity of food by age

We surveyed the amount of feed taken by some serows of various ages in February, May and August. (See Table 31.2 and Figures 31.3 and 31.4.) The others were given food for 7 days in succession and the mean amount per day was measured. In February OMC-11 (6 months old) ate 362.1 g (d.wt) a day, OMC-6 (11 months old), 458.2 g (d.wt) a day, and OMC-4

Figure 31.2: Changes of weight in each month

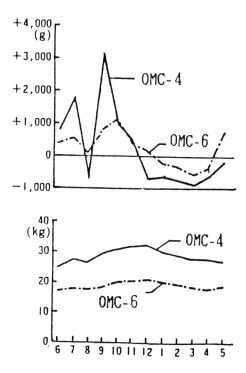

Table 31.2: Quantity of food by age

Month	Age	Mark	Sex	Quantity per day (d.wt), g	Quantity per kg body weight (d.wt)	Live weight, kg
Feb.	6 months	OMC-11	♀	362.1	35.50	10.20
Feb.	11 months	OMC-6	♂	458.2	23.70	19.35
Feb.	2 years	OMC-4	♂	657.2	23.10	28.40
May	11 months	OMC-6	♂	369.0	19.47	18.95
May	2 years	OMC-4	♂	482.0	18.18	26.50
May	7 years	OMC-11	♀	768.1	20.55	37.40
May	20 years	OMC-1	♀	421.0	12.02	35.00
Aug.	11 months	OMC-6	♂	635.6	36.63	17.35
Aug.	2 years	OMC-4	♂	778.4	29.74	26.10
Aug.	5 years	OMC-236	♂	693.9	23.15	29.97
Aug.	13 years	OMC-11	♀	529.8	18.83	28.81

352

Figure 31.3: Quantity of food per day in each month (d.wt)

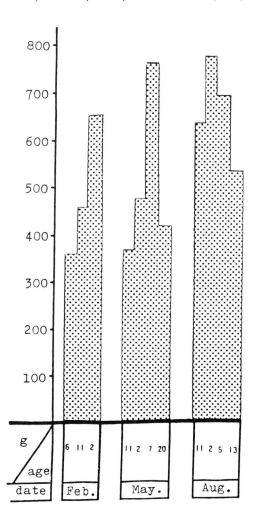

(2 years old), 657.2 g (d.wt) a day. The amount of food per kilogram of body weight is 35.5 g (OMC-11), 23.7 g (OMC-6) and 23.1 g (OMC-4). The amount is almost the same between OMC-6 and OMC-4, but OMC-11 took a lot in comparison with the others. In May OMC-6 ate 369.0 g (d.wt) per day; OMC-4, 482.0 g; OMC-11 (7 years old) 768.1 g; and OMC-1 (20 years old), 421.0 g. The amount per kilogram of body weight was 19.47 g (OMC-6), 18.18 g (OMC-4), 20.55 g (OMC-11), and 12.02 g (OMC-1).

From the above, the amount of feed shows a tendency to increase until

Figure 31.4: Quantity/kg in weight (d.wt)

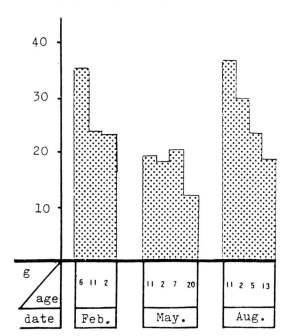

an animal is 7 years old, and to dwindle at the age of 20. The amount per kilogram of body weight was nearly the same in 11-month-old, 2-year-old and 7-year-old serows, but the amount of feed taken by a 20-year-old serow was about 58 per cent of that of a 7-year-old one. In August OMC-6 ate 635.6 g (d.wt) a day; OMC-4, 776.4 g; OMC-236 (5 years old), 693.9 g; and OMC-11 (13 years old), 529.8 g. The amount of feed per kilogram of body weight was 36.63 g (d.wt) a day (OMC-6), 29.74 g (OMC-4), 23.15 g (OMC-236), and 18.83 g (OMC-11). A 13-year-old serow ate less than younger ones. The following conclusions on the quantity of food can be made. It changed in each season: they ate most in autumn, less in winter, more in spring and again less in summer. Their weight also changed in proportion to the quantity of their feed. Each season formed an important factor in this change. The change in quantity of food by age was recognised: growing serows (6 months, 11 months) took more feed, and though their pace of eating slowed down, the amount of food increased until they reached the age of 7 years. Judging from reports of their pregnancy rate from the age of 5 to 10 years the peak is at the age of 5½ years; the pregnancy rate from 7 to 10 will show parallelism or a slight downward curve. From the tenth year onwards serows tend to eat less.

32

Nutritional estimation of Japanese serow by faecal analysis

Yuriko Hazumi, Naoki Maruyama and Keiko Ozawa

Department of Nature Conservation, Tokyo Noko University, School of Agriculture.
3-5-8 Saiwai-cho, Fuchu City, Tokyo 183, Japan

INTRODUCTION

Faecal nutrient contents have been shown to be positively correlated with dietary nutrients in various wild and domestic ruminants (Bredon *et al.* 1963, Nieminen *et al.* 1980, Gates and Hudson 1981, Leslie and Starkey 1985, Renecker and Hudson 1985). In such animals, faeces could be used as a parameter of the nutritional conditions of their diet and habitat, and population qualities. This relation has also been found in the Japanese serow (Hazumi 1986). This species is strictly protected in Japan, so it is very difficult to get material for a study of their nutritional conditions and the related problems. Thus, faecal analysis was chosen as a suitable method.

In this study we examined the nutritional conditions of the Japanese serow populations in Nibetsu, Akita Prefecture, and Kiso and Ina, Nagano Prefecture (Figure 32.1). For the Nibetsu population, seasonal changes of the faecal contents were assessed on four nutritional components: nitrogen, neutral detergent fibre (NDF), acid detergent fibre (ADF) and acid detergent fibre-lignin (ADF-lignin), in 1983–1984. The nitrogen and crude fibre contents of the Ina and Kiso populations were examined in the winters between 1982 and 1984. In addition the nutrient conditions of these populations were compared with each other.

MATERIALS AND METHODS

In Nibetsu, collection of the faeces was carried out every month, except during December and February, from November 1984 to October 1985. A total of 91 faecal pellet groups were obtained. For the Ina and Kiso populations, the faeces were obtained directly from 211 culled serows in the winter (November–February) of 1982 and 1984. Immediately after collection, the faeces were stored in a freezer at $-20°C$ and thawed before

Figure 32.1: Study area. 1: Nibetsu, Akita Prefecture. 2: Kiso, Nagano Prefecture. 3: Ina, Nagano Prefecture

examination. Total nitrogen contents were measured by C–N auto-analyser. Percentage dietary nitrogen content (y) was convertible from the faecal one (x) using the following formula (Hazumi 1986):

$$y = 0.635 + 0.426\ x \qquad (1)$$

The faecal nitrogen content (x) could also be converted to the Riney's kidney fat index, RKFI (y) using the following formula (Hazumi 1986):

$$y = 9.59 + 18.42\ x \qquad (2)$$

NDF, ADF and ADF-lignin were measured according to Horii's method (1975), and crude fibre was estimated according to Tamura's method (1975). NDF was composed of cellulose, hemicellulose and lignin. ADF was composed of cellulose and lignin.

Nibetsu district is located on the western foot of Mt. Taiheisan, Akita Prefecture, northern Honshu, Japan, Japanese cedar (sugi) (*Cryptomeria japonica*) plantations cover 49.6 per cent of its slopes and natural sugi forests 34.6 per cent; the rest is covered by deciduous broad-leaved forests of Japanese chestnut (*Castanea crenata*), Mongolian oak (*Quercus mongo-*

lica var. *grosserrata*), beech (*Fagus crenata*), etc. The serow density was 21.98/km² (R. Kishimoto, pers. comm.). The Kiso and Ina districts were located in the mountainous zone of Chuoh Alps, central Honshu, Japan, below 1500 m in altitude. Ina district is covered mainly with plantations of larch (*Larix leptolepis*) and Japanese red pine (*Pinus densiflora*). Kiso district, on the other hand, is covered mostly with natural forests of two Japanese cypress species (*Chamaecyparis obtusa* and *C. pisifera*) and plantations of *C. obtusa*. The serow densities were not significantly different between these areas (*P*>0.05): a combined value for both areas was estimated at 9.83±5.18 serows/km² in 1982 and 7.41±4.98 serows/km² in 1983, respectively (Tokida and Iwano 1983).

RESULTS

Seasonal changes in faecal nutrients of the Nibetsu serow population

Figure 32.2 shows seasonal changes in percentage nitrogen contents of the faeces collected in Nibetsu. Percentage faecal nitrogen was not significantly

Figure 32.2: Seasonal changes in faecal nitrogen contents of Japanese serows in Nibetsu, Akita Prefecture, 1984–1985. OPL, Optimal growth level; ML. maintenance level. Vertical lines represent range, and boxes represent standard deviation. Numbers are sample sizes

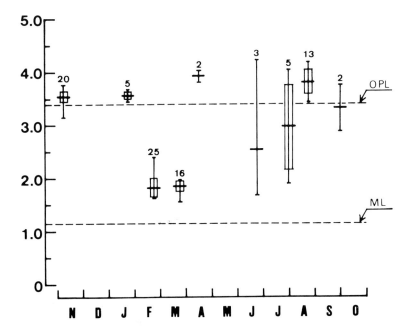

different between November and February (*t*-test, $P > 0.05$), but significantly decreased to the lowest value of 1.9 per cent from early to late March during a winter with abundant snow ($P < 0.05$). However, it increased to 4.0 per cent in the plant-developing season of April 1985, and it decreased again in June. It gradually increased to a value of 3.8 per cent in late August of the same year.

Figure 32.3 shows seasonal changes in percentage NDF, ADF and ADF-lignin in the faeces collected in Nibetsu. Both NDF and ADF showed the same trend throughout the year. NDF and ADF increased respectively

Figure 32.3: Seasonal changes of faecal NDF contents, faecal ADF, and faecal ADF–lignin of Japanese serows in Nibetsu, Akita Prefecture, 1984–1985. Vertical lines represent standard deviation. Numbers are sample sizes

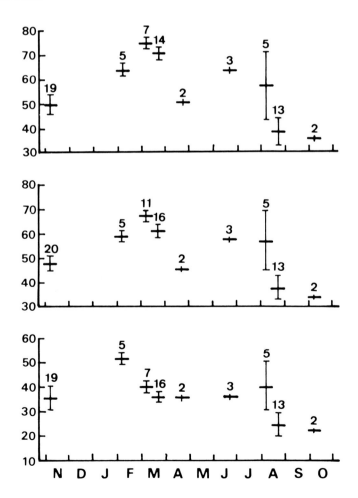

from values of 49.6 per cent and 48.0 per cent in November, to the highest values of 74.9 per cent and 66.9 per cent in early March, but decreased in April. Then, they again increased slightly in June, but abruptly decreased in late August. In late October, both NDF and ADF showed the lowest values, 34.6 per cent and 33.9 per cent, respectively. ADF-lignin reached the highest value (51.7 per cent) in February, and decreased (36.0 per cent) in late March. No significant fluctuation was then found until early August. However, it decreased to a striking 25.4 per cent in late August, but it showed the lowest value (23.1 per cent) in early October.

Faecal nutrient contents of Ina and Kiso populations in winter

Table 32.1 shows the faecal nitrogen contents of the Ina and Kiso populations by age and sex in the two winters of 1982–1984. No significant difference was found between these values. Accordingly, we could not find clear differences in faecal nitrogen content by sex, age and winter for the Ina and Kiso populations.

Figure 32.4 shows the seasonal changes in faecal nitrogen contents for the Ina and Kiso serow populations in the winters of 1982–1983 and 1983–1984. No significant difference was found between both populations

Table 32.1: Percentage nitrogen contents of Japanese serow faeces by age and sex in Ina and Kiso in two winters of 1982–1984

Month	Sex/Age	1982–1983 (x+SD)	1983–1984 (x+SD)
Nov.	Adult male	3.43 ± 0.45 (9)	3.06 (3)
	Pregnant female	3.22 ± 0.27 (4)	3.01 (3)
	Female	2.86 (2)	2.74 ± 0.35 (10)
	Yearling	2.84 (1)	2.47 (1)
	Fawn	3.04 (2)	3.81 (1)
Dec.	Adult male	2.50 ± 0.67 (22)	2.62 ± 0.71 (10)
	Pregnant female	2.28 ± 0.33 (8)	2.35 ± 0.25 (5)
	Female	2.67 ± 0.43 (4)	2.44 ± 0.20 (8)
	Yearling	3.66 (1)	2.35 ± 0.57 (4)
	Fawn	2.62 ± 0.62 (4)	2.60 ± 0.39 (4)
Jan.	Adult male	2.44 ± 0.52 (10)	2.45 ± 0.71 (7)
	Pregnant female	2.22 ± 0.50 (4)	2.29 ± 0.33 (9)
	Female	2.26 ± 0.37 (4)	2.15 ± 0.22 (4)
	Yearling	2.73 ± 0.83 (9)	2.20 ±± 0.10 (4)
	Fawn	2.34 ± 0.28 (5)	2.06 (1)
Feb.	Adult male	2.30 ± 0.20 (8)	1.98 ± 0.32 (9)
	Pregnant female	2.04 ± 0.25 (8)	1.92 ± 0.29 (8)
	Female	2.14 (3)	1.77 ± 0.16 (4)
	Yearling	2.28 ± 0.40 (4)	2.08 (3)
	Fawn	2.53 ± 0.70 (6)	

Figure 32.4: Seasonal changes in faecal nitrogen contents of Japanese serows killed in Ina (solid circle) and Kiso (white circle) in two winters of 1982–1983 and 1983–1984. Vertical lines represent standard deviation. Numbers are sample sizes

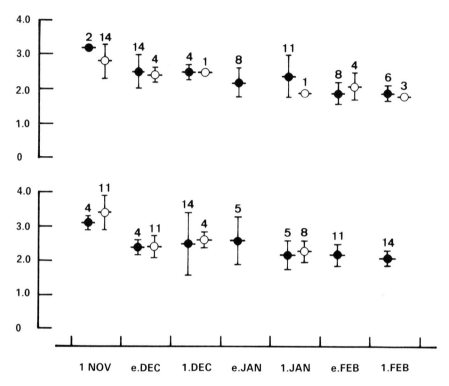

and between the winters. The combined values in both populations and both winters gradually decreased from a mean of 3.1 ± 0.20 (SD)% ($n=31$) in late November to a mean of 1.9 ± 0.20 (SD)% ($n=15$) in late February. Furthermore, this seasonal decrease of faecal nitrogen corresponded well to the decrease of the kidney fat index value, shown by Maruyama and Hazumi (1985) (Figure 32.5).

Figure 32.6 shows the seasonal changes in faecal crude fibre contents for the Ina and Kiso serow populations in the winter of 1983–1984. This figure was plotted using the combined data of both populations, because of their small sample sizes. It increased gradually from 21.8 ± 3.21(SD)% ($n=5$) in late November to 30.9 ± 1.31(SD)% ($n=5$) in late February. Such a seasonal trend was obviously inverse to that of the nitrogen content.

Figure 32.5: Seasonal changes in faecal crude fibre of Japanese serows killed in Ina and Kiso in winter of 1983–1984. Vertical lines represent standard deviation. Numbers are sample sizes

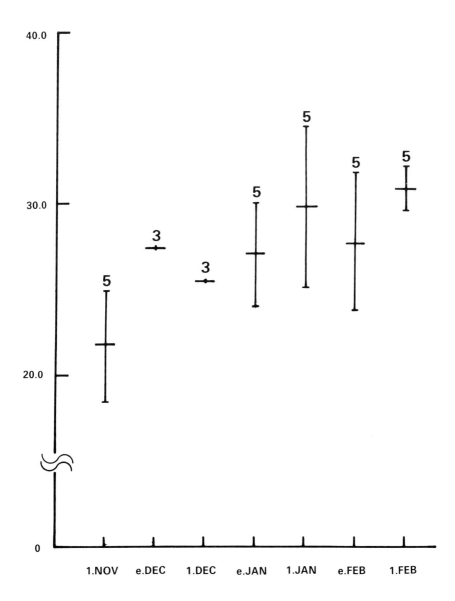

Figure 32.6: Seasonal changes in faecal nitrogen contents and RKFI of Japanese serows killed in Ina and Kiso in the winters of 1982–1983 (solid circle) and 1983–1984 (white circle). Vertical lines represent standard deviation. Numbers are sample sizes

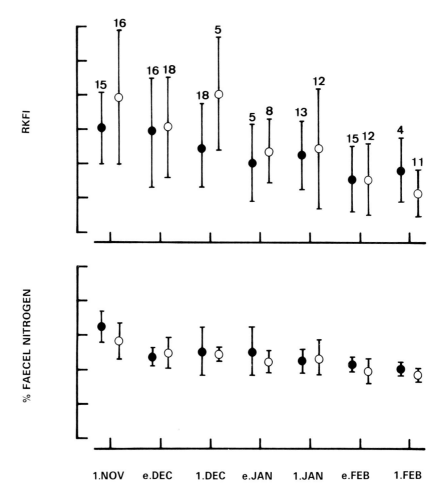

DISCUSSION

From the seasonal changes in faecal nitrogen and fibre contents, we conclude that Nibetsu serows took different plants as food according to the season: highly digestible and nutritious plants were eaten in April and August–November, highly nutritious but hardly digestible ones were eaten in February, and poorly nutritious and hardly digestible in March. This might be a consequence of their mainly taking leaves of broad-leaved deciduous trees, grasses and herbaceous plants in April and August–

362

November, evergreen tree leaves in February, and extremely low quality plants in March. The diet quality decrease in June and early August when food plants were sufficiently available might be caused by extremely small sample sizes, because the nitrogen content of leaves is usually higher than that of the faeces in summer (Katagiri 1978).

In Kiso and Ina, as the winter progresses, available plant matter becomes limited through plant withering and snow cover, and the parts eaten by the animal shift from leaves to twigs and bark (Takatsuki and Suzuki 1985). Decrease of the faecal nitrogen contents and increase of the faecal crude fibre probably reflected the above phenomenon.

Generally speaking, ungulates require a 0.96–1.12 per cent nitrogen content in their diets (0.76–1.14 per cent nitrogen content in faeces converted by (1) as the maintenance level, and 2.08–3.02 per cent (3.39–5.60 per cent faecal nitrogen, converted by formula 1 as the optimal growth level. Using formula 2, the RKFI value of the maintenance level was calculated as 23.0–30.6, the onset level of the decrease in femur marrow fat (Maruyama and Hazumi 1985), and the RKFI value of the optimal growth level was, in the same way, calculated as 72.0–112.7. Nibetsu serows took food with nitrogen contents above the maintenance level throughout the year, and their diets included higher nitrogen contents than the optimal growth level every month except March, June and early August. In February, it prominently exceeded the maintenance level by 2.4 per cent in nitrogen (54 in RKFI), and even in early and mid-March, with the lowest values through the year, they exceeded the maintenance level by 0.7 per cent in nitrogen (23 in RKFI). However, in Kiso and Ina, it was exceeded by 2.1 per cent in nitrogen (48 in RKFI) in late November, and only by 0.8 per cent in nitrogen (24 in RKFI) in late February, which was almost the same as the Nibetsu value in March. Accordingly, those values in Kiso and Ina in the middle and latter part of the winter were obviously lower than the values in Nibetsu in the same seasons. The converted RKFI values of three populations did not decrease below the critical level, at which the femur marrow fat began to move, even in middle and late winter.

Such nutritional differences in the diets may affect the differences in densities of the three serow populations; availability of highly nutritious winter foods might be one of the factors supporting a high-density population in Nibetsu.

ACKNOWLEDGEMENTS

We thank Messrs N. Sakurai and M. Hanai, Culture Agency, Japan; Mr K. Tokida, Japan Wildlife Research Center, Tokyo; Mr K. Maita, Government of Akita Prefecture, Japan; and R. Kishimoto, Osaka City University, for their help in collecting the materials.

REFERENCES

Bredon, R.M., K.W. Harker and B. Marshall. (1963) The nutritive value of grasses grown in Uganda when fed to Zebu cattle. *J. Agr. Sci. 61*, 101–4

Gates, C.C. and R.J. Hudson. (1981) Weight dynamics of wapiti in the boreal forest. *Acta theriol. 26, 27*, 407–18

Hazumi, Y. (1986) The feces of Japanese serows as nutritional indices. MS Thesis, Tokyo Noko University, Tokyo (in Japanese)

Horii, S. (1975) Quality estimation method of forages and crops. In Committee of Crop Chemical Analysis (eds) *Quality estimation method of forages and crops.* Yokendo, Tokyo, pp. 482–500

Katagiri, S. (1978) Studies on mineral cycling in a deciduous broad-leaved forest of Sanbe Forest of Shimane University (IV). Concentration of nutrient of trees. *Res. Bull. Fac. Agr., Shimane Univ. 11*, 60–72

Leslie, D.M. and E.D. Starkey. (1985) Faecal indices to dietary of cervids in old-growth forests. *J. Wildl. Mgmt 49*(1) 142–6

Maruyama, N. and Y. Hazumi. (1985) Changes in fat reserves of Japanese serow in Nagano and Gifu Prefectures. In M. Sugimura (ed.) *Fundamental study on reproduction, morphology, pathology and population ecology of Japanese serows.* Final report of Special Working Group on Serow Biology, Gifu University, Japan (in Japanese)

Nieminen, M., S. Kellokump, P. Vayrynen and H. Hyvarynen. (1980) Rumen function of the reindeer, *Proc. 2nd Int. Reindeer/Caribou Symp.*, Roros, Norway, 1979, pp. 213–23

Renecker, L.A. and R.J. Hudson. (1985) Estimation of dry matter of free ranging moose. *J. Wildl. Mgmt 49*, 785–92

Takatsuki, S. and K. Suzuki. (1985) Analysis of stomach contents of Japanese serows in central Japan in winter. In M. Sugimura (ed.) *Fundamental study on reproduction, morphology, pathology and population ecology of Japanese serows.* Final Report of Special Working Group on Serow Biology, Gifu University, Japan (in Japanese)

Tamura, T. (1975) Analysis and measurements of carbohydrates through specific chemical response. In Committee of Crop Chemical Analysis (eds) *Measurement methods of cultivated plants for nutritional evaluation.* Yokendo, Tokyo, pp. 308–13 (in Japanese)

Tokida, K. and T. Iwano. (1983) Changes in density of Japanese serows in Central Alps, Nagano Prefecture. In Education Committee, Government of Nagano Prefecture and Japan Wildlife Research Center (eds) *Monitoring report on special project of counterplan for damages of serow, special natural monument, to forestry* (in Japanese)

33

Heavy metal accumulation in wild Japanese serow

Katsuhisa Honda,[1] Ryo Tatsukawa[1] and Shingo Miura[2]

[1]Department of Environment Conservation, Ehime University, Tarumi 3-5-7, Matsuyama 790, Japan, and [2]Department of Biology, Hyogo College of Medicine, Nishinomiya 663, Japan

INTRODUCTION

Considering the present status of environmental pollutants such as heavy metals and organochlorine pesticides, their levels are decreasing, at least in the developed countries, thanks to a wider awareness of the problem. At the same time the number of chemicals monitored has increased with the improvement of methods of detection. However, the chronic and complex effects on the biota of these micro-pollutants over a long period of time are not fully understood.

Plankton and fish show a relatively fast uptake and accumulation of micro-pollutants, and their accumulation processes are approximated by a simple mathematical model (Phillips 1980). As regards mammals and birds at a high trophic level, their pollutant accumulations vary widely according to species-specific processes (Honda and Tatsukawa 1985). For example, birds excrete a fairly large amount of certain metals during egg-laying and moulting. Such detailed information on the uptake and accumulation of micro-pollutants by animals, especially those with a long lifespan, is necessary for an understanding of the dynamics of the pollutants and their long-term effects. It is also necessary for the protection and management of wildlife.

The Japanese serow, *Capricornis crispus*, is the only bovine ruminant living wild in Japan. Since the serow is a long-living animal species, which occupies a relatively high trophic level in the food web, this animal is useful as an indicator species for terrestrial environmental pollution of metals and other chemicals, and also for understanding the processes of bioaccumulation of pollutant chemicals. Hitherto, although the growth, reproduction, feeding habits and population dynamics of the serow have been studied, information on accumulation of pollutant chemicals such as heavy metals and organochlorine pesticides has been sparse (Ito 1971, Komori 1975, Sugimura *et al.* 1981, Maruyama and Nakama 1983, Miura and Yasui 1985). This lack is due mainly to the protection of this animal as a special natural monument. As a result, in recent years, the population density of serows has increased. Consequently, during the winter season, saplings are eaten, resulting in failure of afforestation of some areas. The Agency of Cultural Affairs therefore permitted persons engaged in forestry to catch limited numbers of serows, providing us with a rare opportunity to collect the animals.

The purpose of this study, therefore, is to supply information on bio-accumulation processes of heavy metals (iron, manganese, zinc, copper, cobalt, nickel, lead, cadmium and mercury) in wild Japanese serows, focusing on the following five items.

(1) Growth processes such as body weight, body length and the weight of organs and tissues in the Japanese serow.
(2) Organ and tissue distribution of the heavy metals, and their variations with biological processes such as growth stage or age, sex, sexual status, etc.
(3) Transplacental transfer of heavy metals and the relationship between mother and fetus, and uptake and excretion of heavy metals via milk and food.
(4) Variations in heavy metal accumulation in the Japanese serow between habitats and feeding habits.
(5) Heavy metal dynamics *in vivo* of the Japanese serow, and its species-specific accumulation with biological processes.

MATERIALS AND METHODS

Seventy-seven serows were caught with the permission of the Agency of Cultural Affairs during December to January, 1981 to 1983, in Gifu and Nagano Prefectures, Japan (Figure 33.1). The animals consisted of 12 fawns, 6 yearlings and 59 adults (34 males and 25 females including 13 pregnant individuals). These specimens were in good health with no macroscopic pathological symptoms. The animals were frozen at $-20°C$ until autopsy and measurement. The specimens were dissected in the laboratory, and the body weight, body length, sex, basic morphometric data, and weights of organs and tissues were recorded. The age of the animals was determined by counting the growth layers in horn following the method of Miura (1985).

Representative samples for metal analysis were bone, muscle, skin, fleece and ten viscera including the brain, ovary, testis, liver, kidney, spleen, stomach, intestine, heart and lung. The bone samples were taken from the femur and 7th thoracic vertebra, and the adhering muscle and

Figure 33.1: Map showing the collection sites of Japanese serow

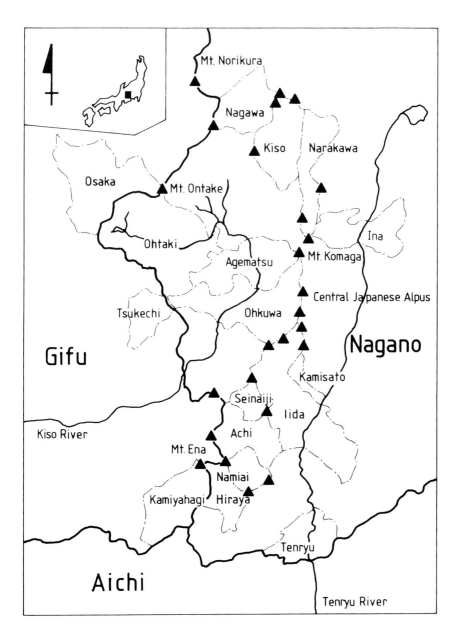

ligament were carefully removed from the bones. The surface of the bone samples was gently washed with distilled water, dried with filter paper and weighed before analysis. The muscle samples were excised from six regions of the body, as shown in Figure 33.2, and weighed separately. The samples of skin and fleece were cut from over the abdominal muscle, *M. obliquus abdominis externus.* All the fleece samples were rinsed thoroughly in tap water, distilled water and acetone, and dried at room temperature. Stomach samples were taken from the four compartments, i.e. the rumen, reticulum, omasum and abomasum, after their contents had been removed. The intestine samples were excised from the medial region of the organ after its contents had been removed. The kidney samples were cut from the left organ and then separated into three parts, namely cortex, medulla and remainder. The samples of liver, spleen and heart were cut from the mediolateral lobe; the samples of lung, testis and ovary were from the medial region of the left organ; the brain samples were from the superficial medial region, including tissue from both hemispheres. All the samples were stored in polyethylene bags at $-20°C$ until analysis.

For an analysis of the heavy metals, homogenised samples of tissue (1–

Figure 33.2: Anatomical region of muscles analysed in the Japanese serow

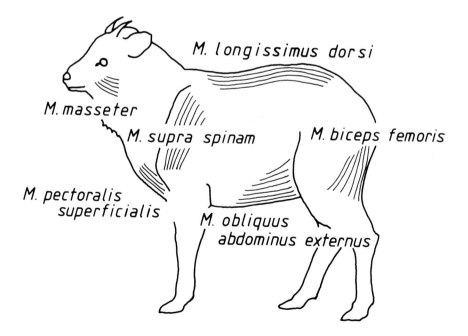

10 g wet weight) were digested to a transparent solution with a mixture of nitric, perchloric and sulphuric acids in a Kjeldahl flask (200 ml volume). The bone samples were digested in a nitric, hydrochloric and perchloric acid mixture. The resultant solutions were diluted to a known volume with deionised water and transferred to acid-washed test tubes with Teflon screw caps. The concentrations of Fe, Mn and Zn were determined by flame atomic absorption spectrophotometer (Model Shimazu AA-670). Cu, Ni, Co, Pb and Cd were measured after methyl isobutyl ketone−diethyl dithiocarbamate treatment by flame atomic absorption spectrophotometry (Honda *et al.* 1982). Corrections for background interference were made by running appropriate blanks. The accuracy of this method has been determined in this laboratory by spiking bone and other tissues of dolphin with known amounts of Fe, Mn, Zn, Cu, Ni, Co, Pb and Cd, with recoveries of 90.0 to 99.9 per cent.

Mercury was determined by cold-vapour atomic absorption spectrophotometry (Honda *et al.* 1983). The procedure consisted of the mineralisation of samples with a nitric and sulphuric acid mixture in a flask equipped with a Liebig condenser, followed by potassium permanganate ($KMnO_4$) digestion. The excess of $KMnO_4$ was reduced with a 20% hydroxylamine hydrochloride solution and the mercury to $Hg°$ with tin(II) chloride. Determinations were made with a Shimazu AA-670 spectrophotometer.

RESULTS AND DISCUSSION

Body and organ weight in Japanese serow

Information on body and organ weight is necessary for understanding the growth and physiological condition of the animals. In the study of the absorption and accumulation of heavy metals, such information has been proved to be particularly useful in estimating the amount of total residues deposited in the organs and tissues using the concentration data. Here we consider the body and organ weights of serows and discuss the relationship between organ weight and age or growth stage.

Body weights and body lengths of the Japanese serows examined ranged between 10.0 kg and 49.5 kg and between 62.0 cm and 84.0 cm, respectively, and their age-related changes are shown in Figure 33.3. The results indicate that the weight and length increased with age until about 2.5 years, which corresponds to the age of sexual maturity of this species as reported by Ito (1971) and Komori (1975). The mean growth rates between 6 months and $1\frac{1}{2}$ years were 10.5 kg/year and 11.2 cm/year, and their values were about two and six times higher than those between $1\frac{1}{2}$ years

Figure 33.3: Growth curves of body weight and body length in the Japanese serow. The numbers on the graph are the numbers of samples analysed, and the vertical and horizontal lines indicate the range and mean, respectively. The data for both males and females were combined, and the body weight did not include the contents of stomach and intestine

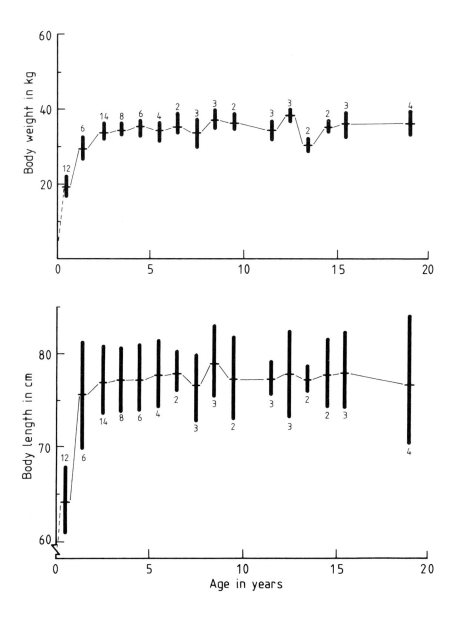

Figure 33.4: Relationship between body weight and body length in the Japanese serow. The length on the abscissa of the graph is shown in the crown–rump length for the fetus and in the body length for the fawn, yearling and adult. The fitted linear line equations are as follows: (1) log W (kg) = 4.94 log BL (cm) −4.80, (2) log W (kg) = 2.76 log CRL (cm) −1.12. The fetal linear equation (2) is cited after Sugimura *et al.* (1983), and the dots in the graph show the present data

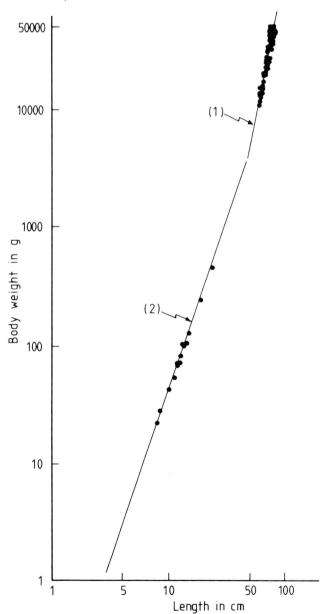

and $2^1/_2$ years respectively. When the body weight and body length were plotted on a logarithmic scale, its relationship was linear (Figure 33.4). From the fitted straight-line equation the relationship between body weight and body length in this species can be drawn as log W (kg) = 4.94 log BL (cm) -4.80, which is different from the relation drawn between crown–rump length and weight in fetal serows (log W (kg) = 2.76 log CRL (cm) -1.12) as reported by Sugimura *et al.* (1983). From the point of intersection of these two lines, the weight and length of the serow at birth may be expected to be 3.53 kg and 48.5 cm respectively. These sizes approximate to 3.507 kg and 47.0–51.0 cm at birth, which were estimated from the relationship between crown–rump length and gestation age by Sugimura *et al.* (1983).

Organ weights of the serows increased with age, as did the growth curve of body weight with age. However, their growth patterns were organ specific. A concept of the relative growth is useful for understanding the difference in the growth rates of different organs. The muscle of adult serows comprised about 50 per cent of the body weight, followed by the bone, viscera, fur and adipose tissue (Table 33.1). Although there was no significant difference in percentage of the organ weights between sexes, their percentages varied organ-specifically with growth stage, especially during the fetal and fawn periods. When the relative growth rates of organs and tissues were plotted against age, they could be classified conveniently into three patterns (Figure 33.5). Relatively high percentages in weight of the brain, liver, lungs and bone were in the fetal period, and their values decreased with age during the fawn and yearling periods. The relative growth rate of liver weight with age is shown in Figure 33.5 as an example. This indicates that the growth of these four organs in the fetus is very fast. In contrast, the percentage in weight of the muscle increased with age during the fawn and yearling periods, and thereafter remained constant. Similar trends were also observed with stomach, intestine, fur and adipose tissue (not shown). In kidney, spleen and heart, weight percentages were constant throughout the lifespan, indicating that the growth of these organs was approximated by the age-dependent increase of body weight.

These observations indicate that consideration of organ weights with age or growth stage is necessary in order to understand the phenomenon of bioaccumulation of heavy metals. In particular, taking into consideration the rapid and wide variation of the organ weights in the fetal and fawn stages, these data are essential for an understanding of the complex processes of accumulation of heavy metals and also the toxicological effect of these substances on this animal. Furthermore, the growth rates of organs and tissues in the Japanese serow with age are different from those of other animals such as marine mammals, the rat and humans, and so a consideration of the species-specific process of growth is also needed for an understanding of bioaccumulation of heavy metals in different animals.

Table 33.1: Percentages (mean ± standard deviation) of tissue and organ weights to body weight of Japanese serow

	Fetus $N = 13$	Fawn + yearling $N = 18$	Adult Male, $N = 34$	Adult Female, $N = 25$
Muscle	40.2 ± 5.35	46.0 ± 2.97	47.0 ± 2.66	49.5 ± 2.57
Viscera	15.6 ± 1.72	17.2 ± 2.88	16.2 ± 1.79	14.1 ± 1.97
Brain	5.25 ± 0.33	0.65 ± 0.13	0.45 ± 0.06	0.38 ± 0.04
Liver	2.48 ± 0.89	2.33 ± 0.54	2.10 ± 0.45	1.75 ± 0.48
Kidney	0.42 ± 0.09	0.40 ± 0.08	0.45 ± 0.13	0.34 ± 0.08
Spleen	0.40 ± 0.02	0.45 ± 0.06	0.55 ± 0.06	0.40 ± 0.12
Lungs	3.52 ± 0.95	1.35 ± 0.24	1.50 ± 0.45	1.30 ± 0.43
Heart	0.78 ± 0.09	0.75 ± 0.06	0.85 ± 0.13	0.75 ± 0.10
Stomach	1.10 ± 0.48	3.70 ± 0.63	3.63 ± 0.85	3.15 ± 0.64
Intestine	1.60 ± 0.95	7.58 ± 1.95	6.65 ± 0.85	6.03 ± 0.43
Fur (skin + fleece)	15.7 ± 2.00	14.9 ± 1.97	13.4 ± 0.47	12.1 ± 1.73
Abdominal fat	0	2.13 ± 0.86	4.30 ± 2.42	4.72 ± 2.28
Others	1.00 ± 0.03	2.03 ± 0.78	2.80 ± 1.68	2.43 ± 1.30
Content[a]	—	14.4 ± 1.27	13.7 ± 2.47	11.6 ± 3.32
Body weight[b]	0.121 ± 0.117	23.2 ± 11.5	32.6 ± 4.67	33.2 ± 5.91

[a]Contents of stomach and intestine: (content/body weight) × 100
[b]Body weight does not include the contents of stomach and intestine (kg)

Figure 33.5: Age-related changes in the relative growth rate (%) of liver, muscle and kidney for the Japanese serow. The fetal age was estimated from the regression equation between crown–rump length and gestational age, as reported by Sugimura *et al.* (1983), and shown in the period of −1 to 0 year on the abscissa of the graphs. The data for both males and females were combined. The vertical line and horizontal line indicate the range and mean, respectively

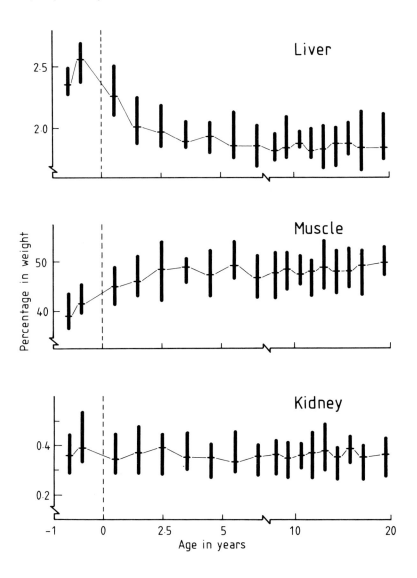

Organ and tissue distribution of heavy metals

The metal concentrations were generally high in the liver and kidney and low in the brain and muscle. This is particularly so in the case of Cd (Figure 33.6). However, there are some organ- and metal-specific accumulations. The concentrations of Ni and Pb in the liver and kidney were relatively low; Cu in the brain, and Zn in the muscle showed comparatively high accumulations; Fe and Mn in the fleece, Fe in the lung, Mn and Ni in the stomach and intestine, and Cu in the liver were present in very high

Figure 33.6: Organ and tissue distribution of heavy metal concentrations in the Japanese serow. Br, Brain; Li, liver; Ki, kidney; Sp, spleen; St, stomach; In, intestine; He, heart; Lu, lung; Mu, muscle; Sk, skin; Fl, fleece; Bo, bone; Go, gonads. □ adult male, ■ adult female

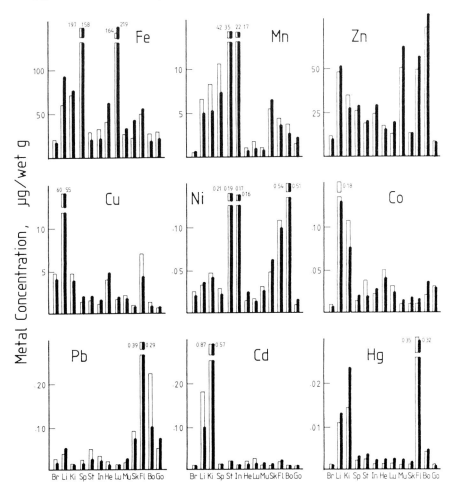

375

concentrations; higher accumulations of Zn, Ni, Pb, and Hg were in the hard tissues such as fleece and bone, especially Hg, which in the fleece showed the highest concentration.

Such distribution patterns are similar to those in other animals such as goat, sheep and marine mammals (Hidiroglou 1979, Honda *et al.* 1982, 1983, 1984, 1985, Allen and Masters 1985). However, a very high accumulation of Cu in the liver is species-specific as it is found only in ruminants and in certain other animals such as duck and frog (Underwood 1971). Furthermore, the concentrations of Mn and Ni in the gastro-intestinal organs of the serows were very high, whereas they were low or normal in the case of other animals ($<$ 2 µg/g w.wt). The serows also showed higher concentrations of Mn and Ni in the kidney than in the liver: again, different from other animals. Usually, Mn and Ni are accumulated in liver, kidney and bone, and their differences in concentration between animal species are relatively small. Also, Mn and Ni in the body are excreted mainly through bile and a little via urine, so their body concentrations remain relatively unchanged, even though the animals consume foods that contain high concentrations of Mn and Ni (Hall and Symonds 1981, Hall *et al.* 1982, Symonds and Hall 1983). In the case of Japanese serow, although the concentrations of Mn and Ni in the bile (2.53–6.87 µg/ml for Mn and 0.05–0.06 µg/ml for Ni) were relatively high compared with those in the goat and sheep (0.35–2.28 µg/ml for Mn and 0.02–0.05 µg/ml for Ni), the bile tract was very small, i.e. below 3 g in weight. Consequently, we assume that very high accumulations of Mn and Ni in the gastrointestinal organs of this animal are species-specific, which might be due to be an undeveloped excretory system for Mn and Ni by bile. Also, higher concentrations of Mn and Ni in the kidney than in the liver may be explained by an overload of these metals due to the absence of an effective bile excretion. However, further data are needed to elucidate the ability to excrete metals in the bile of the serow.

Significant differences in the metal concentrations in the regions of the muscle, kidney and bone were observed (Table 33.2). In muscle, most of the metals showed higher concentrations in *M. masseter* than in the other muscles, but the opposite was observed with Zn. Relatively high concentrations of Mn, Ni, Pb and Cd were also found in *M. obliquus abdominus externus*, especially for Ni and Cd. In kidney, the metal concentrations were highest in the cortex, followed by the medulla and remainder. But such a difference was not noticeable for Cu. In bone, the concentrations of Fe, Cu, Co, Cd and Hg were higher in the 7th vertebra than in the femur, and their accumulations depended upon the amounts of blood and red marrow in the bones. On the other hand, the distribution patterns of Mn, Zn, Ni and Pb concentrations in the bone were similar to that of Ca, which indicates that these metals are accumulated in the bone with ossification, probably as metal-binding hydroxyapatite.

Table 33.2: Metal concentrations (μg/g wt.wet) in various parts of muscle, stomach, kidney and bone for the Japanese serow

Tissue parts analysed	Fe m	Fe f	Mn m	Mn f	Zn m	Zn f	Cu m	Cu f	Ni m	Ni f	Co[b] m	Co[b] f	Pb m	Pb f	Cd[b] m	Cd[b] f	Hg[b] m	Hg[b] f
Muscle[a]																		
M m.	57	45	0.81	0.57	15.6	19.9	2.20	2.31	0.04	0.04	16	17	0.08	0.08	3.2	4.6	15.2	13.0
M. s. s.	25	28	0.22	0.20	38.7	46.6	1.46	1.42	0.01	0.02	7	9	0.03	0.04	0.8	1.4	7.4	6.7
M. p. s.	28	29	0.49	0.26	47.5	66.7	1.17	1.25	0.01	0.01	10	13	0.04	0.02	1.4	2.2	6.2	4.1
M. l. d.	32	36	0.46	0.36	63.3	54.2	1.40	1.40	0.01	0.01	12	8	0.03	0.04	1.2	1.6	3.4	5.4
M. o. a. e.	33	30	0.62	0.48	68.9	52.8	0.98	1.07	0.03	0.04	7	9	0.04	0.04	2.4	3.8	4.3	7.3
M. b. f.	29	31	0.40	0.28	63.1	63.7	1.42	0.02	0.02	0.02	8	10	0.03	0.03	0.5	1.4	4.4	4.4
Stomach																		
Rumen	28	19	42.2	34.6	17.3	17.7	1.41	1.66	0.21	0.19	26	18	0.03	0.03	40.1	20.2	8.2	7.4
Reticulum	34	15	36.9	35.8	16.9	13.4	1.34	1.59	0.18	0.22	23	17	0.02	0.03	32.3	21.4	8.1	7.2
Omasum	20	25	47.4	44.6	18.8	16.9	1.53	1.73	0.27	0.31	25	16	0.02	0.03	36.2	29.1	8.4	8.3
Abomasum	33	18	37.2	25.1	15.4	11.4	1.60	1.42	0.17	0.16	28	23	0.02	0.03	40.4	28.2	7.4	7.4
Kidney																		
Cortex	106	121	2.54	1.95	33.4	29.2	2.90	2.73	0.07	0.05	60	53	0.19	0.10	990	650	12.4	10.3
Medulla	59	55	1.03	0.78	17.9	15.2	2.41	2.30	0.03	0.03	31	33	0.09	0.05	390	350	4.4	6.2
Remainder	23	25	0.50	0.36	8.3	7.2	2.15	2.19	0.01	0.02	16	14	0.04	0.02	182	186	2.1	3.0
Bone																		
Femur	29	14	1.05	0.96	74.3	88.9	0.54	0.35	0.54	0.51	46	33	0.24	0.26	0.6	0.4	6.2	5.3
7th vertebra	106	78	0.75	0.81	68.4	75.9	0.83	0.66	0.25	0.34	71	60	0.22	0.22	0.8	0.6	17.1	13.1

[a] M.m., M. masseter; M.s.s., M. supra spinum; M.p.s., M. pectoralis superficialis; M.l.d., M. longissimus dorsi; M.o.a.e., M. obliquus abdominus externus; M.b.f., M. biceps femoris

[b] ng/g wet weight

Information on the tissue distribution of the heavy metal burden is useful for understanding the uptake and excretion of heavy metals in animals. Figure 33.7 shows the tissue distribution of the metal burdens as a percentage of the tissue burden to the total body burden. Here, the heavy metal burdens were calculated from the weights of tissues and their metal concentrations.

In general, relatively high percentages of the metal burdens in the whole body were in the muscle, being 50 per cent or more with Hg, Ni, Pb, Zn and Fe. However, the burdens of Hg, Ni, Pb and Zn were comparatively high in the hard tissues such as fleece and bone. In particular, Hg in the fleece was about 40 per cent of the body burden, indicating that an excretion of Hg by moulting is significant. Cu in the liver, Mn in the gastrointestinal organs, and Cd in the liver and kidney were also very high.

In summary, higher accumulations of Fe, Cu, Co and Cd were observed in soft tissues such as liver and kidney than in hard tissues. A very high accumulation of Mn was detected in the gastrointestinal organs, and Zn and Hg showed high concentrations in fleece, with relatively high concentrations in liver and kidney. In contrast, higher accumulations of Zn, Ni and Pb were observed in hard tissues such as fleece and bone than in soft tissues. Consequently, when the accumulation processes of these metals in the serow are discussed, analyses of liver, kidney, fleece and bone are essential. If the tissue of the highest accumulation is needed, liver and kidney for Fe, Co and Cd, liver for Cu, gastrointestinal organs for Mn, bone and fleece for Zn, Pb and Ni, and fleece for Hg should be selected and analysed. The accumulations of toxic metals such as Hg, Pb and Cd in muscle can be used for a comparative study of residue levels between animal species, and also for metal pollution in some cases. However, when the heavy metal accumulations are considered with muscle, kidney and bone, the significant regional variation in concentration within the tissues should be kept in mind. Usually, femoral muscle, femur and kidney cortex are recommended for comparison.

Variations in heavy metal accumulation with growth stage, age and sex

Age-related accumulation of heavy metals

The average concentrations of heavy metals in whole bodies were in the order of Fe · Zn > Mn > Cu > Ni · Pb > Co > Cd · Hg, and this order agreed with that of the food plants, indicating that accumulation of heavy metals is influenced by the metal contents of the dietary foods. Their concentrations, however, varied widely with growth stage or age, and also with metal species. Figure 33.8 shows schematically the age-related accumulations of heavy metals in the whole body.

The whole-body concentration of Fe increased with age and reached a

Figure 33.7: Organ and tissue distribution of heavy metal burdens in the Japanese serow. The heavy metal burdens, which were calculated from the weights of organs and tissues and their metal concentrations, are shown as percentages of the tissue burden to the total body burden

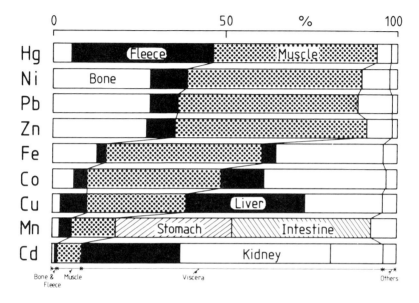

Figure 33.8: Schematic diagram of the age-related accumulations of heavy metals in the whole body of Japanese serow. The concentration on the ordinate of the graph is shown in the appropriate unit, so comparison of concentrations among the metals is not meaningful

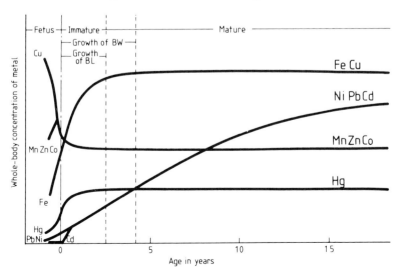

379

plateau at age $2^{1}/_{2}$ years. This age-related accumulation depended mainly on the muscular Fe accumulation as myoglobin. The concentration of Cu was very high in the first stage of the second trimester of gestation, and after birth its concentration increased with age until about $2^{1}/_{2}$ years of age, as a result of the hepatic accumulation. The concentrations of Mn, Zn and Co rapidly changed during the fetal period, and after birth their concentrations remained constant. The concentrations of Ni, Pb, Cd and Hg were very low in the fetus, and in particular the transplacental transfer of Cd was negligible. After birth, their concentrations increased year by year. However, the concentration of Hg remained constant after about $2^{1}/_{2}$ years of age, indicating the different age-dependent increase of Hg in comparison with other mammals such as marine mammals.

Such an unusual accumulation process of Hg is interesting in view of the absorption and excretion of Hg by this animal. Usually, Hg concentration in the body increases with age, as shown in marine mammals, whereas a discrepant age pattern is shown in birds (Figure 33.9). For example, in the

Figure 33.9: Schematic diagram of the age-related accumulations of Hg in the whole bodies of marine mammals and birds (from Honda and Tatsukawa 1985)

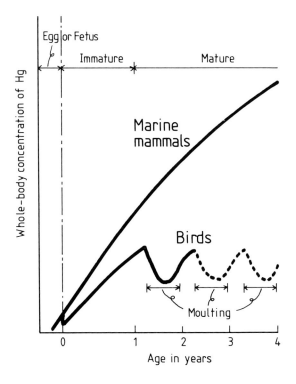

case of black-eared kite, the whole-body concentration of Hg increased with age throughout the chick period, whereas in adults its concentration changed seasonally as a result of excretion of Hg by moulting. This mechanism can be applied to the accumulation process of Hg in the serow, because of a relatively high burden of Hg in the fleece.

The age-related accumulations of heavy metals also showed organ-specific variations. For example, the muscular Fe concentration increased with age until about 2½ years of sexual maturity (Figure 33.10), and its age pattern approximated to that of the body weight shown in Figure 33.3.

Figure 33.10: Age-related accumulations of Fe in the muscle, liver, kidney and fleece of Japanese serow. ●, male , ○, female

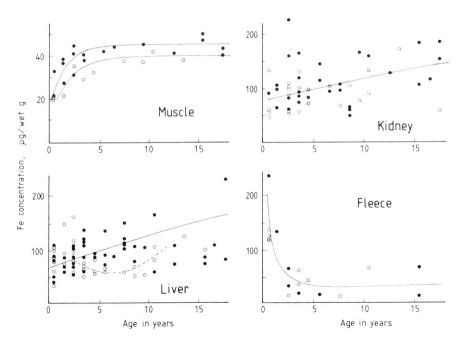

In contrast, Fe in the fleece exhibited the highest concentration in the fawn stage, and its concentration decreased with age until about 2½ years. The hepatic and renal Fe concentrations increased slightly with age, with wide variations during the puberty period of 1½–3½ years. As for Ni, Pb, Cd and Hg in the tissues, although these varied widely during the puberty stage, their age-related accumulations were generally similar to those in the whole body (see Figure 33.8). However, different trends were found with Ni, Pb, Cd and Hg in the fleece. The age-related accumulations of Hg and Pb in the liver and Pb in the fleece are shown in Figure 33.11 as examples.

Figure 33.11: Age-related accumulations of Hg and Pb in the liver and Pb in the fleece of Japanese serow. ●, male; ○, female

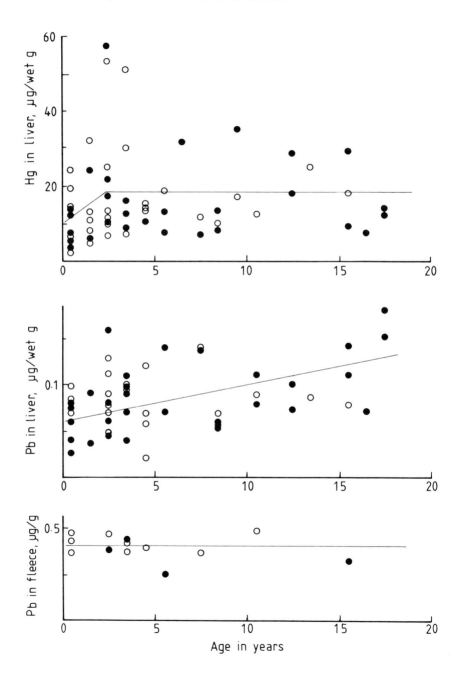

Furthermore, these age trends showed some sexual differences, which are discussed in the next section.

Consequently, these results indicate that accumulation of essential metals such as Fe, Mn, Zn, Cu and Co depends on the metabolic turnover. Also, their variations are generally small, except for the wide variations that occur during the fetal, lamb and puberty periods. In contrast, Ni, Pb, Cd and Hg are accumulated with age, indicating that age or exposure time is a dominant factor for their accumulation. However, Hg in the body is excreted by moulting and its concentration remains constant after about $2^{1}/_{2}$ years of age.

Sexual difference in metal concentrations

The heavy metal burdens of the fetus were very low compared with those of their mothers, and so parturition does not much change the whole-body concentrations of metals in mothers. Sexual differences of the metal concentrations, however, were observed in some tissues. Fe in the muscle and liver, and Cu in the liver showed significantly higher concentrations in the males than in the females. In the case of the hepatic Fe, such sexual difference appeared only in the mature stage, as shown in Figure 33.10: matured male serows showed age-dependent increase of Fe concentration, whereas the Fe concentration in matured females decreased with age until about 10 years, thereafter increasing up to the levels in the matured males. The hepatic Fe is accumulated as storage Fe, i.e. haemosiderin and ferritin, so the accumulated Fe in the liver is largely used up during gestation, parturition and lactation (Underwood 1971). This mechanism accounts for the lower concentration of Fe in the matured females during the period of 5–10 years. Also, an increase in the Fe concentration in females after 10 years of age might be due to decrease of the hepatic Fe consumption, suggesting that reproductive activities of the serow are inactive after 10 years of age.

Variations in heavy metal accumulation between habitats and feeding habits

Variations in heavy metal accumulation in Japanese serow between habitats were investigated in muscle, liver and kidney tissue of the serows, which were collected at different locations as shown in Figure 33.1.

Among the metals examined, only in the case of Cd was there a significant difference in concentration between habitats. Figure 33.12 shows the age-related accumulation of Cd in the liver of serow, as separately plotted for each sampling area. With a few exceptions, the five locations of Osaka, Ohkuwa, Tsukechi, Ohtaki and Agematsu, found in the north-west of the central Japanese Alpus mountain belt, showed a relatively high concentra-

Figure 33.12: Age-related accumulation of Cd in the liver of Japanese serows collected at different locations as shown in Figure 33.1. ●: Osaka, Ohkuwa, Tsukechi, Ohtaki and Agematsu; ○: Nagawa, Kiso, Narakawa, Ina, Kamisato, Seinaiji, Iida, Achi, Namiai, Kamiyahagi, Hiraya and Tenryu

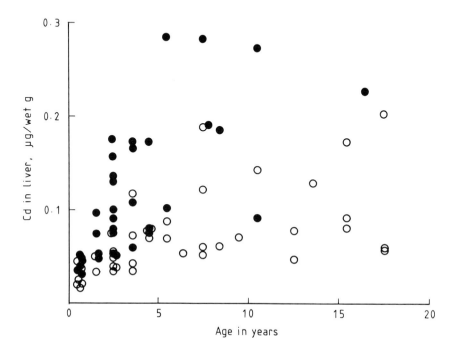

tion of Cd when compared with the other locations. Similar trends were also observed with the muscular and renal Cd.

Taking into consideration the locational difference of the Cd residue level in the serow, the concentration of Cd in food plants and faeces were determined. The results are shown in Table 33.3. Whereas the concentration of Cd in food plants varied widely with plant species, its locational difference was not significant. In contrast, the concentration of Cd in the faeces was significantly higher in the five locations mentioned above than in the other locations.

In this connection, absorption efficiency of metals by animals is generally very low, i.e. most of the metals taken are excreted via the faeces. It is also known that the serow occasionally eats soil to compensate for the deficiency of minerals such as Co. The locational differences of Cd concentration in the faeces are therefore considered to be due to variation in the amount of the direct intake from soil and/or in the concentration of Cd in soil. This situation may also account for the locational difference of the Cd

Table 33.3: Cd concentration (μg/g d.wt) in the faeces of Japanese serows collected at different locations

Location[a]	N	Range	Mean ± SD
Osaka	25	(0.442–1.944)	0.850 ± 0.415
Tsukechi	13	(0.475–2.194)	0.875 ± 0.481
Ohtaki	12	(0.783–4.355)	1.350 ± 0.990
Agematsu	2	(0.903–2.077)	1.530
Okhuwa	1	(0.862)	0.862
Kamiyahagi	22	(0.113–1.467)	0.701 ± 0.357
Iida	17	(0.191–2.524)	0.693 ± 0.576
Ina	16	(0.212–1.387)	0.501 ± 0.266
Seinaiji	1	(0.659)	0.659
Narakawa	1	(0.511)	0.511

[a]See Figure 33.1

residue level in the Japanese serow. In other words, this indicates that faeces are useful as a biological indicator for estimating the residue and intake levels of some metals in this animal.

SUMMARY AND CONCLUSIONS

Organ and tissue accumulations of nine heavy metals (Fe, Mn, Zn, Cu, Co, Ni, Pb, Cd and Hg), and their variations with biological processes, habitats and food habits in the Japanese serow were investigated, and the results are as follows:

(1) The whole-body concentrations of heavy metals were in the order Fe · Zn > Mn > Cu > Ni · Pb > Co > Cd · Hg, and this order agreed well with that of the food plants, indicating that accumulation of heavy metals in the body is influenced by the metal concentrations in the food plants.

(2) The distribution of heavy metals in the body was organ- and metal-specific. In particular, Hg in the fleece was about 40 per cent of the body burden, indicating that Hg in the body is excreted by moulting. Also, high accumulations of Mn and Ni in the gastrointestinal organs are species-specific in this animal, which might be due to an un-developed bile excretory system for Mn and Ni.

(3) Metal concentrations varied widely with metal species, and also with growth stage or age, especially during the fetal, fawn and puberty periods. Accumulations of Fe, Mn, Zn, Cu and Co depended upon the metabolic turnover, whereas age or exposure time is a dominant factor for Ni, Pb, Cd and Hg accumulation. However, an unchanged Hg

concentration with age after $2^{1}/_{2}$ years was due to excretion of Hg by moulting of the fleece.

(4) The heavy metal burdens of the fetus were very low compared with those of their mothers, and so parturition does not much change the whole-body concentrations of metals in mothers. In particular the residue level of Cd in the fetus was very low, indicating that trans-placental transfer of Cd is negligible. Sexual differences of the metal concentrations, however, were observed: Fe in the muscle and liver, and Cu in the liver showed significantly higher concentrations in males than in females. Also, the hepatic Fe concentration in matured females decreased with age through $2^{1}/_{2}$ to 10 years of age, which corresponded to the period of reproductive activity of this animal, and after 10 years its concentration increased.

(5) Among the metals examined, only in the case of Cd was there a significant difference in concentration between the living locations of the serow, which might be due to variations in amount of direct intake from soil and/or in concentration of Cd in dietary soil. Furthermore, analysis of the faeces was found to be useful in explaining the residue level of Cd in the body.

Wide variations of the heavy metal concentrations mentioned above indicate that consideration of age and biological processes in this species is necessary for understanding the phenomena of bioaccumulation of heavy metals and also their effects on them. In particular, taking into consideration a rapid and wide variation of the metal concentrations during the young and puberty stages, detailed research on the processes of bioaccumulation at these stages is needed.

ACKNOWLEDGEMENTS

We would like to thank Prof. M. Sugimura, Gifu University; Dr N. Maruyama, Tokyo University of Agriculture and Technology; Dr H. Saitoh, Research Center for Japanese Wildlife; Mr N. Sakurai, Agency of Cultural Affairs; and Mr O. Takeshita, Gifu Prefectural Government, for their help in the collection of the materials. This work was supported in part by Grant-in-Aid for Scientific Research from the Ministry of Education, Science and Culture of Japan (Project Nos 58030020, 59030053 and 60030060).

REFERENCES

Allen, J.G. and H.G. Masters. (1985) Renal lesions and tissue concentrations of zinc, copper, iron and manganese in experimentally zinc-intoxicated sheep. *Res. Vet. Sci. 39*, 249

Hall, E.D. and H.W. Symonds. (1981) The maximum capacity of the bovine liver to excrete manganese in bile, and the effect of a manganese load on the rate of excretion of copper, iron and zinc in bile. *Brit. J. Nutr. 45*, 605

Hall, E.D., H.W. Symonds and C.B. Mallinson. (1982) Maximum capacity of the bovine liver to remove manganese from portal plasma and the effect of the route of entry of manganese on its rate of removal. *Res. Vet. Sci. 33*, 89

Hidiroglou, M. (1979) Manganese in ruminant nutrition. *Can. J. Anim. Sci. 59*, 217

Honda, K., Y. Fujise, K. Itano and R. Tatsukawa. (1984) Composition of chemical components in bone of striped dolphin, *Stenella coeruleoalba*: distribution characteristics of heavy metals in various bones. *Agr. Biol. Chem. 48*, 677

Honda, K., B.Y. Min and R. Tatsukawa. (1985) Heavy metal distribution in organs and tissues of the eastern great white egret, *Egretta alba modesta*. *Bull. Environ. Contam. Toxicol. 35*, 781

Honda, K. and R. Tatsukawa. (1985) Comparative biology of heavy metal accumulations in marine mammals, birds and some other vertebrates. In T.D. Lekkas (ed.) *Heavy metals in the environment.* CEP Consultants Ltd, Edinburgh, pp. 706–8

Honda, K., R. Tatsukawa and T. Fujiyama. (1982) Distribution characteristics of heavy metals in the organs and tissues of striped dolphin, *Stenella coeruleoalba*. *Agr. Biol. Chem. 46*, 3011

Honda, K., R. Tatsukawa, K. Itano, N. Miyazaki and T. Fujiyama. (1983) Heavy metal concentrations in muscle, liver and kidney tissue of striped dolphin, *Stenella coeruleoalba*, and their variations with body length, weight, age and sex. *Agr. Biol. Chem. 47*, 1219

Ito, T. (1971) On the oestrous cycle and gestation period of the Japanese serow, *Capricornis crispus. J. Mammal. Sci. 5*, 104

Komori, A. (1975) Survey on the breeding of Japanese serows, *Capricornis crispus*, in captivity. *J. Japan. Assoc. Zool. Gard. Aq. 7*, 53

Maruyama, N. and S. Nakama. (1983) Blood count method for estimating serow populations. *Japan. J. Ecol. 33*, 243

Miura, S. (1985) Horn and cementum annulation as age criteria in Japanese serow. *J. Wildl. Mgmt. 49*, 152

Miura, S. and K. Yasui. (1985) Validity of tooth eruption-wear patterns as age criteria in Japanese serow, *Capricornis crispus. J. Mammal. Soc. 10*, 169

Phillips, D.J.H. (1980) *Quantitative aquatic biological indicators.* Applied Science, London, p. 488

Sugimura, M., Y. Suzuki, S. Kamiya and T. Fujita. (1981) Reproduction and pre-natal growth in the wild Japanese serow, *Capricornis crispus. Japan. J. Vet. Sci. 43*, 553

Sugimura, M., Y. Suzuki, I. Kita, Y. Ide, S. Kodera and M. Yoshizawa. (1983) Prenatal development of Japanese serow, *Capricornis crispus*, and reproduction in females. *J. Mammal. 64*, 302

Symonds, H.W. and E.D. Hall. (1983) Acute manganese toxicity and the absorption and biliary excretion of manganese in cattle. *Res. Vet. Sci. 35*, 5

Underwood, E.J. (1971) *Trace elements in human and animal nutrition*, 3rd edn, Academic Press, New York, p. 491

Index

Tierpark Berlin 44-7, 66-8, 193, 195, 197
Tooth 270
Toxoplasma gondii 300, 301, 307
Tracking duration 120
Trauma 47-9
Tur 34
Twin birth 188, 197, 200

Urials 7, 12-14, 21, 32, 56
Urogenital organ 229
Uterine infection 151

Veterinary medical care 188

Visual count method 111, 112
VTR 112
Vulva 150, 151

Wapiti 7, 17
Weaning age 149
Weapon differentiation 22
Weasel 130
Wild boar 130
Wild rabbit 200
Willow 187
Wood pecker 115

Yokohama Zoological Garden 312